Indian Hotspots

Chandrakasan Sivaperuman
Krishnamoorthy Venkataraman
Editors

Indian Hotspots

Vertebrate Faunal Diversity, Conservation and Management
Volume 1

Springer

Editors
Chandrakasan Sivaperuman
Zoological Survey of India, Andaman & Nicobar Regional Centre
Port Blair, Andaman & Nicobar Islands
India

Krishnamoorthy Venkataraman
National Centre for Sustainable Coastal Management
Anna University Campus
Chennai, Tamil Nadu, India

ISBN 978-981-10-6604-7 ISBN 978-981-10-6605-4 (eBook)
https://doi.org/10.1007/978-981-10-6605-4

Library of Congress Control Number: 2017964256

Printed on acid-free paper

This Springer imprint is published by Springer Nature
The registered company is Springer Nature Singapore Pte Ltd.
The registered company address is: 152 Beach Road, #21-01/04 Gateway East, Singapore 189721, Singapore

Foreword

India is 1 of the 17 mega-biodiversity countries of the world. Out of a total of 35 biodiversity hotspots, India has 4, namely, the Eastern Himalaya, the Indo-Burma, the Western Ghats and Sri Lanka and Sundaland. There are ten biogeographical zones in India; these are Trans-Himalayan, Himalayan, desert, semiarid, Western Ghats, Deccan Plateau, Gangetic Plain, coasts, north-east and islands. The varied edaphic, climatic and topographic conditions and years of geological stability have resulted in a wide range of ecosystems and habitats such as forests, grasslands, wetlands, deserts and coastal and marine ecosystem. The location of India is at the confluence of three major biogeographic realms, namely, the Indo-Malayan, the Eurasian and the Afro-tropical, which enabled it to have an assemblage of diverse elements from all these regions.

The tropical forest ecosystem is one of the most diverse and rich in biological diversity on Earth. India has a diverse range of forests from the rainforest of Kerala in the south to the alpine pastures of Ladakh in the north, from the deserts of Rajasthan in the west to the evergreen forests in the north-east. The forests in India are spread over an area of 692,027 km^2, covering 23.39% of the geographical area of the country. The tropical rainforests support the greatest diversity of living organisms on Earth. India ranks tenth in the world and fourth in Asia in plant diversity, and 11% of the world flora is reported from India.

The present book entitled *Indian Hotspots: Vertebrate Faunal Diversity, Conservation and Management* (Volumes I and II) embodies 39 research chapters of high standard grouped into *vertebrate faunal diversity*, *conservation* and *management*. This is a welcome step on the conservation of vertebrate faunal communities in the tropical forest ecosystem especially in Indian hotspots. I congratulate the editors Dr. C. Sivaperuman and Dr. K. Venkataraman for their earnest effort to bring this volume with a treasure of knowledge to the public domain.

Amita Prasad

Ministry of Environment, Forest and Climate Change,
Government of India
New Delhi, India
29 August 2017

Preface

The tropical forest ecosystems are one of the most diverse and rich in biological diversity on Earth. The tropical rainforests of India are found in the Western Ghats, north-eastern India and the Andaman and Nicobar Islands. The forests in India are spread over an area of 692,027 km^2, covering 21.05% of the geographical area of the country. The location of India is at the confluence of three major biogeographic realms, namely, the Indo-Malayan, the Eurasian and the Afro-tropical, which enabled it to have an assemblage of diverse elements from all these regions. The country is exceptionally rich in biodiversity and considered as 1 of the 17 mega-biodiversity countries in the world. Of the 35 biodiversity hotspots identified in the world, India has 4 biodiversity hotspots, i.e. the Eastern Himalaya, Indo-Burma, Western Ghats and Sri Lanka and Sundaland, with unique assemblage of plant and animal communities. There are ten biogeographical zones in India; these are Trans-Himalayan, Himalayan, desert, semiarid, Western Ghats, Deccan Plateau, Gangetic Plain, coasts, north-east and islands. The varied edaphic, climatic and topographic conditions and years of geological stability have resulted in a wide range of ecosystems and habitats such as forests, grasslands, wetlands, deserts and coastal and marine ecosystem.

India occupies about 2.4% of the world's land area and 4% of freshwater and supports about 8% of the world's total species. This region is home to a rich endemic assemblage of plants, birds, reptiles and amphibians, as well as important populations of Asian elephants, Indian tigers and the endangered lion-tailed macaque. India ranks tenth in the world and fourth in Asia in plant diversity and tenth in the number of endemic species of higher vertebrates in the world. In terms of species richness, India ranks eighth in mammals, ninth in birds, fifth in reptiles and fifteenth in Amphibians. Eleven percent of the world flora is reported from India.

This volume is a culmination of detailed studies carried out by reputed researchers working in the field of biodiversity conservation. This book contains a collection of different chapters, and concerted effort has been made by the professionals in their respective group. This book is divided into two volumes and covers different faunal communities from the Indian biodiversity hotspots, e.g. biodiversity in Indian hotspots; endemic, endangered and threatened vertebrates; chiropteran fauna; herpetofaunal diversity; group size composition of gaur; grizzled giant squirrel; Nilgiri tahr; elephant from the Western Ghats; freshwater fishes; birds; mammals of the

Andaman and Nicobar Islands; avifauna of the north-west Himalaya; bat fauna of the north-east and Western Ghats; and golden langur from the north-east.

This book depicts unique information on vertebrate faunal diversity of Indian hotspots. We sincerely hope that this book will be of great help to the researchers and field scientists in the area of biodiversity conservation not only in India but also in neighbouring countries.

Port Blair, Andaman & Nicobar Island, India Chandrakasan Sivaperuman
Chennai, Tamil Nadu, India Krishnamoorthy Venkataraman

Acknowledgements

We express our heartfelt gratitude to all those who helped in different ways to complete this work. Our sincere thanks to the officials of the Ministry of Environment, Forest and Climate Change, Government of India, for their support and encouragement. We also thank all the authors who have contributed the various articles for this book.

Contents

About the Editors

Chandrakasan Sivaperuman is currently a Scientist D at the Zoological Survey of India (ZSI), Port Blair. He received his master's degree in wildlife biology from Bharathidasan University, Tamil Nadu, and his doctorate degree in ecology of wetland birds in the Vembanad-Kole Ramsar site from the Kerala Forest Research Institute, Kerala, and Forest Research Institute Deemed University, Dehra Dun. He has been extensively involved in field surveys in different parts of the country. He has published more than 200 research papers in national and international journals and newsletters. He also authored/edited more than 25 books published by respected national and international publishers. He has participated in the 36th Indian Scientific Expedition to Antarctica during 2016–2017 and carried out studies on the species abundance and distribution of birds and mammals in Antarctica. He has participated and presented research papers at 60 national and international seminars and symposia. He is life member of various scientific societies in India and abroad.

Krishnamoorthy Venkataraman former director of Zoological Survey of India, is presently working as a senior scientific consultant at the National Centre for Sustainable Coastal Management, Anna University, Campus, Chennai.

His research areas include coastal and marine biodiversity assessment; coral reef restoration; integrated management of coastal, marine and wetland ecosystems; and biodiversity laws and policies. He is an expert on strategy formulation, climate change resilience in the management of marine ecosystem, meiofaunal assemblage in the Antarctic ecosystem and training and organizing awareness programmes on faunal resources conservation. He has 37 years of research experience in faunal diversity studies and has served at Madurai Kamaraj University, the Bombay Natural History Society (BNHS), the National Biodiversity Authority, the Loss of Ecology Authority and the Zoological Survey of India (ZSI) in various capacities.

He was the director of the Zoological Survey of India from 2010 to 2015. At BNHS he worked with internationally renowned ornithologist, Dr. Salim Ali, in the management of Keoladeo National Park, Bharatpur, Rajasthan. He has contributed over 458 publications in the field of marine, wetland, terrestrial and freshwater ecosystems, of which 41 have been published as books. He is a member of the editorial board of leading scientific journals in India and has reviewed several articles for internationally respected journals and various Government of India committees. He has participated in over 500 national and 50 international meetings, chairing sessions and panel discussions and delivered lead talks.

He was the founder Member Secretary of the National Biodiversity Authority of the Ministry of Environment, Forests and Climate Change, Government of India, and coordinated the preparation of the thematic National Biodiversity Strategy and Action Plan and the action plan for biophysical monitoring of coral reefs in the Andaman and Nicobar Islands and Gulf of Mannar Biosphere Reserve. His input and involvement in the enactment and implementation of the Biological Diversity Act, 2002 and establishment of State Biodiversity Boards are worth mentioning.

He was instrumental in the Great Nicobar Biosphere Reserve (GNBR), Andaman and Nicobar Islands, becoming recognized in the UNESCO-MAB World Network of Biosphere Reserves.

Biodiversity Hotspots in India

1

Krishnamoorthy Venkataraman
and Chandrakasan Sivaperuman

Abstract

India is one of the 17 megadiverse countries in the world, and there are four biodiversity hotspots found in India. These are Indo-Burma, Himalaya, Western Ghats-Sri Lanka, and Sundaland. For biological diversity point of view, India is very rich in resources due to its diversified habitat and climatic conditions. India also supports 7.5% of the total animal species of the world. In this chapter, an effort made to compile and provide the detailed account on the vertebrate faunal diversity in the Indian biodiversity hotspots.

Keywords

Biodiversity · Faunal · Hotspots · India · Threatened

1.1 Introduction

Biodiversity is defined in the Convention on Biological Diversity as "the variability among living organisms from all sources including, inter alia, terrestrial, marine and other aquatic ecosystems and the ecological complexes of which they are part; this includes diversity within species, between species and of ecosystems" (Article 2).

K. Venkataraman
National Centre for Sustainable Coastal Management, Anna University Campus, Chennai, Tamil Nadu, India

C. Sivaperuman (✉)
Zoological Survey of India, Andaman & Nicobar Regional Centre, Port Blair, Andaman & Nicobar Island, India
e-mail: c_sivaperuman1@rediffmail.com

C. Sivaperuman, K. Venkataraman (eds.), *Indian Hotspots*,
https://doi.org/10.1007/978-981-10-6605-4_1

Myers (1988) defined "hotspots" as high concentrations of endemic species with high habitat loss. This hotspot approach can be applied at any geographical scale and both in terrestrial and marine environments. However, hotspots represent conservation priorities in terrestrial ecosystems but remain largely unexplored in marine habitats, where the amount of data is still poor (Mittermeier et al. 2011). A hotspot must contain at least 1500 species of vascular plants as endemics, and it has to have lost at least 70% of its original habitat (Myers et al. 2000). India is composed of a diversity of ecological habitats like forests, grasslands, wetlands, deserts, coastal, and marine ecosystems. India lies between 8° 04′ and 37° 06′ N latitude and 68° 07′ and 97° 25′ E longitude with a total geographical area of 329 million ha. India is one of the megadiverse countries due to its rich biological diversity, and there were 8.4 million species reported. According to the publication of Venkataraman (2006), India holds three major biological realms, namely, Indo-Malayan, Eurasian, and Afrotropical.

1.2 Biogeographic Zones of India

Biogeography is the study of distribution of plants and animals over their evolutionary history. The "biogeographic classification" for conservation planning divided the country into 10 zones and 26 provinces (WII 2009; Table 1.1). There are four levels of biogeographic classification.

Biogeographic zone is the large distinctive units of similar ecology, biome representation, community, and species (e.g., the Himalaya, the Western Ghats).

Biotic province is the secondary units within a zone, giving weight to particular communities separated by dispersal barriers or gradual change in environmental factors (e.g., North West and West Himalaya either side of the Sutlej River).

Land region is a tertiary set of units within a province, indicating different landforms (e.g., Aravalli Mountains and Malwa Plateau in Gujarat Rajwada Province).

Biome is an ecological unit, not a biogeographic unit. A biome such as swamp/wetland or temperate broad-leaved forests could be found in several biogeographic zones or provinces.

Table 1.1 List of biogeographic zone with provinces

Sl. no.	Name of the biogeographic zone	Number of provinces
1.	Trans-Himalaya	3
2.	Himalaya	4
3.	Indian desert	2
4.	Semiarid zone	2
5.	Western Ghats	2
6.	Deccan Peninsula	3
7.	Gangetic Peninsula	2
8.	Coasts	3
9.	North East India	2
10.	Islands	2

1.3 Biodiversity in India

India is very rich in terms of biological diversity due to its diversified habitat and climatic conditions. More than 50% of the world's plant diversity and 42% of terrestrial vertebrate diversity are endemic within the 35 biodiversity hotspots of the world. Overall, 7.5% of the total animal species of the world are found in India, though the Indian landmass is about 2%. India is known to have nearly 100,693 animal species, of which insect alone is 65,047 (Table 1.2). The inventories of reptiles, amphibians, fish, birds, and mammals are fairly complete. More than 5150 species of plants, 20,765 insect species, 46 mammal species, 176 bird species, 214 reptile species, 138 amphibian species, and 435 fish species are endemic in India (Ravindranath et al. 2006; Ramakrishna and Alfred 2007; ZSI 2016).

Table 1.2 Total number of animal species recorded in India

		Number of species			
Kingdom	Phylum	World (living and fossil)	World (living)	India	Percentage
Protista	Protozoa	36,400 (excluding fossil)	34,400	3510	9.64
Animalia	Mesozoa	122	122	10	8.02
	Porifera	11,055	8,838	545	6.16
	Cnidaria	17,702	11,522	1396	12.12
	Ctenophora	199	199	19	9.55
	Platyhelminthes	29,488	29,487	1738	5.89
	Rotifera	2049	2049	466	2.24
	Gastrotricha	828	828	162	19.56
	Kinorhyncha	196	196	10	5.10
	Nematoda	25,043	25,033	2914	11.63
	Acanthocephala	1461	1330	301	22.63
	Sipuncula	156	156	41	26.28
	Echiura	198	198	47	23.73
	Annelida	17,426	17,388	1024	5.89
	Onychophora	187	183	1	0.53
	Arthropoda	1,302,809	1,257,040	75,528	6.00
	Subphylum: Chelicerata	1,15,992	1,13,773	5945	5.23
	Class: Arachnida	1,14,275	1,12,442	5907	5.25
	Class: Merostomata	103	4	2	50.00
	Class: Pycnogonida	1346	1335	36	2.69
	Subphylum: Crustacea	73,141	67,735	3796	5.61
	Subphylum: Hexapoda	1,080,760	1,063,533	65,409	6.15
	Class: Collembola	8187	8162	324	3.97
	Class: Diplura	976	975	18	1.85
	Class: Protura	816	816	20	2.45
	Class: Insecta	1,070,781	1,053,578	65,047	6.17
	Subphylum: Myriapoda	12,010	11,999	378	3.15

(continued)

Table 1.2 (continued)

Kingdom	Phylum	Number of species World (living and fossil)	World (living)	India	Percentage
	Class: Chilopoda	3118	3112	101	3.25
	Class: Diplopoda	7842	7837	270	3.45
	Class: Symphyla	204	204	7	3.43
	Phoronida	16	16	3	18.75
	Bryozoa (Ectoprocta)	11,652	6186	327	5.29
	Entoprocta	186	186	10	5.37
	Brachiopoda	7390	392	8	2.04
	Chaetognatha	186	170	44	25.88
	Tardigrada	1335	1167	30	2.57
	Mollusca	118,062	84,978	5189	6.11
	Nemertea	1368	1368	6	0.43
	Echinodermata	20,550	7550	777	10.29
	Hemichordata	162	139	14	10.07
	Chordata	89,955	71,526	6573	9.08
	Subphylum: Cephalochordata	33	33	6	18.18
	Subphylum: Urochordata	2804	2804	516	18.40
	Subphylum: Vertebrata	88,512	68,689	6051	6.85
	Class: Pisces	37,172	34,362	3324	9.70
	Class: Amphibia	8007	7667	388	5.06
	Class: Reptilia	16,123	10,357	527	5.47
	Class: Aves	11,241	10,357	1340	12.93
	Class: Mammalia	15,969	5853	427	7.29
	Total	**1,664,289**	**1,529,953**	**97,183**	
Grand total (Protista + Animalia		**1,700,689**	**1,566,353**	**100,693**	

Source: ZSI (2017)

The forest cover of the country constitutes about 692,027 km^2 (21.05%) of India's total geographical area (FSI 2011). Champion and Seth (1968) have classified the Indian forests into 16 major forest types and 221 subtypes. Wetland is another important habitat in India, and the extent of wetlands in India is about 4.1 million hectares, these excluding paddy fields and mangroves. The mangrove forest constitutes 7% of the world's mangroves with an extent of about 6700 km^2. The coral reef ecosystem is another important habitat in the marine environment; these are found in Andaman and Nicobar Islands, Lakshadweep Islands, Gulf of Kutch, and Gulf of Mannar. The Great Indian Desert covers about 2% of the total landmass which covers the states of Rajasthan, Gujarat, Punjab, and Haryana. The cold desert is also found in India, which generally lies in Ladakh, Jammu and Kashmir, and

Table 1.3 Total number of plant species recorded in India

Sl. no.	Type	Number of species World	India	% in India	No. of endemic species	No. of threatened species
Flowering plants						
1.	Gymnosperms	1021	74	7.35	8	7
2.	Angiosperms	268,600	18,043	6.72	4036	1700
Non-flowering plants						
1.	Bryophytes	16,236	2523	15.54	629	ca. 80
2.	Pteridophytes	12,000	1267	10.57	47	414
Others						
1.	Virus and bacteria	11,813	986	8.77	NA	NA
2.	Algae	40,000	7284	18.21	1924	NA
3.	Fungi	98,998	14,883	15.09	4100	ca. 580
4.	Lichens	17,000	2401	14.12	520	NA
Total		**465,668**	**47,513**	–	**11,273**	**2781**

Source: Chapman (2009) and Singh and Dash (2014)
NA not available

Lahaul-Spiti in Himachal Pradesh covering an area of about 109,990 km^2. The India's landmass is only 2.4% of the world; however, it supports 47,513 species of plant (Singh and Dash 2014). Of the total recorded flora of the country, 28% are endemic. The detail comparative account of major plant species reported from India is provided in Table 1.3.

1.4 The Biodiversity Hotspots Concept

The first published biodiversity hotspot concept thesis was by the British Ecologist Norman Myers in the year 1988. Myers concept on biodiversity hotspots is the only relying sources, though he has used only qualitative criteria to assess the habitat loss and the presence of the highest number of plant endemism (Mittermeier et al. 2011). Later, eight more hotspots were analyzed and identified by Myers (1990), which includes four in Mediterranean regions. The Conservation International also adopted the Myers' hotspot concept and thereafter worked systematically to update the global biodiversity hotspots. Myers, Conservation International, and collaborators later revised estimates of the remaining primary habitat and defined the hotspots formally as biogeographic regions with more than 1500 endemic vascular plant species and ≤30% of original primary habitat (Myers et al. 2000). Based on this collaboration, an extensive global review has been made and the scientific publications on the hotspots also expand in greater number (Mittermeier et al. 1999; Myers et al.

2000). During the year 2004, a second major revision carried out and updated the biodiversity hotspots without changing the criteria; however they were redefining several hotspot boundaries. Based on the results, a total of 34 biodiversity hotspots were classified by Mittermeier et al. (2011). The Forests of East Australia is added as the 35th biodiversity hotspot by Williams et al. (2011). Overall the hotspots are maintaining the 77% of endemic plant species, 43% of vertebrates, and 80% of all threatened amphibians (Mittermeier et al. 2011; Williams et al. 2011).

1.5 Indian Biodiversity Hotspots

India is one of the world's most biodiverse countries. India's political boundaries encompass a wide range of ecozones, namely, the desert, the high mountains, the highlands, the tropical and temperate forests, the swamplands, the plains, the grasslands, and the islands. Four global biodiversity hotspots are found in India: the Western Ghats-Sri Lanka, the Himalaya, the Indo-Burma, and the Sundaland (Table 1.4). India is situated in the confluence of Oriental, Palaearctic, and Ethiopian biogeographical regions (Mani 1974). There are a total of 668 protected areas including 102 national parks, 515 wildlife sanctuaries, 47 conservation reserves, and 4 community reserves covering a total of 161,221.57 km^2 of the country. Also, there are 47 tiger reserves, 18 biosphere reserves, 25 elephant reserves, 5 natural world heritage sites, and 25 Ramsar sites designated in India (Anon. 2015).

Table 1.4 Biodiversity hotspots in India

Sl. no.	Name of the biodiversity hotspots	Province
1	Himalaya	These hotspots cover the Indian Himalayan region (and that falling in Pakistan, Tibet, Nepal, Bhutan, China, and Myanmar)
2	Indo-Burma	It covers the northeastern India, except Assam and Andaman group of Islands (and Myanmar, Thailand, Vietnam, Laos, Cambodia, and southern China)
3	Western Ghats-Sri Lanka	It includes the Western Ghats and Sri Lanka
4	Sundaland	Entire Nicobar group of islands is included in this hotspot (and Indonesia, Malaysia, Singapore, Brunei, and Philippines)

www.conservation.org

1.6 Hotspot: The Himalaya

The world's highest mountains and Mount Everest are found in the Indian Himalayan Biodiversity Hotspot. The forests of these mountains are subtropical broadleaf forest to alpine. Several vascular plants have also been reported at the height of 6000 m. This hotspot supports several important animals and bird species, which include vultures, tigers, elephants, rhinoceros, and buffalo. The Himalaya hotspot includes several of the world's deepest rivers. The mountain range of this hotspot covers about 750,000 km^2 and is divided into two major regions, i.e., the Eastern Himalaya (Nepal, Bhutan, the northeast Indian states of West Bengal, Sikkim, Assam, Arunachal Pradesh, southeast Tibet, and northern Myanmar) and the Western Himalaya (Kumaon-Garhwal, northwest Kashmir, and northern Pakistan) (Table 1.5).

1.7 Biodiversity

The Himalayan Biodiversity Hotspot supports about 163 globally threatened species which include one-horned rhinoceros *Rhinoceros unicornis*, wild Asian water buffalo *Bubalus bubalis*, and above 45 species of mammals, 50 species of birds, 17 species of reptiles, 12 species amphibians, 3 species invertebrates, and 36 species of plant (Table 1.6). The endangered species of the relict dragonfly *Epiophlebia*

Table 1.5 Details of the Himalayan Biodiversity Hotspot

Original extent (km^2)	741,706
Remaining vegetation (km^2)	185,427
Number of plant species are endemic	3160
Number of threatened and endemic birds	8
Number of endemic and threatened mammals	4
Number of endemic and threatened amphibians	4
Extent of protected area (km^2)	112,578

Source: www.conservation.org

Table 1.6 Biodiversity and endemic species

Group	Total number of species	Number of endemic species	Percentage of endemism
Plants	10,000	3160	31.6
Mammals	300	12	4.0
Birds	977	15	1.5
Reptiles	176	48	27.3
Amphibians	105	42	40.0
Freshwater fishes	269	33	12.3

Source: www.conservation.org

Table 1.7 Description of the Indo-Burma

Original extent (km^2)	2,373,057
Remaining vegetation (km^2)	118,653
Number of plant species are endemic	7000
Number of threatened and endemic birds	18
Number of endemic and threatened mammals	25
Number of endemic and threatened amphibians	35
Extent of protected area (km^2)	235,758

Source: www.conservation.org

laidlawi is also found in this hotspot. This region is also home to the salamander species Himalayan newts, *Tylototriton verrucosus*. More than 10,000 species of plants in the Himalayas were reported from here, of these one-third of the species are endemic. Five families, namely, Tetracentraceae, Hamamelidaceae, Circaesteraceae, Butomaceae, and Stachyuraceae, are completely endemic to this region. Many threatened and endemic bird species are also found in this hotspot, namely, the Himalayan quail, cheer pheasant, western tragopan Himalayan vulture, and white-bellied heron.

1.8 Hotspot: Indo-Burma

The Indo-Burma Biodiversity Hotspot is one of the top ten hotspots in the world, which is impossible to replace the original habitat and also under the threat is in the five. Only 5% of its natural habitat is remaining and with more human population than any other hotspots (Mittermeier et al. 2004). This hotspot encompasses several countries, which spread out from eastern Bangladesh to Malaysia, the south of Brahmaputra River of the northeastern India, the southern part of China's Yunnan province, Lao People's Democratic Republic, Cambodia, Vietnam, and Thailand. An extensive variety of diversities is represented in this hotspot, which includes mixed wet evergreen, dry evergreen, deciduous, and montane forests. Several patches of shrublands and woodlands on karst limestone outcrops and, in some coastal areas, scattered heath forests are also found here. Besides, a large variety of distinctive localized vegetation formations occur in Indo-Burma; these consist of lowland floodplain swamps, mangroves, and seasonally inundated grasslands (Table 1.7).

1.9 Biodiversity

A large amount of area in this hotspot, which is still untouched, however, has been declining rapidly in the past few decades. In recent years, about six species of mammals were discovered from this hotspot, namely, large-antlered muntjac, Annamite muntjac, gray-shanked douc, Annamite striped rabbit, leaf deer, and the Saola; this

Table 1.8 Biodiversity and endemism

Taxonomic group	Total number of species	Number of endemic species	Percentage of endemism
Plants	13,500	7000	51.9
Mammals	433	73	16.9
Birds	1266	64	5.1
Reptiles	522	204	39.1
Amphibians	286	154	53.8
Freshwater fishes	1262	553	43.8

Source: www.conservation.org

Table 1.9 Description of the Western Ghats and Sri Lanka hotspots

Original extent (km^2)	189,611
Remaining vegetation (km^2)	43,611
Number of plant species are endemic	3049
Number of threatened and endemic birds	10
Number of endemic and threatened mammals	14
Number of endemic and threatened amphibians	87
Number of extinct species	20
Extent of protected area (km^2)	26,130

Source: www.conservation.org
Recorded extinctions since 1500

is also home for many species of primates. Several species of freshwater turtle are endemic in this region. More than 1300 species of birds can be found, which include the threatened white-eared night heron, the gray-crowned crocias, and the orange-necked partridge. There is also an estimated 13,500 species of plants, of which more than 50%, are endemic (Table 1.8).

1.10 Hotspot: Western Ghats and Sri Lanka

The Western Ghats area chain of hills that run along the southwestern coast of India lies the Western Ghats Mountain Range (Table 1.9). This also known by the name "Sahyadri" constitutes a 1600-km-long mountain range, and this is originating from south of the Tapti River and extending up to Kanyakumari at Southern India. The Western Ghats Mountains are ranges recognized for their high biodiversity and natural heritage. The average elevation is 900–1500 m, and the highest mountain peak is having an altitude of 2969 m. The western slopes of Western Ghats receive very high rainfall ranging from 2000 to 6000 mm per year (Nair 1991). The hotspots support more species diversity and endemism. About 77% of amphibians and 62% of the reptile species found only in this hotspot which is not found anywhere.

1.11 Biodiversity

About 6000 species of vascular plant that belongs to 2500 genera were reported from this hotspot. Of these, 3000 species are endemic to this region. The world's spices, namely, the black pepper and cardamom, were originated from Western Ghats Mountain. The Agasthyamalai Hills support highest concentration of species in the Western Ghats. This hotspot is also home to special habitat of fresh water swamps known as Myristica swamps. The Western Ghats is home for more than 510 species of birds, 140 species of mammals, 260 species of reptiles, and 181 species of amphibians (Tables 1.10 and 1.11).

Table 1.10 Biodiversity and endemism

Taxonomic group	Total number of species	Number of endemic species	Percentage of endemism
Plants	5916	3049	51.5
Mammals	140	18	12.9
Birds	510	35	7.6
Reptiles	267	174	65.2
Amphibians	181	130	73.0
Freshwater fishes	191	139	72.8

Source: www.conservation.org

Table 1.11 Animal groups recorded from Western Ghats

Animal group	No. of species	Endemism%
Mammals	137	11.7
Birds	508	0.3
Reptiles	203	61.8
Amphibians	181	87.8
Fishes	290	65.0
Land snails	269	76.0
Freshwater snails	77	36.0
Butterflies	332	11.0
Odonata	174	39.6

1.12 Hotspot: Sundaland

The Sundaland Biodiversity Hotspot is located in Southeast Asia (Myers et al. 2000). This hotspot comprises the landmasses of the Malay Peninsula, Sumatra, Java, Bali, and Borneo besides several smaller islands. The Nicobar group of islands, which are jurisdictionally controlled by India, form part of this hotspot, and the fauna and flora have close affinities (Davis et al. 1995). The extent of the Sundaland hotspot is about 1.5 million km^2 and covering half of the Indo-Malayan archipelago. This hotspot also includes more than 17,000 islands, of which Borneo covers about 725,500 km^2 and Sumatra, 427,300 km^2. The dividing boundary between the Sundaland hotspot and the Mainland Southeast Asia hotspot to the northwest is here taken as the Kangar-Pattani Line, which lies near the Thailand-Malaysia border (van Steenis 1950; Whitmore 1984). Sundaland is rich in biodiversity on earth, supporting more than 25,000 species of vascular plants, of which 117 species are endemic. About 770 species of birds were reported from this region, which include 150 endemic species. High concentration of mammals was also reported; 380 mammal species are found in Sundaland, and 170 species are endemic (Tables 1.12, 1.13 and 1.14).

Table 1.12 Species diversity and endemism in Sundaland hotspots

Taxonomic group	Total number of species	Number of endemic species	Percentage of endemism
Plants	25,000	15,000	60.0
Mammals	380	172	45.3
Birds	769	142	18.5
Reptiles	452	243	53.8
Amphibians	244	196	80.3
Freshwater fishes	950	350	36.8

Source: Conservation International: www.conservation.org

Table 1.13 Description of the Sundaland

Original extent (km^2)	1,501,063
Remaining vegetation (km^2)	100,571
Number of plant species are endemic	15,000
Number of threatened and endemic birds	43
Number of endemic and threatened mammals	60
Number of endemic and threatened amphibians	59
Number of extinct species	4
Extent of protected area (km^2)	179,723

Source: www.conservation.org

Table 1.14 Biodiversity diversity and endemism

Taxonomic group	Total number of species	Number of endemic species	Percentage of endemism
Plants	25,000	15,000	60.0
Mammals	380	172	45.3
Birds	769	142	18.5
Reptiles	452	243	53.8
Amphibians	244	196	80.3
Freshwater fishes	950	350	36.8

Source: www.conservation.org

The Andaman and Nicobar archipelagos, also often referred to as the Emerald Islands, comprise 572 islands, islets, and rocky outcrops and extending over 800 km. The Andaman and Nicobar Islands running between 6° 45′ N and 13° 30′ N latitudes and 90° 20′ E and 93° 56′ E longitudes with extent of 8249 km^2 are broadly divided into two groups of islands, namely, the Andaman and the Nicobar. These two groups are separated by the Ten Degree Channel which is about 150-km-wide, 400 fathoms deep. The Andaman group consisting of 550 islands covers a land area of 6408 km^2, and the Nicobar group comprising 22 islands has an area of 1841 km^2. The Nicobar Islands are located in Southeast Asia, 150 km north of Aceh on Sumatra, and separated from Thailand to the east by the Andaman Sea. Located 1300 km southeast of the Indian subcontinent, across the Bay of Bengal, they form part of the Union Territory of Andaman and Nicobar Islands, India. The Nicobar groups of islands are further divided into three distinct subgroups, namely, Great Nicobar, Nancowry, and Car Nicobar. The protected areas in the Nicobar groups are Great Nicobar Biosphere Reserve, Campbell Bay National Park, Galathea National Park, Galathea Bay Wildlife Sanctuary, Megapode Island Wildlife Sanctuary, Tillangchong Island Wildlife Sanctuary, and Batimaliv Island Wildlife Sanctuary. The Tillangchong Island, Camorta Island, Katchal Island, Nancowry Island, and Trinkat Island are the important bird areas identified by the BirdLife International.

1.13 Important Major Fauna of Sundaland Hotspots (Nicobar Islands)

1.13.1 Coconut Crab, *Birgus latro* (Linnaeus, 1767)

The coconut crab, *Birgus latro* (Linnaeus), has wide distribution ranging from Eastern Africa, through the Indian Ocean islands, to the Pacific Ocean islands. This crab is the largest living terrestrial arthropod and weighs up to 4 kg and measures 200 mm in carapace width (Lavery et al. 1996a). *Birgus latro* is considered T4 terrestrial species, according to the dependence level it has on the aquatic environments. These species within the grade do not require immersion in standing water but are dependent on water for the pelagic larvae (Powers and Bliss 1983; Hartnoll 1988; Greenaway 2003). During the larval phases, the *Birgus latro* spends 3–4 weeks in the sea before undertaking migration to terrestrial habitats. It is widely distributed in the tropical islands of the Indian and Pacific Oceans (Reyne 1939; Robertson 1991; Lavery et al. 1996b). The coconut crab is most commonly found on island habitats and generally within 4 km from the sea (Fletcher and Amos 1994). They mainly inhabit dense forest regions but can be abundant in sandy coconut groves (Grubb 1971). The populations of *Birgus latro* are declining throughout their range, with loss of habitat, and harvesting for human consumption is thought to be the key drivers of the declines. However, populations in general are poorly studied, and consequently the species is considered data deficient under the International Union for Conservation of Nature (IUCN) red list (Eldredge 1996). In India, the coconut crab occurs in Nicobar group of islands and North Sentinel Island in Andaman group (Hume 1874; Alcock 1905; Altevogt and Davis 1975; Bhaskar and Rao 1992). During the tsunami in 2004, the favored habitats of these animals have been severely affected (Ramachandran et al. 2005). After the tsunami, the sighting of the species is very less in most of the Nicobar group of islands.

1.13.2 Nicobar Megapode *Megapodius nicobariensis* (Blyth, 1846)

Megapodes are medium-sized to large terrestrial birds with large legs and feet with sharp claws, which are endemic to the Nicobar group of islands of India. They split into two subspecies, i.e., *Megapodius nicobariensis abbotti* and *Megapodius nicobariensis nicobariensis*. The *M. n. abbotti* is distributed in Great Nicobar, Little Nicobar, Kondul, Menchal, Treis, and Meroe, and *M. n. nicobariensis* is found in the islands of Camorta, Trinkat, Nancowry, Katchal, Teressa, Bompoka, and Tillangchong. It generally inhabits forests and secondary growth, with the greatest concentrations in coastal forests. It incubates its eggs in nest mounds close to the shore which are built from sand, loam, and humus. The species is primarily monogamous, although extra-pair copulations have been observed. In a pair, both the male and female contribute to the mound maintenance. The key threat is the loss of coastal forest through conversion to agriculture (coconut, banana, and cashew plantations, and rice-paddy cultivation), road development projects, which threaten to fragment habitat blocks, particularly on Great Nicobar, and settlement expansion. Megapode builds a large mound nest with soil and vegetation, with the eggs hatched by the heat produced by decomposition, and it is also called "thermometer birds."

1.13.3 Nicobar Tree Shrew *Tupaia nicobarica* (Zelebor, 1869)

Tree shrews are a group of tropical small mammals found in South and Southeast Asia. Tree shrews have been previously classified in different orders which include Primates and Insectivora. They are considered by some to resemble primitive mammals. Currently they are classified under the order Scandentia and belong to the family Tupaiidae. Nineteen species of tree shrews were distributed under five genera (Anderson and Jones 1984). The Nicobar tree shrew is a small tupaiid and found only on two islands (Great Nicobar and Little Nicobar islands) in the Sundaland hotspots.

1.13.4 Nicobar Long-Tailed Macaque *Macaca fascicularis umbrosa* (Miller, 1902)

India is well known for its rich primate diversity with as many as 21 living primate species. The highest primate diversity in India is localized toward the northeastern states of India, where as many as ten species occur in sympatry. A critically endangered species of Nicobar long-tailed macaque *(Macaca fascicularis umbrosa)* inhabits the Nicobar Islands. Their preferred habitats are mangroves, coastal forests, and riverine; it is also found in inland forest at altitude of up to 600 m above sea level. The Nicobar long-tailed macaque is one of the endangered primates in India. This species has been listed in Schedule I of Wildlife Protection Act 1972. This species occurs only in Nicobar Islands. According to the IUCN, their status is *near threatened*, having been amended in 2004 from the taxon's previous status as *data deficient* following some basic surveys. This island faced maximum ecological damage during the December 2004 tsunami. The vegetation structure in this island except that one of two major food, *Pandanus*, has become rare due to flooding of coastal area and flushing of seawater in river beds. But the other fruits such as coconut, banana, etc. are available and abundant in unguarded agricultural fields. According to Velankar et al. (2016), the population of Nicobar long-tailed macaque has been recovered from the severe decline caused by tsunami 2004. The threats reported by Umapathy et al. (2003) still exist such as domestic dogs escape the tsunami disaster. The main causes of habitat destruction of long-tailed macaques were found to be the construction of new settlements for rehabilitation of local people and the new road alignment from Campbell Bay to India Point. In addition, if the proposed marine jetty at Galathea Bay is materialized, the existing population from Galathea Bay to India Point will face serious threats.

Nicobar long-tailed macaques also face pressure in Great Nicobar Islands, due to habitat loss and other anthropogenic pressure like other primates in the world. The local tribes and the settlers subsist on coconut, and they have converted coastal areas near their villages into coconut, banana, and tuber-bearing plants. *Pandanus* fruit is the staple diet of long-tailed macaques and most of the places this habitat has been destructed due to tsunami. In addition, the existing habitats were also converted for construction of new settlements, roads, and development of other infrastructures. In view of the fact, an intensive long-term research study is needed for quantitative information on the status, distribution, demographic, and habitat of this species to develop appropriate conservation and management plan.

1.13.5 Leatherback Turtle *Dermochelys coriacea*

Four species of marine turtles found in Andaman, namely, leatherback *Dermochelys coriacea*, hawksbill *Eretmochelys imbricata*, green *Chelonia mydas*, and olive ridley *Lepidochelys olivacea* turtles. The survey shows that the Andaman and Nicobar Islands have the largest nesting populations of leatherback, hawksbill, and green turtles (Andrews et al. 2006; Bhaskar 1979a, b, 1993; Kar and Bhaskar 1982; Fatima et al. 2011). The leatherback nesting population in the Nicobar Islands is the largest in South Asian region (Andrews and Shanker 2002; Fatima et al. 2011). Leatherback turtle is the largest turtle species, crossing both the Atlantic and Pacific Oceans during their migration. Pacific leatherbacks migrate from nesting beaches in the Coral Triangle to the California coast to feed on the abundant jellyfish every summer and fall. Also they are distributed in small group in British Columbia, Newfoundland, and the British Isles, and Australia, Cape of Good Hope, and Argentina. The leatherback turtle is known to have wide nesting distribution in the tropical Atlantic and Pacific oceans, especially in the mainland shores. In the Indian Ocean, they are nesting in South Africa, Andaman and Nicobar Islands, Sri Lanka, and Malaysia. The highest number of nesting was reported from Great Nicobar Island followed by Middle Andaman and South Andaman (Andrews and Shanker 2002; Bhaskar 1993; Andrews et al. 2006).

1.14 Major Threats to the Biodiversity

Like other ecosystems, the mountains obtain the major negative impact due to various unplanned developmental activities such as construction of roads and degradation which result in landslides and erosion. The mountain ecosystems in the Himalayas and Western Ghats have been considered fragile and have attracted special attention. The grassland ecosystem is one of the highly threatened ecosystems in India. This ecosystem is also under severe pressures from grazing, fire, pollution, development project, conversion for agriculture, and plantations. The other habitats like lakes, marshes, and river system are threatened due to pollution, sewage, and toxic effluents. The mangroves are also facing threats due to their reclamation for urban development, waste disposal, oil spillage, etc. Coral reef ecosystems are threatened because of mining, blasting, dredging, collection of reef biota, coastal clearance for development, sewage disposal, discharge of effluents from industries and thermal power plants, chemical pollution, and oil spillage. The desert of Western India, which is one of the high-density populated deserts in the world with more livestock, is under heavy biotic pressure.

1.15 Threats to Species and Genetic Diversity

India is also facing threats to the species and genetic diversity like other parts of the world, and these threats are directly affecting the ecosystem. Other major facts are habitat destruction, over exploitation, floods, droughts, and cyclones. In the past century, the following species are reported to have become extinct, e.g., the Indian cheetah, the lesser Indian rhino, the pink-headed duck, the forest owlet, and the Himalayan mountain quail. Among the reported animal species in India, 32 mammal species, 11 bird species, 17 reptiles, 3 amphibians, 4 fishes, and good number of butterflies, moths, and beetles are listed under various threatened categories of IUCN (Ramakrishna and Alfred 2007).

1.16 Government of India Efforts for Conservation of Biodiversity

1.16.1 Protected Areas

A total of 103 national parks covering 39,155 km^2, 537 wildlife sanctuaries, 67 conservation reserves, and 26 community reserves were found in India with extent of 160,901.77 km^2 (Table 1.15).

1.16.2 Project Tiger

The Government of India launched the Project Tiger during the year 1973 in order to conserve their natural habitats. Initially, nine reserves were designated during 1973–1974; however at present, the number of tiger reserves is raised up to 50 with extent of 71,027.10 (Table 1.16).

Table 1.15 National parks and wildlife sanctuaries in different biogeographic zones (km^2)

Sl. no.	Name of the biogeographic zone	National parks	Extent of area	Wildlife sanctuaries	Extent of area
1	Trans-Himalaya	3	5809.00	4	10,438.56
2	Himalaya	12	7366.92	65	16,065.85
3	Desert	1	3162.00	5	12,914.09
4	Semiarid	10	1505.78	81	12,410.66
5	Western Ghats	16	3673.52	47	10,018.86
6	Deccan Peninsula	24	9712.24	127	44,329.08
7	Gangetic Plain	6	2363.62	32	5473.24
8	Coasts	5	1731.18	20	2959.45
9	North East	13	2674.00	36	3428.62
10	Island	9	1156.91	96	389.39
Grand total		99	39,155.00	513	118,417.00

Source: wiienvis.nic.in

Table 1.16 List of tiger reserves in India

Sl. no.	Name of the tiger reserve	States	Core area (km²)	Butter area (km²)	Total (km²)
1	Bandipur	Karnataka	872.24	584.06	1456.3
2	Corbett	Uttarakhand	821.99	466.32	1288.31
	Amangarh (buffer of Corbett TR)	Uttar Pradesh		80.60	80.60
3	Kanha	Madhya Pradesh	917.43	1134.361	2051.791
4	Manas	Assam	840.04	2310.88	3150.92
5	Melghat	Maharashtra	1500.49	1268.03	2768.52
6	Palamau	Jharkhand	414.08	715.85	1129.93
7	Ranthambore	Rajasthan	1113.364	297.9265	1411.291
8	Simlipal	Odisha	1194.75	1555.25	2750.00
9	Sunderbans	West Bengal	1699.62	885.27	2584.89
10	Periyar	Kerala	881.00	44.00	925.00
11	Sariska	Rajasthan	881.1124	332.23	1213.342
12	Buxa	West Bengal	390.5813	367.3225	757.9038
13	Indravati	Chhattisgarh	1258.37	1540.70	2799.07
14	Namdapha	Arunachal Pradesh	1807.82	245.00	2052.82
15	Dudhwa	Uttar Pradesh	1093.79	1107.9848	2201.7748
16	Kalakkad-Mundanthurai	Tamil Nadu	895.00	706.542	1601.542
17	Valmiki	Bihar	598.45	300.93	899.38
18	Pench	Madhya Pradesh	411.33	768.30225	1179.63225
19	Tadoba-Andhari	Maharashtra	625.82	1101.7711	1727.5911
20	Bandhavgarh	Madhya Pradesh	716.903	820.03509	1598.10
21	Panna	Madhya Pradesh	576.13	1021.97	1578.55
22	Dampa	Mizoram	500.00	488.00	988.00
23	Bhadra	Karnataka	492.46	571.83	1064.29
24	Pench	Maharashtra	257.26	483.96	741.22
25	Pakke	Arunachal Pradesh	683.45	515.00	1198.45
26	Nameri	Assam	200.00	144.00	344.00
27	Satpura	Madhya Pradesh	1339.264	794.04397	2133.30797
28	Anamalai	Tamil Nadu	958.59	521.28	1479.87
29	Udanti-Sitanadi	Chhattisgarh	851.09	991.45	1842.54
30	Satkosia	Odisha	523.61	440.26	963.87
31	Kaziranga	Assam	625.58	548.00	1173.58
32	Achanakmar	Chhattisgarh	626.195	287.822	914.017

(continued)

Table 1.16 (continued)

Sl. no.	Name of the tiger reserve	States	Core area (km²)	Butter area (km²)	Total (km²)
33	Dandeli-Anshi	Karnataka	814.884	282.63	1097.514
34	Sanjay-Dubri	Madhya Pradesh	812.571	861.931	1674.502
35	Mudumalai	Tamil Nadu	321.00	367.59	688.59
36	Nagarahole	Karnataka	643.35	562.41	1205.76
37	Parambikulam	Kerala	390.89	252.772	643.662
38	Sahyadri	Maharashtra	600.12	565.45	1165.57
39	Biligiri Ranganatha Temple	Karnataka	359.10	215.72	574.82
40	Kawal	Telangana	893.23	1125.89	2019.12
41	Sathyamangalam	Tamil Nadu	793.49	614.91	1408.40
42	Mukundra Hills	Rajasthan	417.17	342.82	759.99
43	Navegaon-Nagzira	Maharashtra	653.674	–	653.674
44	Nagarjunsagar-Srisailam (part)	Andhra Pradesh	2595.72	700.59	3296.31
45	Amrabad	Telangana	2166.37	445.02	2611.39
46	Pilibhit	Uttar Pradesh	602.7980	127.4518	730.2498
47	Bor	Maharashtra	138.12	–	138.12
48	Rajaji	Uttarakhand	819.54	255.63	1075.17
49	Orang	Assam	79.28	413.18	492.46
50	Kamlang	Arunachal Pradesh	671.00	112.00	783.00
Total			40,340.12	30,686.98	71,027.10

Source: wiienvis.nic.in

1.16.3 Project Elephant

The Indian elephant *Elephas maximus* occurs in the 16 states of the country, and their population is showing an increasing trend in different distributional ranges. The Indian elephant is listed in the Schedule I of the Indian Wildlife (Protection) Act, 1972, and Appendix I of the Convention on International Trade in Endangered Species of Flora and Fauna (CITES). The Project Elephant was launched by the Government of India during the year 1992 under the Centrally Sponsored Scheme (CSS) in 16 different states and union territories, namely, Andhra Pradesh, Arunachal Pradesh, Assam, Chhattisgarh, Jharkhand, Karnataka, Kerala, Maharashtra, Meghalaya, Nagaland, Orissa, Tamil Nadu, Tripura, Uttarakhand, Uttar Pradesh, and West Bengal. The Ministry of Environment, Forest, and Climate Change provides the financial and technical support to major elephant range states in the country through Project Elephant. There are 32 elephant reserves in India (Table 1.17).

Table 1.17 List of elephant reserves in India

Sl. no.	Elephant reserve (ER)	Elephant range	Name of state	Date of notification	Total area (Km^2)
1	Mayurjharna		West Bengal	24/10/2002	414
2	Singhbhum	East-Central	Jharkhand	26/09/2001	4530
3	Mayurbhanj	Landscape	Orissa	29/09/2001	3214
4	Mahanadi	(Southwest Bengal-Jharkhand – Orissa)	Orissa	20/07/2002	1038
5	Sambalpur		Orissa	27/03/2002	427
6	Baitami		Orissa		1755
7	South Orissa		Orissa		4216
8	Lemru ER		Chhattisgarh		450
9	Badalkhol-Tamor Pingla		Chhattisgarh		1048.3
				Total	**17,092.3**
10	Kameng	Kameng-Sonitpur Landscape	Arunachal	19/06/2003	1892
11	Sonitpur	Arunachal – Assam	Assam	06/03/2003	1420
				Total	**3312**
12	Dehing-Patkai	Eastern-South Bank Landscape	Assam	17/04/2003	937
13	South Arunachal	Assam-Arunachal	Arunachal Pradesh	29/02/2008	1957.5
				Total	**2894.5**
14	Kaziranga-Karbi Anglong	Kaziranga-Karbi	Assam	17/04/2003	3270
15	Dhansiri-Lungding	Anglong-Intanki Landscape	Assam	19/04/2003	2740
16	Intanki	Assam-Nagaland	Nagaland	28/02/2005	202
				Total	**6212**
17	Chirang-Ripu	North Bengal- Greater Manas Landscape	Assam	07/03/2003	2600
18	Eastern Dooars	Assam-West Bengal	West Bengal	28/08/2002	978
				Total	**3578**
19	Garo Hills	Meghalaya Landscape	Meghalaya	31/10/2001	3500
20	Khasi Hills	Meghalaya	Meghalaya		1331
				Total	**4831**
21	Mysore	Brahmagiri-Nilgiri-Eastern	Karnataka	25/11/2002	6724
22	Wayanad	Ghat Landscape	Kerala	02/04/2002	1200
23	Nilgiri	Karnataka-Kerala	Tamil Nadu	19/09/2003	4663
24	Rayala	Tamil Nadu- Andhra	Andhra Pradesh	09/12/2003	766
25	Nilambur		Kerala	02/04/2002	1419
26	Coimbatore		Tamil Nadu	19/09/2003	566
				Total	**15,335**
27	Anamalai	Anamalai-Nelliyampathy – High-Range Landscape	Tamil Nadu	19/09/2003	1457

(continued)

Table 1.17 (continued)

Sl. no.	Elephant reserve (ER)	Elephant range	Name of state	Date of notification	Total area (Km^2)
28	Anamudi	Tamil Nadu-Kerala	Kerala	02/04/2002	3728
				Total	**5185**
29	Periyar	Periyar-Agasthyamalai Landscape	Kerala	02/04/2002	3742
30	Srivilliputhur	Kerala-Tamil Nadu	Tamil Nadu	19/09/2003	1249
				Total	**4991**
31	Shivalik	Northwestern Landscape	Uttarakhand	28/10/2002	5405
32	Uttar Pradesh	Uttarakhand-Uttar Pradesh	Uttar Pradesh	09/09/2009	744
				Total	**6149**
				Grand total	**69,582.80**

Source: wiienvis.nic.in

1.17 Biosphere Reserves of India

The Man and the Biosphere Programme (MAB) was initiated by the UNESCO in the year 1971, and the purpose of the formation is to conserve the biodiversity under in situ program. A total of 651 biosphere reserves were designated in 120 countries, and 18 biosphere reserves exist in India (Table 1.18).

1.18 Marine Protected Areas in India

At present 27 coastal and marine protected areas are found in India. Among these, 18 areas are offshore or away from the Indian mainland, which is protecting or conserving exclusively marine life forms, and these protected areas are considered as exclusive marine protected areas (Tables 1.19 and 1.20).

1.19 The Ramsar Convention

The Ramsar Convention is an international treaty for the conservation and sustainable utilization of wetlands, recognizing the fundamental ecological functions of wetlands and their economic, cultural, scientific, and recreational value. There are over 2000 Ramsar sites on the territories of over 160 Ramsar Contracting Parties across the world. In India the Ramsar Convention came into force in 1975, and 26 wetlands were identified and designated with extent of 12,119 km^2 (Table 1.21).

Table 1.18 Biosphere reserves in India

Sl. no.	Name of the reserve	Extent of area (km^2)	Date of declaration
1.	Nilgiri	5520	01.08.1986
2.	Nanda Devi	5860.69	18.01.1988
3.	Nokrek	820	01.09.1988
4.	Manas	2837	14.03.1989
5	Sunderbans	9640	29.03.1989
6.	Gulf of Mannar	10,500	18.02.1989
7.	Great Nicobar	885	06.01.1989
8.	Simlipal	4374	21.06.1994
9.	Dibru-Saikhowa	765	28.07.1997
10.	Dehang-Debang	5111.5	02.09.1998
11.	Pachmarhi	4981.72	03.03.1999
12.	Khangchendzonga	2612.92	07.02.2000
13.	Agasthyamalai	3500.36	12.11.2001
14.	Achanakmar-Amarkantak	3835.51	30.03.2005
15.	Kachchh	12,454	29.01.2008
16.	Cold desert	7770	28.08.2009
17.	Seshachalam	4655.997	20.09.2010
18.	Panna	2998.98	25.08.2011

Source: wiienvis.nic.in

Table 1.19 Marine protected area (MPA) in India (Mainland)

States	MPA	Year of establishment	Area (ha)	Conservation importance
Gujarat	Gulf of Kachchh Marine Sanctuary	1980	29,503	Corals, dugongs, turtles
	Gulf of Kachchh Marine National Park	1982	16,289	Corals, mangroves
Maharashtra	Malvan Marine Sanctuary	1987	2912	Corals, mangroves, and marine life
Orissa	Gahirmatha Wildlife Sanctuary	1997	143,500	Turtles, dolphins
Tamil Nadu	Gulf of Mannar Marine National Park	1980	623	Corals, mangroves, seagrass

Source: wiienvis.nic.in

Table 1.20 Marine protected area (MPA) in India (islands)

States	MPA	Year of establishment	Area (ha)	Conservation importance
Lakshadweep	Pitti		1	Birds
Andaman and Nicobar	Mahatma Gandhi Marine National Park	1983	28,150	Corals, dolphins, turtles, and other marine life forms
	Rani Jhansi Marine National Park	1996	25,614	Corals, dolphins, turtles, and other marine life forms
	Cinque Marine Sanctuary	1987	991	Corals and other marine life forms
	North Button National Park	1987	44	Corals, turtles, and other marine life forms
	Middle Button National Park	1987	44	Corals, turtles, and other marine life forms
	South Button National Park	1987	3	Corals,, turtles, and other marine life forms
	Cuthbert Bay Sanctuary	1997	582	Sea turtles
	Galathea Bay Sanctuary	1992	1144	Sea turtles
	North-Reef Marine Sanctuary	1987	348	Corals and other marine life forms
	South-Reef Marine Sanctuary	1987	117	Corals and other marine life forms

Source: wiienvis.nic.in

Table 1.21 List of designated Ramsar sites in India

Sl. no.	States	Name of the site	Date designation	Extent of area (km^2)
1.	Andhra Pradesh	Kolleru Lake	19.8.2002	673
2.	Assam	Deepor Beel	19.8.2002	4.14
3.	Gujarat	Nalsarovar Bird Sanctuary	24/09/12	120
4.	Himachal Pradesh	Chandertal Wetland	8.11.2005	38.56
5.		Pong Dam Lake	19.8.2002	307.29
6.		Renuka Wetland	8.11.2005	*
7.	Jammu and Kashmir	Wular Lake	23.3.1990	173
8.		Hokera Wetland	8.11.2005	13.75
9.		Surinsar-Mansar Lakes	8.11.2005	3.50
10.		Tso Moriri Lake	19.8.2002	120
11.	Kerala	Ashtamudi Wetland	19.8.2002	1860
12.		Sasthamkotta Lake	19.8.2002	11.3
13.		Vembanad-Kol Wetland	19.8.2002	4583
14.	Madhya Pradesh	Bhoj Wetlands	19.8.2002	31

(continued)

Table 1.21 (continued)

Sl. no.	States	Name of the site	Date designation	Extent of area (km²)
15.	Manipur	Loktak Lake	23.3.1990	945
16.	Orissa	Bhitarkanika Mangroves	19.8.2002	525
17.		Chilka Lake	1.10.1981	1140
18.	Punjab	Harike Lake	23.3.1990	86
19.		Kanjli Lake	22.1.2002	14.84
20.		Ropar Lake	22.1.2002	41.36
21.	Rajasthan	Keoladeo Ghana NP	1.10.1981	28.73
22.		Sambhar Lake	23.3.1990	736
23.	Tamil Nadu	Point Calimere	19.8.2002	17.26
24.	Tripura	Rudrasagar Lake	8.11.2005	2.40
25.	Uttar Pradesh	Upper Ganga River (Brijghat to Narora stretch)	8.11.2005	265.90
26.	West Bengal	East Calcutta Wetlands	19.8.2002	378
		Total		**12,119.03**

Source: wiienvis.nic.in
* = 0.2

1.20 Species Recovery Program for Saving Critically Endangered Animals

The country's flagship and charismatic species face a variety of threats, ranging from habitat destruction and illegal wildlife trade to reduction in forest cover outside protected areas. Significant populations of these species exist outside protected areas moving for dispersal from their natal habitats or for seasonal migrations. The Government of India has identified 16 terrestrial and 7 aquatic critically endangered species/ecosystems in order to provide protection outside protected areas in different habitat and landscapes. The objective of this program was saving the critically endangered species in their native habitat. The following programs are initially proposed under this component, namely, Asian wild buffalo, Asiatic lion, brow-antlered deer or sangai, dugong, edible-nest swiftlet, Gangetic river dolphin, great Indian bustard, Hangul, Indian rhino or great one-horned rhinoceros, Jerdon's courser, Malabar civet, marine turtles, Nicobar megapode, Nilgiri tahr, snow leopard, swamp deer, and vultures.

1.21 Legislative and Policy Framework

The Central and State Government legislates and formulates policies and program on the subject. At present, the major Central Acts having direct bearing on biodiversity issues are the following:

- The Indian Forest Act, 1942
- The Forest (Conservation) Act, 1980

- The Wildlife (Protection) Act, 1972
- The Environment (Protection) Act, 1986

The National Forest Policy, as amended in 1988, stresses the sustainable use of forests and the need for greater attention to ecologically fragile but biologically rich mountain and island ecosystem. The National Wildlife Action Plan (1973) identified broad goals of establishing a network of representative protected areas and developing appropriate management systems. One of the major considerations in the environment impact assessment of development project carried out by the Ministry of Environment and Forests is the protection of habitat and valuable ecosystem. The National Afforestation and Eco-Development Board of the ministry undertakes large-scale rehabilitation of degraded forest lands in the country. India is an active participant in the following International Conventions and agreement relevant to biodiversity: the Convention on International Trade in Wildlife Species of Endangered Fauna and Flora (CITES), the Ramsar Convention on Wetlands of International Importance especially as Water Fowl Habitat, the World Heritage Convention, the Bonn Convention on Conservation of Migratory Species of Wild Animals, the FAO commission on Plant Genetic Resources, and the UN Law of the Seas (UNCLOS).

1.22 National Biodiversity Authority

India is a party to the Convention on Biological Diversity (CBD) 1992 which recognizes the sovereign rights of states to use their own biological resources. In order to help in realizing the objectives of CBD, India has enacted an umbrella legislation called the Biological Diversity Act, 2002 (No.18 of 2003) aimed at conservation of biological resources and associated knowledge as well as facilitating access to them in a sustainable manner. In the exercise of the powers conferred by Subsection (1) (4) of Section 8 of the Biological Diversity Act, 2002 (18 of 2003), the Central Government has established a statutory body called the National Biodiversity Authority, on and from the 1st day of October, 2003. The main functions of the Authority are:

1. To lay down procedures and guidelines to govern the activities provided under Section 3, 4, and 6 (permission to foreigners/NRI's foreign companies)
2. Regulate activities, approve, and advice the government of India on research, commercial, bio-survey, and bio-utilization
3. Grant approval to Section 3, 4, and 6
   - Certain persons not to undertake biodiversity-related activities without approval of National Biodiversity Authority (Section 3) (access to biological resources or associated knowledge)
   - Results of research not to be transferred to certain persons without approval of National Biodiversity Authority (Section 4) (Transfer of Research Results)
   - Application of IPR rights not to be made without approval of National Biodiversity Authority (Section 6) (Seeking IPR)

4. Certain persons not to transfer of biological resource or knowledge without approval of National Biodiversity Authority (Section 20) (Third Party Transfer)
5. Determination of equitable benefit sharing arising out of the use of accessed biological resources (Section 21)

1.23 Other Important Central Acts Relevant to Biodiversity

- Fisheries Act, 1987
- Destructive Insect and Pest Act, 1914
- Indian Coffee Act, 1942
- Agricultural Produce (Grading and Marketing) Act, 1937
- Import and Export (Control) Act, 1947
- Rubber (Production and Marketing) Act, 1947
- Tea Act, 1953
- Prevention of Cruelty to Animals Act, 1960
- Customs Act, 1962
- Cardamom Act, 1965
- Seeds Act, 1966
- Marine Products Exports Development Authority Act, 1972
- Water (prevention and Control of Pollution) Act, 1974
- Tobacco Board Act, 1975
- Territorial Water, Continental Shelf, Exclusive Economic Zone, and Other Maritime Zones Act, 1976
- Water (Prevention and Control of Pollution) Cess Act, 1977
- Coconut Development Board Act, 1979
- Maritime Zones of India (Regulation and Fishing by Foreign Vessels) Act, 1980
- Air (Prevention and Control of Pollution) Act, 1981
- National Oilseeds and Vegetable Oils Development Board, 1983
- Agricultural and Processed Food Products Export Development Authority Act, 1985/1986
- Spices Board Act, 1986
- National Dairy Development Board Act, 1987
- New Seed Development Policy, 1988
- Foreign Trade (Development and Regulation) Act, 1992

References

Alcock A (1905) Catalogue of the Indian decapod crustacea (Pt 2, Fasc. I). Calcutta

Altevogt R, Davis TA (1975) *Birgus latro,* India's monstrous crab: a study and an appeal. Bull Dept Mar Sci Univ Cochin 7:11–24

Anderson S, Jones JK Jr (1984) Orders and families of recent mammals of the world. Wiley, New York, xii +686 pp

Andrews HV, Shanker K (2002) A significant population of leatherback turtles in the Indian ocean. Kachhapa 6:17

Andrews HV, Krishnan S, Biswas P (2006) Distribution and status of marine turtles in the Andaman and Nicobar Islands. In: Shanker K, Choudhury BC (eds) Marine turtles of the Indian subcontinent. University Press, Hyderabad, pp 33–57

Anon. (2015) Annual report (2014–2015), ministry of environment and forests. Government of India, New Delhi

Bhaskar S (1979a) Sea turtles in the south Andaman Islands. Hamadryad 4:3–5

Bhaskar S (1979b) Sea turtle survey in the Andaman and Nicobars. Hamadryad 4:2–26

Bhaskar S (1993) The status and ecology of sea turtles in the Andaman and Nicobar Islands. ST 1/93. Centre for Herpetology, Madras Crocodile Bank Trust, Post Bag 4, Mamallapuram, Tamil Nadu- 603104, India

Bhaskar S, Rao GC (1992) Present status of some endangered animals in Nicobar Islands. J Andaman Sci Assoc 8:181–186

Champion HG, Seth SK (1968) A revised survey of the forest types of India. Govt. of India Press, Nasik

Chapman AD (2009) Numbers of living species in Australia and the world, 2nd edn. Australian Government, Department of the Environment, Water, Heritage and the Arts, Australia

Davis SD, Heywood VH, Hamilton AC (1995) Centers of plant diversity. A guide and strategy for their conservation, Vol. 2: Asia, Australasia and the Pacific. WWF/IUCN-The World Conservation Union. IUCN Publications Unit, Cambridge

Eldredge LG (1996) *Birgus latro*. The IUCN red list of threatened species 1996: e.T2811A9484078.. Downloaded on 22 June 2016

Fatima E, Andrews H, John S, Shanker K (2011) Status of marine turtles in Cuthbert Bay, middle Andaman Islands. Mar Turt Newsl 130:6–9

Fletcher WJ, Amos M (1994) Stock assessment of coconut crabs. AICAR monograph No.29, 32 p

FSI (2011) Forest Survey of India (MInistry of Environment & Forests) Kaulagarh Road, P.O -IPE Dehradun, 286 p

Greenaway P (2003) Terrestrial adaptations in the Anomura (Crustacea: Decapoda). Mem Mus Vic 60(1):13–26

Grubb P (1971) The growth, ecology and population structure of giant tortoises on Aldabra. Phil Trans R Soc Lond B 260:327–372

Hartnoll RG (1988) Biology of the land crabs: an introduction. In: WW Burggren, BR McMahon (eds) Evolution, systematics, and geographical distribution. Cambridge University Press, pp 6–54

Hume AO (1874) The islands of the Bay of Bengal. Stray Feathers 2:91

Kar CS, Bhaskar S (1982) Status of sea turtles in the Eastern Indian ocean. In: Bjorndal KA (ed) Biology and conservation of sea turtles. Smithsonian Institution Press, Washington, DC, pp 365–372

Lavery S, Moritz C, Fielder DR (1996a) Indo-Pacific population structure and evolutionary history of the coconut crab *Birgus latro*. Mol Ecol 5:557–570

Lavery S, Moritz C, Fielder DR (1996b) Indo-Pacific population structure and evolutionary history of the coconut crab *Birgus latro*. Mol Ecol 5(4):557–570

Mani MS (1974) Ecology and biogeography in India. Dr. W. Junk Publishers, The Hague, p 773

Miller GS (1902) The mammals of the Andaman and Nicobar Islands. Proc U S Nat Mus 24(7S):1–795

Mittermeier RA, Myers N, Robles-Gil P, Mittermeier CG (1999) Hotspots: Earth's biologically richest and most endangered terrestrial ecoregions. CEMEX/Agrupación Sierra Madre, Mexico City

Mittermeier RA, Turner WR, Larsen FW, Brooks TM, Gascon C (2011) Global biodiversity conservation: the critical role of hotspots. In: Zachos FE, Habel JC (eds) Biodiversity hotspots. Springer Publishers, London, pp 3–22

Myers N (1988) Threatened biotas: "hotspots" in tropical forests. Environmentalist 8:1–20

Myers N (1990) The biodiversity challenge: expanded hotspots analysis. Environmentalist 10:243–256

Myers N, Mittermeier RA, Mittermeier CG, da Fonseca GAB, Kent J (2000) Biodiversity hotspots for conservation priorities. Nature 403:853–858

Nair SC (1991) The southern Western Ghats: a biodiversity conservation plan. INTACH, New Delhi, 92 p

Powers LW, Bliss DE (1983) Terrestrial adaptations. In: Vernberg FJ, Vernberg WB (eds) The biology of crustacean. Environmental Adaptations, vol 8. Academic, New York, pp 272–333

Ramachandran S, Anitha S, Balamurugan V, Dharanirajan K, Vendhan EK, Divien MIP (2005) Ecological impact of tsunami on Nicobar Islands(Camorta, Katchal, Nancowry and Trinkat). Curr Sci 89:195–200

Ramakrishna, Alfred JRB (2007) Faunal resources in India. Zoological Survey of India, Kolkata

Ravindranath NH, Sukumar R, Joshi NV, Murthy IK (2006) Impact of climate changes on biodiversity in India and adaptation. DD Verma, S Arora, RK Rai (eds). Ministry of Environments and Forests, Government of India, pp 21–40

Reyne A (1939) On the food habits of the coconut crab (*Birgus latro* L.), with notes on its distribution. *Arch. Neerland.* De Zool 3:283–320

Robertson M (1991) Husbandry and moulting behavior of the robber or coconut crab *Birgus latro*: at London zoo. Int Zoo Year Book 30(1):60–67

Singh P, Dash SS (2014) Plant discoveries 2013 – new genera, species and new records. Botanical Survey of India, Kolkata

Umapathy G, Singh M, Mohnot SM (2003) Status and distribution of *Macaca fascicularis umbrosa* in the Nicobar Islands, India. Int J Primatol 24(2):281–293

van Steenis CGGJ (1950) The delimitation of Malesia and its main plant geographical divisions. Flora Malesiana I(1):lxx–lxxv

Velankar AD, Kumara HN, Pal A, Mishra PS, Sing M (2016) Population recovery of Nicobar long-tailed macaque Macaca fascicularis umbrosus following a tsunami in the Nicobar Islands, India. PLoS One 11(2):e0148205. https://doi.org/10.1371/journal.pone.0148205

Venkataraman K (2006) Biodiversity legislations in likeminded mega diversity countries (eds) DD Verma, S Arora, RK Rai. Ministry of Environment and Forests, pp 79–92

Whitmore TC (ed) (1984) Tropical rain forests of the far East, 2nd edn. Oxford University Press, Oxford

WII (2009) India's green book: forests and wildlife. Wildlife Institute of India, Dehra Dun. wiienvis.nic.in – Wildlife Institute of India, ENVIS Centre on Wildlife and Protected Areaswiienvis.nic.in/Database/br_8225.aspx

Williams KJ, Ford A, Rosauer DF, De Silva N, Mittermeier R, Bruce C, Larsen FW, Margules C (2011) Forests of east Australia: the 35th biodiversity hotspot. In: Zachos FE, Habel JC (eds) Biodiversity hotspots. Springer Publishers, London, pp 295–310

ZSI (2016). Animal discoveries of 2015 - New species and new records. Zoological Survey of India, Kolkata. 152 p

ZSI (2016) Animal discoveries of 2015 – new species and new records. Zoological Survey of India, Kolkata, p 152

Endemic, Endangered and Threatened Vertebrates in the Indian Hotspots

2

Chandrakasan Sivaperuman, Krishnamoorthy Venkataraman, and Anil Kumar

Abstract

In this chapter we presented the status and distribution of the endemic, threatened vertebrate fauna of the four Indian hotpots. Also detailed information on the threatened animals is provided.

Keywords

Faunal diversity · Endemism · Hotspots · India

2.1 Introduction

The Convention of Biological Diversity (CBD) defines biodiversity as "the variability among living organisms from which all sources including inter alia, terrestrial, marine and other aquatic ecosystems and the complexes of which they are part; this includes diversity within species, between species and of ecosystems themselves". Endemism is rich in the biodiversity hotspots especially in the terrestrial ecosystem. Total of 35 biodiversity hotspots were identified, and each of these hotspots has already lost at least 70 % of its original natural vegetation. These hotspots support

C. Sivaperuman (✉)
Zoological Survey of India, Andaman & Nicobar Regional Centre,
Port Blair, Andaman & Nicobar Island, India
e-mail: c_sivaperuman1@rediffmail.com

K. Venkataraman
National Centre for Sustainable Coastal Management, Anna University Campus,
Chennai, Tamil Nadu, India

A. Kumar
Northern Regional Centre, Zoological Survey of India, Dehra Dun, Uttarakhand, India

C. Sivaperuman, K. Venkataraman (eds.), *Indian Hotspots*,
https://doi.org/10.1007/978-981-10-6605-4_2

more than 50% of the plant species and 42% endemic terrestrial vertebrate species (Mittermeier et al. 2005). India is also one among the 17 Like-Minded Megadiverse Countries of the world. Among the 35 hotspots identified, four hotspots are extended within Indian subcontinent. These are Himalayan, Indo-Burma, Western Ghats-Sri Lanka and Sundaland. India is home to 7.6% of all mammals, 12.6% of avifauna, 6.2% of reptiles, 4.4% of amphibians, 11.7% of fishes and 6.0% of flowering plant species. These hotspots are also home to many large mammals, e.g. the Asian elephant, the Indian rhinoceros, the Royal Bengal tiger and the endangered Asiatic lion. In this chapter, we made an attempt to present endemic, endangered and threatened vertebrate fauna of Indian hotspots based on the fieldwork carried out by the authors and available literature.

2.2 Endemism

The term "endemism" was coined by A. P. de Candolle (1855) for the distribution of an organism in a limited geographical area. Endemism is defined as the species being unique to a distinct geographic location, e.g. an island, nation, country or other defined zone, or habitat type; organisms that are indigenous to a place are not endemic to it if they are also found elsewhere. The cosmopolitan distribution is opposite to the endemism.

2.3 Threatened Species

A particular species may become threatened due to various human interventions or natural incident; the former is affecting many more species. The introductions of exotic species, habitat loss and over-exploitation are the major anthropogenic threats to the biodiversity. These species are at significant risk of extinction due to the small size of remaining populations. The most threatened species are, therefore, of the highest importance for the conservation of biodiversity.

2.4 Distribution of Amphibians of India

Three hundred and eighty four species of amphibians were reported from India under 3 orders and 15 families (Table 2.1). More number of species of amphibians were reported from Rhacophoridae (109 species) followed by Dicroglossidae (50 species) and Ranidae (39 species).

2.5 Distribution of Reptiles of India

A total of 555 species or subspecies of reptiles were reported from India; these include crocodiles (3 species), turtles and tortoises (34 species), lizards (202 species) and snakes (316 species) (Table 2.2).

Table 2.1 Order- and family-wise diversity of amphibians of India

Order	Family	Number of species
Anura	Bufonidae	28
	Dicroglossidae	60
	Hylidae	1
	Megophryidae	22
	Micrixalidae	23
	Microhylidae	23
	Nasikabatrachidae	1
	Nyctibatrachidae	27
	Ranidae	39
	Ranixalidae	11
	Rhacophoridae	109
Caudata	Salamandridae	1
Gymnophiona	Chikilidae	4
	Ichthyophiidae	22
	Indotyphlidae	13
	Total	**384**

Source: Dinesh et al. (2015)

Table 2.2 Order- and family-wise diversity of reptiles in India

Order	Family	Number of species
Crocodylia	Crocodylidae	2
	Gavialidae	1
Testudines	Bataguridae	17
	Cheloniidae	4
	Dermochelyidae	1
	Testudinidae	4
	Trionychidae	8
Squamata	Agamidae	47
	Anguidae	1
	Chamaeleonidae	1
	Dibamidae	1
	Eublepharidae	3
	Gekkonidae	77
	Lacertidae	10
	Scincidae	57
	Uromastycidae	1
	Varanidae	4
	Typhlopidae	18
	Leptotyphlopidae	2
	Uropeltidae	39
	Xenopeltidae	1
	Pythonidae	4
	Boidae	5
	Acrochordidae	1

(continued)

Table 2.2 (continued)

Order	Family	Number of species
	Colubridae	175
	Elapidae	22
	Hydrophiidae	23
	Viperidae	26
	Total	**555**

Source: Aengals et al. (2011)

2.6 Birds of India

One thousand three hundred and eight five species of birds have been reported from India (Table 2.3). Highest number of species recorded from the order Passeriformes (746 species), followed by Charadriiformes (126 species) and Accipitriformes (82 species).

2.7 Mammals of India

Four hundred and twenty six species of mammals have been reported from India under 14 orders and 52 families (Table 2.4). Of these, 395 species were terrestrial mammals (belongs to 12 orders, 43 families and 176 genera) and 31 species were marine mammals (belongs to 2 orders, 9 families and 23 genera).

2.8 Few Threatened Animals of India

2.8.1 Jerdon's Courser *Rhinoptilus bitorquatus* (Blyth, 1848)

The Jerdon's courser, found only in the northern part of Andhra Pradesh, is a nocturnal bird. This is a flagship species found in the highly threatened scrub jungle. This species was rediscovered in year 1986 and the area was subsequently declared as the Sri Lankamalleswara Wildlife Sanctuary; however, till 1986 this species was considered under the extinct category.

2.8.2 Forest Owlet *Heteroglaux blewitti* (Hume, 1873)

The forest owlet was rediscovered in the year 1997 after the gap of 113 years and reappeared in the Indian bird list. The very interesting history of the species is that Dr. Salim Ali, the father of Indian ornithology, made a poster and appealed to the public to search out this species when it was not sighted for decades.

Table 2.3 Order- and family-wise diversity of birds in India

Order	Family	Number of species
Gaviiformes	Gaviidae	1
Podicipediformes	Podicipedidae	5
Procellariiformes	Procellariidae	9
	Hydrobatidae	4
Phaethontiformes	Phaethontidae	3
Pelecaniformes	Pelecanidae	3
	Threskiornithidae	4
Suliformes	Sulidae	3
	Phalacrocoracidae	3
	Anhingidae	1
	Fregatidae	3
Ciconiiformes	Ardeidae	20
	Ciconiidae	8
Phoenicopteriformes	Phoenicopteridae	1
Anseriformes	Anatidae	45
Accipitriformes	Accipitridae	67
	Pandionidae	1
	Falconidae	14
Galliformes	Megapodiidae	2
	Phasianidae	46
Gruiformes	Turnicidae	4
	Gruidae	6
	Rallidae	21
	Heliornithidae	1
Otidiformes	Otididae	4
Charadriiformes	Jacanidae	2
	Rostratulidae	1
	Haematopodidae	1
	Charadriidae	19
	Scolopacidae	41
	Ibidorhynchidae	1
	Recurvirostridae	2
	Phalaropidae	2
	Dromadidae	1
	Burhinidae	3
	Glareolidae	6
	Stercorariidae	5
	Laridae	33
	Rynchopidae	1
	Pteroclididae	8
Columbiformes	Columbidae	37
Psittaciformes	Psittacidae	17
Cuculiformes	Cuculidae	25
Strigiformes	Tytonidae	4
	Strigidae	33
Caprimulgiformes	Podargidae	2
	Caprimulgidae	10
Apodiformes	Apodidae	17
	Hemiprocnidae	1

(continued)

Table 2.3 (continued)

Order	Family	Number of species
Trogoniformes	Trogonidae	3
Coraciiformes	Alcedinidae	20
	Meropidae	7
	Coraciidae	4
	Upupidae	1
	Bucerotidae	9
Piciformes	Capitonidae	9
	Indicatoridae	1
	Picidae	34
Passeriformes	Eurylaimidae	2
	Pittidae	5
	Alaudidae	21
	Hirundinidae	17
	Motacillidae	22
	Campephagidae	19
	Pycnonotidae	23
	Irenidae	5
	Lanidae	11
	Bombycillidae	1
	Cinclidae	2
	Troglodytidae	1
	Prunellidae	8
	Turdidae	99
	Timaliidae	120
	Panurinae	10
	Sylviidae	107
	Muscicapidae	36
	Monarchidae	5
	Rhipiduridae	3
	Pachycephalidae	1
	Aegithalidae	4
	Remizidae	2
	Paridae	15
	Sittidae	10
	Certhiidae	5
	Dicaeidae	10
	Nectariniidae	18
	Zosteropidae	2
	Emberizidae	20
	Fringillidae	45
	Estrildidae	10
	Passeridae	15
	Ploceidae	4
	Sturnidae	24
	Oriolidae	7
	Dicruridae	12
	Artamidae	2
	Corvidae	23
	Total	**1385**

Table 2.4 Order- and family-wise diversity of mammals in India

Order	Family	Number of species
Terrestrial mammals		
Proboscidea	Elephantidae	1
Scandentia	Tupaiidae	3
Primates	Lorisidae	2
	Cercopithecidae	21
	Hylobatidae	2
Rodentia	Sciuridae	26
	Dipodidae	1
	Platacanthomyidae	1
	Spalacidae	2
	Cricetidae	14
	Muridae	55
	Hystricidae	3
Lagomorpha	Ochotonidae	7
	Leporidae	4
Erinaceomorpha	Erinaceidae	3
Soricomorpha	Soricidae	30
	Talpidae	2
Chiroptera	Pteropodidae	13
	Rhinolophidae	17
	Hipposideridae	14
	Megadermatidae	2
	Rhinopomatidae	2
	Emballonuridae	6
	Molossidae	4
	Vespertilionidae	57
	Miniopteridae	3
Pholidota	Manidae	2
Carnivora	Canidae	6
	Mustelidae	15
	Ailuridae	1
	Ursidae	4
	Felidae	15
	Prionodontidae	1
	Hyaenidae	1
	Herpestidae	7
	Viverridae	8
Perissodactyla	Equidae	2
	Rhinocerotidae	2
Artiodactyla	Suidae	2
	Tragulidae	1
	Moschidae	4
	Cervidae	8
	Bovidae	21
	Total	**395**

(continued)

Table 2.4 (continued)

Order	Family	Number of species
Aquatic mammals		
Sirenia	Dugongidae	1
Cetacea	Balaenidae	1
	Balaenopteridae	5
	Delphinidae	16
	Phocoenidae	1
	Physeteridae	1
	Kogiidae	2
	Platanistidae	1
	Ziphiidae	3
	Total	**31**

Source: Alfred et al. (2002), Anon. (2008), Srinivasulu and Srinivasult (2012), and Sharma et al. (2013)

2.8.3 White-Bellied Heron *Ardea insignis* (Hume, 1878)

The white-bellied heron is highly restricted-range species and found only in five or six locations in Assam and Arunachal Pradesh and few sites in Bhutan and Myanmar. It is naturally very rare and the populations have never been known to be very high. Recently, the nesting site of this species was discovered in Assam in 2015.

2.8.4 Vultures

Nine species of vultures are found in India; of these 99% of population has declined for the three species, viz. white-backed vulture *Gyps bengalensis*, slender-billed vulture *Gyps tenuirostris* and long-billed vulture *Gyps indicus*. Similarly, the red-headed vulture *Sarcogyps calvus* has also suffered a rapid decline in the recent past. Vultures keep the environment clean, by scavenging on animal carcasses. Efforts are being made to captive breed the vultures by the Bombay Natural History Society in two places in India.

2.8.5 Bengal Florican *Houbaropsis bengalensis* (Gmelin, 1789)

This is one of the grasslands species, and due to habitat loss and other anthropogenic pressures, this is also facing severe threat. This species is known for its mating dance.

2.8.6 Himalayan Quail *Ophrysia superciliosa* (Gray, JE 1846)

The Himalayan quail *Ophrysia superciliosa* is another species considered extinct. However, extensive and intensive surveys are required to find this species.

2.8.7 Pink-Headed Duck *Rhodonessa caryophyllacea* (Latham, 1790)

This species has not been recorded since 1949, and the name of the species has been derived from the deep pink head and neck colour of the males.

2.8.8 Spoon-Billed Sandpiper *Eurynorhynchus pygmeus* (Linnaeus, 1758)

This species is found in the coastal wetlands and requires highly specialized habitat for breeding. This is a winter visitor to India and sighting of the species is very rare.

2.8.9 Siberian Crane *Grus leucogeranus* (Pallas, 1773)

The Siberian crane is a regular winter visitor to Keoladeo National Park, Rajasthan. However, after 2002, this has not been sighted from this region.

2.8.10 Sociable Lapwing *Vanellus gregarius* (Pallas, 1771)

This species is also a winter visitor to India, and the population has suddenly declined due to various anthropogenic activities.

2.8.11 Pygmy Hog *Porcula salvania* (Hodgson, 1847)

This is the smallest wild pig and breeds throughout the year. The adult average weight is about 8.0 kg. The pygmy hog is the indicator species of the management status of grassland habitats. At present, efforts are being made by the Department of Forests, Government of West Bengal, to reintroduce this species in its habitat.

2.8.12 Andaman and Nicobar Shrew

Andaman white-toothed shrew *Crocidura andamanensis* (Miller, 1902); Jenkin's Andaman spiny shrew *Crocidura jenkinsi* (Chakraborty, 1978) and the Nicobar white-tailed shrew *Crocidura nicobarica* (Miller, 1902) are endemic to India. They are usually active by twilight or in the night and have specialized habitat requirements.

2.8.13 Malabar Large-Spotted Civet *Viverra civettina* (Blyth, 1862)

This species is found only in the Western Ghats and considered to be one of the world's rarest species. This is endemic to India and first reported from Travancore, Kerala.

2.9 Endemic Birds Species of India

Seventy four species of birds are endemic to India. Of these, the Himalayan quail, great Indian bustard, Jerdon's courser, forest owlet and Bugun liocichla were listed as critically endangered in the IUCN Red List. The Narcondam hornbill, Banasura laughing thrush, Nilgiri laughing thrush, Nilgiri blue robin and white-bellied blue robin were listed as endangered (Table 2.5).

2.10 Hotspot: Western Ghats

The Western Ghats is also known as Sahyadri hills. This mountain chain is running along the Arabian Sea of the west coast of India from south to north. The Western Ghats extends from the mouth of the Tapti River (21°N) to the tip of southern India (8o N). The Western Ghats is spread over six Indian states, namely, Gujarat, Maharashtra, Goa, Karnataka, Kerala and Tamil Nadu (Pascal 1988). The average elevation is 900–1500 m with highest peak having an altitude of 2969 m. They have different vegetation types such as scrub jungles and grasslands at low altitudes, dry and moist deciduous forests, montane grasslands and shoals and precious tropical evergreen and semievergreen forests. Complex topography, high rain fall and relative inaccessibility have helped the region retain its biodiversity. Of the 15,000 flowering plant species in India, there are an estimated 4780 species in the Western Ghats region. There is also a great diversity of traditional crop plants and an equal diversity of animal life (Myers et al. 2000). The Western Ghats is the second largest endemic centre in India with 1550 endemics out of the estimated 4250 species of vascular plants. The South Western Ghats consisting Southern Karnataka, Kerala and part of Tamil Nadu are considered as the most species-rich region with respect to endemism (Nayar 1996).

Table 2.5 Endemic birds of India

Sl. no	Family	Common name	Scientific name	IUCN status
1.	Phasianidae	Himalayan quail	*Ophrysia superciliosa*	CR
2.	Otididae	Great Indian bustard	*Ardeotis nigriceps*	CR
3.	Glareolidae	Jerdon's courser	*Rhinoptilus bitorquatus*	CR
4.	Strigidae	Forest owlet	*Heteroglaux blewitti*	CR
5.	Leiothrichidae	Bugun liocichla	*Liocichla bugunorum*	CR
6.	Bucerotidae	Narcondam hornbill	*Rhyticeros narcondami*	EN
7.	Leiothrichidae	Banasura laughing thrush	*Trochalopteron jerdoni*	EN
8.	Leiothrichidae	Nilgiri laughing thrush	*Trochalopteron cachinnans*	EN
9.	Muscicapidae	Nilgiri blue robin	*Myiomela major*	EN
10.	Muscicapidae	White-bellied blue robin	*Myiomela albiventris*	EN
11.	Columbidae	Andaman wood pigeon	*Columba palumboides*	NT
12.	Columbidae	Andaman cuckoo-dove	*Macropygia rufipennis*	NT
13.	Columbidae	Andaman green pigeon	*Treron chloropterus*	NT
14.	Rallidae	Andaman crake	*Rallina canningi*	NT
15.	Strigidae	Andaman boobook	*Ninox affinis*	NT
16.	Strigidae	Andaman scops-owl	*Otus balli*	NT
17.	Accipitridae	Great Nicobar serpent eagle	*Spilornis klossi*	NT
18.	Accipitridae	Andaman serpent eagle	*Spilornis elgini*	NT
19.	Picidae	Andaman woodpecker	*Dryocopus hodgei*	NT
20.	Psittacidae	Nicobar parakeet	*Psittacula caniceps*	NT
21.	Campephagidae	Andaman cuckooshrike	*Coracina dobsoni*	NT
22.	Corvidae	Andaman treepie	*Dendrocitta bayleyii*	NT
23.	Pycnonotidae	Nicobar bulbul	*Ixos nicobariensis*	NT
34.	Pycnonotidae	Andaman bulbul	*Brachypodius fuscoflavescens*	NT
25.	Pycnonotidae	Grey-headed bulbul	*Brachypodius priocephalus*	NT
26.	Timaliidae	Naga wren-babbler	*Spelaeornis chocolatinus*	NT
27.	Leiothrichidae	Palani laughing thrush	*Trochalopteron fairbanki*	NT
28.	Muscicapidae	Andaman shama	*Kittacincla albiventris*	NT
29.	Muscicapidae	Nilgiri flycatcher	*Eumyias albicaudatus*	NT
30.	Muscicapidae	Nicobar jungle flycatcher	*Cyornis Nicobaricus*	NT
31.	Muscicapidae	Black-and-orange flycatcher	*Ficedula nigrorufa*	NT
32.	Megapodiidae	Nicobar scrubfowl	*Megapodius nicobariensis*	VU
33.	Anatidae	Andaman teal	*Anas albogularis*	VU
34.	Columbidae	Nilgiri wood pigeon	*Columba elphinstonii*	VU
35.	Accipitridae	Nicobar sparrowhawk	*Accipiter butleri*	VU
36.	Paridae	White-naped tit	*Machlolophus nuchalis*	VU
37.	Locustellidae	Broad-tailed grassbird	*Schoenicola platyurus*	VU
38.	Pycnonotidae	Yellow-throated bulbul	*Pycnonotus xantholaemus*	VU

(continued)

Table 2.5 (continued)

Sl. no	Family	Common name	Scientific name	IUCN status
39.	Timaliidae	Rusty-throated wren-babbler	*Spelaeornis badeigularis*	VU
40.	Timaliidae	Tawny-breasted wren-babbler	*Spelaeornis longicaudatus*	VU
41.	Leiothrichidae	Travancore laughing thrush	*Trochalopteron meridionale*	VU
42.	Estrildidae	Green avadavat	*Amandava formosa*	VU
43.	Motacillidae	Nilgiri pipit	*Anthus nilghiriensis*	VU
44.	Strigidae	Nicobar scops-owl	*Otus alius*	DD
45.	Phasianidae	Red spur fowl	*Galloperdix spadicea*	LC
46.	Phasianidae	Painted spur fowl	*Galloperdix lunulata*	LC
47.	Phasianidae	Rock bush-quail	*Perdicula argoondah*	LC
48.	Phasianidae	Painted bush-quail	*Perdicula erythrorhyncha*	LC
49.	Phasianidae	Grey jungle fowl	*Gallus sonneratii*	LC
50.	Columbidae	Grey-fronted green pigeon	*Treron affinis*	LC
51.	Columbidae	Nilgiri imperial pigeon	*Ducula cuprea*	LC
52.	Caprimulgidae	Andaman nightjar	*Caprimulgus andamanicus*	LC
53.	Strigidae	Hume's boobook	*Ninox obscura*	LC
54.	Strigidae	Mottled wood-owl	*Strix ocellata*	LC
55.	Bucerotidae	Malabar grey hornbill	*Ocyceros griseus*	LC
56.	Megalaimidae	Malabar barbet	*Psilopogon malabaricus*	LC
57.	Megalaimidae	White-cheeked barbet	*Psilopogon viridis*	LC
58.	Campephagidae	White-bellied minivet	*Pericrocotus erythropygius*	LC
59.	Corvidae	White-bellied treepie	*Dendrocitta leucogastra*	LC
60.	Alaudidae	Malabar lark	*Galerida malabarica*	LC
61.	Alaudidae	Sykes's lark	*Galerida deva*	LC
62.	Locustellidae	Himalayan grasshopper warbler	*Locustella kashmirensis*	LC
63.	Pycnonotidae	Flame-throated bulbul	*Pycnonotus gularis*	LC
64.	Timaliidae	Indian scimitar-babbler	*Pomatorhinus horsfieldii*	LC
65.	Leiothrichidae	Rufous babbler	*Argya subrufa*	LC
66.	Leiothrichidae	Black-winged babbler	*Turdoides somervillei*	LC
67.	Leiothrichidae	Wynaad laughing thrush	*Garrulax delesserti*	LC
68.	Sittidae	Indian spotted creeper	*Salpornis spilonota*	LC
69.	Sturnidae	White-headed starling	*Sturnia erythropygia*	LC
70.	Muscicapidae	White-bellied blue flycatcher	*Cyornis pallipes*	LC
71.	Muscicapidae	Malabar whistling-thrush	*Myophonus horsfieldii*	LC
72.	Dicaeidae	Nilgiri flower pecker	*Dicaeum concolor*	LC
73.	Nectariniidae	Crimson-backed sunbird	*Leptocoma minima*	LC
74.	Nectariniidae	Sahyadri sunbird	*Aethopyga vigorsii*	LC

Source: BirdLife International (2017)

Table 2.6 Diversity of different species of vertebrates and endemism in the Western Ghats biodiversity hotspots

Group	Number of species	Endemism
Mammals	139	11.7
Birds	508	3.0
Reptiles	203	61.8
Amphibians	181	87.8
Fishes	290	65

Source: Radhakrishnan and Rajmohana (2012)

Table 2.7 Diversity of threatened vertebrate fauna of the Western Ghats biodiversity hotspots

Animal groups	Critically endangered	Endangered	Vulnerable	Total
Mammals	3	7	21	31
Birds	2	1	12	15
Reptiles	0	1	3	4
Amphibians	16	28	15	59
Fishes	12	54	31	97

Source: Radhakrishnan and Rajmohana (2012)

2.11 Vertebrate Fauna of Western Ghats

Among the vertebrates, birds shows highest in species diversity with 508 species in the Western Ghats biodiversity hotspots, followed by fishes (290 species), reptiles (203 species), amphibians (181 species) and mammals (139 species) (Table 2.6). Of the recorded species of vertebrates, highest numbers of fishes were reported under IUCN threatened categories, followed by amphibians, mammals, birds and reptiles (Table 2.7).

2.12 Endemic Birds of Western Ghats

Of the recorded species of birds from Western Ghats, 16 were endemic. Of these, seven species have distribution from the sea level to 1500 m height. These species occur in different habitats especially evergreen and semievergreen rain forests, moist deciduous forest and subtropical hill forest (Stattersfield et al. 1998) (Table 2.8).

2.13 Endemic Mammals of Western Ghats

Western Ghats is very rich for its diversity of mammalian species. The Western Ghats has the representation of about 137 species of mammals. Of these 15 are endemic to Western Ghats (Table 2.9).

Table 2.8 Endemic birds of Western Ghats biodiversity hotspots

Common name	Scientific name	IUCN status
Nilgiri wood pigeon	*Columba elphinstonii* (Sykes, 1832)	VU
Malabar parakeet	*Psittacula columboides* (Vigors, 1830)	LC
Malabar grey hornbill	*Ocyceros griseus* (Latham, 1790)	LC
Nilgiri pipit	*Anthus nilghiriensis* (Sharpe, 1885)	VU
Grey-headed bulbul	*Pycnonotus priocephalus* (Jerdon, 1839)	NT
White-bellied shortwing	*Sholicola major* (Jerdon, 1844)	EN
Wynaad laughing thrush	*Garrulax delesserti* (Jerdon, 1839)	LC
Rufous-breasted laughing thrush	*Montecincla cachinnans* (Jordon, 1839)	EN
Grey-breasted laughing thrush	*Garrulax jerdoni* (Blyth, 1851)	NT
Rufous babbler	*Turdoides subrufa* (Jerdon, 1839)	LC
Broad-tailed grassbird	*Schoenicola platyurus* Jerdon, 1844	VU
Black-and-rufous flycatcher	*Ficedula nigrorufa* (Jerdon, 1839)	NT
Nilgiri flycatcher	*Eumyias albicaudatus* (Jerdon, 1840)	NT
White-bellied blue flycatcher	*Cyornis pallipes* (Jerdon, 1840)	LC
Crimson-backed sunbird	*Leptocoma minima* (Sykes, 1832)	LC
White-bellied treepie	*Dendrocitta leucogastra* (Gould, 1833)	LC

Table 2.9 List of endemic mammals of Western Ghats biodiversity hotspots

Common name	Scientific name
Lion-tailed macaque	*Macaca silenus* (Linnaeus, 1758)
Black-footed grey langur	*Semnopithecus hypoleucos* (Blyth, 1841)
Nilgiri langur	*Semnopithecus johnii* (J. Fischer, 1829)
Nilgiri palm squirrel	*Funambulus sublineatus* (Waterhouse, 1838)
Jungle palm squirrel	*Funambulus tristriatus* (Waterhouse, 1837)
Spiny tree mouse	*Platacanthomys lasiurus* (Blyth, 1859)
Bonhote's mouse	*Mus famulus* (Banhote, 1898)
Ranjini's field rat	*Rattus rnjiniae* (Agarwall and Ghosal, 1969)
Sahyadris forest rat	*Rattus satarae* (Hinton, 1918)
Nilgiri vandeleuria	*Vandeleuria nilagirica* (Jerdeon, 1867)
Day's shrew	*Suncus dayi* (Dobson, 1888)
Hill shrew	*Suncus niger* (Horsfield, 1851)
Nilgiri marten	*Martes gwatkinsii* (Horfield, 1851)
Brown palm civet	*Paradoxurus jerdoni* (Blanford, 1885)
Nilgiri tahr	*Nilgiritragus hylocrius* (Ogilby, 1838)

2.14 Hotspot: Indo-Burma

This biodiversity hotspot includes the Himalayan mountain chain and the associated foothills in Nepal, Bhutan and India. It also covers the coastal lowlands of southern China, many offshore islands, namely, Hainan Island, China in the South China Sea and the Andaman group of Islands in India. This hotspot includes 33 terrestrial ecoregions, such as the tropical and subtropical moist broadleaf forests, tropical and subtropical dry broadleaf forests, tropical and subtropical coniferous forests, temperate broadleaf and mixed forests and mangroves. The Indo-Burma hotspots cover wide variety of ecosystems; these are mixed wet evergreen, dry evergreen, deciduous and montane forests. These hotspots also include portions of eastern India and Andaman and Nicobar Islands, eastern part of Bangladesh, southernmost China, most of Myanmar (excluding the northern tip), most of Thailand (excluding the southern tip) and all of Cambodia, Laos and Vietnam. A total of 1277 bird species were reported from the Indo-Burma hotspot, and of the recorded species, 74 were endemic (Birdlife International 2003; UNDP 1998; Karmakar 2010). The Indo-Burma hotspot is equally rich in faunal diversity. Total of 430 mammal species reported, of these Similarly, 71 are endemic in this hotspot. Three hundred and twenty three species of amphibians were reported, of which 139 are endemic. This region also supports highest diversity of freshwater turtle. Total of 1262 species of freshwater fishes were documented which is accounting for about 10% of the world species and these include 566 endemics (Tordoff et al. 2012; Pande and Arora 2014).

Recently, a new species of leaf deer *Muntiacus putaoensis* was discovered in Myanmar and also reported from the forests of Arunachal Pradesh during the year 2003 (Datta et al. 2003). A Tawang macaque *Macaca munjala* is also reported from this region which is a new record for India (Mishra et al. 2004). Among the Indian primates, 15 species were found in this region (Mohnot 1980; Roonwal and Mohnot 1977). Of the six species of larger cats of the world recorded from India, four of them reported are from here, the tiger *Panthera tigris*, leopard *Panthera pardus*, snow leopard *Uncia uncia* and clouded leopard *Neofelis nebulosa*.

2.15 Endemic and Threatened Birds of Indo-Burma Hotspot

One hundred and forty five species of endemic and threatened birds were reported from Indo-Burma hotspot. Of these eight species are listed as critically endangered, namely, Baer's pochard, white-bellied heron, red-headed vulture, white-rumped vulture, Indian vulture, slender-billed vulture, Bengal florican and sociable lapwing. Eleven species are listed as endangered, namely, white-winged duck, green peafowl, greater adjutant, Egyptian vulture, steppe eagle, masked finfoot, great knot, black-bellied tern, swamp prinia, yellow-breasted bunting and Narcondam hornbill (Table 2.10).

Table 2.10 List of endemic and threatened species of the birds of Indo-Burma hotspot

Sl. no.	Common name	Scientific name	IUCN status
1.	Baer's pochard	*Aythya baeri*	CR
2.	White-bellied heron	*Ardea insignis*	CR
3.	Red-headed vulture	*Sarcogyps calvus*	CR
4.	White-rumped vulture	*Gyps bengalensis*	CR
5.	Indian vulture	*Gyps indicus*	CR
6.	Slender-billed vulture	*Gyps tenuirostris*	CR
7.	Bengal florican	*Houbaropsis bengalensis*	CR
8.	Sociable lapwing	*Vanellus gregarius*	CR
9.	White-winged duck	*Asarcornis scutulata*	EN
10.	Green peafowl	*Pavo muticus*	EN
11.	Greater adjutant	*Leptoptilos dubius*	EN
12.	Egyptian vulture	*Neophron percnopterus*	EN
13.	Steppe eagle	*Aquila nipalensis*	EN
14.	Masked finfoot	*Heliopais personatus*	EN
15.	Great knot	*Calidris tenuirostris*	EN
16.	Black-bellied tern	*Sterna acuticauda*	EN
17.	Swamp prinia	*Prinia cinerascens*	EN
18.	Yellow-breasted bunting	*Emberiza aureola*	EN
19.	Narcondam hornbill	*Rhyticeros narcondami*	EN
20.	Red spurfowl	*Galloperdix spadicea*	Endemic
21.	White-cheeked barbet	*Psilopogon viridis*	Endemic
22.	Vigors's sunbird	*Aethopyga vigorsii*	Endemic
23.	Pink-headed duck	*Rhodonessa caryophyllacea*	Endemic (Possibly extinct)
24.	Manipur bush-quail	*Perdicula manipurensis*	Endemic/EN
25.	Andaman cuckooshrike	*Coracina dobsoni*	Endemic/NT
26.	Dark-rumped swift	*Apus acuticauda*	Endemic/VU
27.	Yellow-throated bulbul	*Pycnonotus xantholaemus*	Endemic/VU
28.	Rusty-throated wren-babbler	*Spelaeornis badeigularis*	Endemic/VU
29.	Tawny-breasted wren-babbler	*Spelaeornis longicaudatus*	Endemic/VU
30.	Snowy-throated babbler	*Stachyris oglei*	Endemic/VU
31.	Marsh babbler	*Pellorneum palustre*	Endemic/VU
32.	Andaman coucal	*Centropus andamanensis*	LC
33.	Andaman barn owl	*Tyto deroepstorffi*	LC
34.	Andaman Hume's hawk owl	*Ninox obscura*	LC
35.	Andaman nightjar	*Caprimulgus andamanicus*	LC
36.	Andaman cuckooshrike	*Coracina dobsoni*	LC
37.	Andaman bulbul	*Microtarsus fuscoflavescens*	LC
38.	Andaman shama	*Copsychus albiventris*	LC
39.	Andaman flowerpecker	*Dicaeum virescens*	LC
40.	Andaman white-headed starling	*Sturnia erythropygius*	LC

(continued)

Table 2.10 (continued)

Sl. no.	Common name	Scientific name	IUCN status
41.	Falcated duck	*Anas falcata*	NT
42.	Ferruginous duck	*Aythya nyroca*	NT
43.	White-cheeked partridge	*Arborophila atrogularis*	NT
44.	Japanese quail	*Coturnix japonica*	NT
45.	Satyr tragopan	*Tragopan satyra*	NT
46.	Hume's pheasant	*Syrmaticus humiae*	NT
47.	Black-necked stork	*Ephippiorhynchus asiaticus*	NT
48.	Painted stork	*Mycteria leucocephala*	NT
49.	Oriental darter	*Anhinga melanogaster*	NT
50.	Spot-billed pelican	*Pelecanus philippensis*	NT
51.	Black-headed ibis	*Threskiornis melanocephalus*	NT
52.	Lammergeier	*Gypaetus barbatus*	NT
53.	Cinereous vulture	*Aegypius monachus*	NT
54.	Himalayan griffon	*Gyps himalayensis*	NT
55.	Pallid harrier	*Circus macrourus*	NT
56.	Lesser fish eagle	*Ichthyophaga humilis*	NT
57.	Grey-headed fish eagle	*Ichthyophaga ichthyaetus*	NT
58.	Great thick-knee	*Esacus recurvirostris*	NT
59.	Northern lapwing	*Vanellus vanellus*	NT
60.	River lapwing	*Vanellus duvaucelii*	NT
61.	Eurasian curlew	*Numenius arquata*	NT
62.	Black-tailed godwit	*Limosa limosa*	NT
63.	Bar-tailed godwit	*Limosa lapponica*	NT
64.	Curlew sandpiper	*Calidris ferruginea*	NT
65.	Red-necked stint	*Calidris ruficollis*	NT
66.	Asian dowitcher	*Limnodromus semipalmatus*	NT
67.	Great snipe	*Gallinago media*	NT
68.	River tern	*Sterna aurantia*	NT
69.	Ashy-headed green pigeon	*Treron phayrei*	NT
70.	Great hornbill	*Buceros bicornis*	NT
71.	Brown hornbill	*Anorrhinus austeni*	NT
72.	Malabar pied-hornbill	*Anthracoceros coronatus*	NT
73.	Blyth's kingfisher	*Alcedo hercules*	NT
74.	Brown-winged kingfisher	*Pelargopsis amauroptera*	NT
75.	Yellow-rumped honeyguide	*Indicator xanthonotus*	NT
76.	Laggar falcon	*Falco jugger*	NT
77.	Alexandrine parakeet	*Psittacula eupatria*	NT
78.	Grey-headed parakeet	*Psittacula finschii*	NT
79.	Blossom-headed parakeet	*Psittacula roseata*	NT
80.	Derbyan parakeet	*Psittacula derbiana*	NT
81.	Red-breasted parakeet	*Psittacula alexandri*	NT

(continued)

Table 2.10 (continued)

Sl. no.	Common name	Scientific name	IUCN status
82.	Long-billed bush-warbler	*Locustella major*	NT
83.	Rufous-vented prinia	*Prinia burnesii*	NT
84.	Rufous-throated wren-babbler	*Spelaeornis caudatus*	NT
85.	Long-tailed wren-babbler	*Spelaeornis chocolatinus*	NT
86.	Chevron-breasted babbler	*Sphenocichla roberti*	NT
87.	Blackish-breasted babbler	*Stachyris humei*	NT
88.	Chestnut-backed laughing thrush	*Ianthocincla nuchalis*	NT
89.	Rusty-bellied shortwing	*Brachypteryx hyperythra*	NT
90.	Firethroat	*Calliope pectardens*	NT
91.	Andaman teal	*Anas gibberifrons*	NT
92.	Andaman serpent eagle	*Spilornis elgini*	NT
93.	Andaman crake	*Rallina canningi*	NT
94.	Andaman wood pigeon	*Columba palumboide*	NT
95.	Andaman cuckoo-dove	*Macropygia rufipennis*	NT
96.	Andaman green pigeon	*Treron chloropterus*	NT
97.	Andaman scops-owl	*Otus balli*	NT
98.	Andaman hawk owl	*Ninox affinis*	NT
99.	Andaman woodpecker	*Dryocopus hodgei*	NT
100.	Andaman drongo	*Dicrurus andamanensis*	NT
101.	Andaman treepie	*Dendrocitta bayleyi*	NT
102.	Taiga bean-goose	*Anser fabalis*	Rare
103.	Mandarin duck	*Aix galericulata*	Rare
104.	Eastern spot-billed duck	*Anas zonorhyncha*	Rare
105.	White-tailed tropicbird	*Phaethon lepturus*	Rare
106.	Goliath heron	*Ardea goliath*	Rare
107.	Eastern marsh-harrier	*Circus spilonotus*	Rare
108.	White-browed crake	*Amaurornis cinerea*	Rare
109.	Little gull	*Hydrocoloeus minutus*	Rare
110.	Salim Ali's swift	*Apus salimalii*	Rare
111.	Buff-throated warbler	*Phylloscopus subaffinis*	Rare
112.	Chestnut-cheeked starling	*Agropsar philippensis*	Rare
113.	Buff-breasted sandpiper	*Calidris subruficollis*	Rare/NT
114.	Red-breasted goose	*Branta ruficollis*	Rare/VU
115.	Long-tailed duck	*Clangula hyemalis*	Rare/VU
116.	Spoon-billed sandpiper	*Calidris pygmaea*	Rare/CR
117.	Red knot	*Calidris canutus*	Rare/NT
118.	Lesser white-fronted goose	*Anser erythropus*	VU
119.	Marbled teal	*Marmaronetta angustirostris*	VU
120.	Common pochard	*Aythya ferina*	VU
121.	Chestnut-breasted partridge	*Arborophila mandellii*	VU
122.	Swamp francolin	*Francolinus gularis*	VU

(continued)

Table 2.10 (continued)

Sl. no.	Common name	Scientific name	IUCN status
123.	Sclater's monal	*Lophophorus sclateri*	VU
124.	Blyth's tragopan	*Tragopan blythii*	VU
125.	Lesser adjutant	*Leptoptilos javanicus*	VU
126.	Dalmatian pelican	*Pelecanus crispus*	VU
127.	Indian spotted eagle	*Clanga hastata*	VU
128.	Greater spotted eagle	*Clanga clanga*	VU
129.	Imperial eagle	*Aquila heliaca*	VU
130.	Pallas's fish eagle	*Haliaeetus leucoryphus*	VU
131.	Sarus crane	*Antigone antigone*	VU
132.	Wood snipe	*Gallinago nemoricola*	VU
133.	Indian skimmer	*Rynchops albicollis*	VU
134.	Pale-capped pigeon	*Columba punicea*	VU
135.	Rufous-necked hornbill	*Aceros nipalensis*	VU
136.	Great slaty woodpecker	*Mulleripicus pulverulentus*	VU
137.	Beautiful nuthatch	*Sitta formosa*	VU
138.	Bristled grassbird	*Chaetornis striata*	VU
139.	Grey-crowned prinia	*Prinia cinereocapilla*	VU
140.	Jerdon's babbler	*Chrysomma altirostre*	VU
141.	Black-breasted parrotbill	*Paradoxornis flavirostris*	VU
142.	Slender-billed babbler	*Turdoides longirostris*	VU
143.	White-throated bushchat	*Saxicola insignis*	VU
144.	Grey-sided thrush	*Turdus feae*	VU
145.	Yellow weaver	*Ploceus megarhynchus*	VU

2.16 Endemic and Threatened Species of Fishes

The Indo-Burma biodiversity hotspot is enriched with diversified fish fauna. Thirty five species of endemic and threatened fishes were reported (Table 2.11). Of these, six species were listed as critically endangered, namely, *Moringua hodgarti*, *Garra litanensis*, *Garra manipurensis*, *Schistura singhi*, *Aborichthys garoensis* and *Kryptopterus indicus*. Eight species were listed as endangered, viz. *Barilius dogarsinghi*, *Acanthocobitis pavonacea*, *Schistura arunachalensis*, *Schistura barapaniensis*, *Schistura nagaensis*, *Mesonoemacheilus reticulofasciatus*, *Aborichthys tikaderi* and *Chaudhuria* (*Garo*) *khajuriai*.

Table 2.11 Endemic and threatened species of fishes of Indo-Burma hotspot

Sl. no.	Name of the species	IUCN status
1.	*Moringua hodgarti* Chaudhuri	CR
2.	*Garra litanensis* Vishwanath	CR
3.	*Garra manipurensis* Vishwanath and Sarojnalini	CR
4.	*Schistura singhi* (Menon)	CR
5.	*Aborichthys garoensis* Hora	CR
6.	*Kryptopterus indicus* Datta, Bannan and Jayaram	CR
7.	*Tor progenius* (McClelland)	DD
8.	*Barilius dogarsinghi* Hora	EN
9.	*Acanthocobitis pavonacea* (McClelland)	EN
10.	*Schistura arunachalensis* (Menon)	EN
11.	*Schistura barapaniensis* (Menon)	EN
12.	*Schistura nagaensis* (Menon)	EN
13.	*Mesonoemacheilus reticulofasciatus* (Singh and Banarescu)	EN
14.	*Aborichthys tikaderi* Bannan	EN
15.	*Chaudhuria (Garo) khajuriai* (Talwar, Yazdani and Kundu)	EN
16.	*Barilius lairokensis* Arunkumar and Singh	NE
17.	*Danio yuensis* Kumar and Singh	NE
18.	*Semiplotus manipurensis* Vishwanath and Kosygin	NE
19.	*Garra elongata* Vishwanath and Kosygin	NE
20.	*Schistura prashadi* (Hora)	NE
21.	*Acanthophthalmus longipinnis* (Menon)	NE
22.	*Nangra assamensis* Sen and Biswas	NE
23.	*Myerglanis jayrami* Vishwanath and Kosygin	NE
24.	*Danio naganensis* Chaudhuri	VU
25.	*Brachydanio acuticephala* (Hora)	VU
26.	*Puntius shalynius* Yazdani and Talukdar	VU
27.	*Garra kempi* Hora	VU
28.	*Garra lissorhynchus* (McClelland)	VU
29.	*Garra naganensis* Hora	VU
30.	*Garra rupecula* (McClelland)	VU
31.	*Schistura manipurensis* (Chaudhuri)	VU
32.	*Mesonoemacheilus sijuensis* (Menon)	VU
33.	*Neoeucirrhichthys maydelli* Banarescu and Nalbant	VU
34.	*Lepidocephalus irrorata* (Hora)	VU
35.	*Chaudhuria (Pillaia) indica* (Yazdani)	VU

Source: Sen (2003)

2.17 Hotspot: Sundaland (Nicobar Islands)

The Sundaland hotspot covers the western half of the Indo-Malayan archipelago (1.5 million km^2), an arc of some 17,000 islands lying north and south of the equator and spanning a distance of 5000 km between the Asian mainland and Australia. The hotspot is dominated by Borneo (725,500 km^2) and Sumatra (427,300 km^2), the third and sixth largest islands on Earth, with the remainder occupied by the Malay Peninsula and the island of Java (126,700 km^2). The hotspot is bordered by three other hotspots. The dividing boundary between the Sundaland hotspot and the Mainland Southeast Asia hotspot to northwest is here taken as the Kangar-Pattani Line, which lies near the Thailand-Malaysia border (van Steenis 1950; Whitmore 1984). Wallacea lies immediately to the east of the Sundaland hotspot, separated by the famous Wallace's Line, while the 7100 islands of the Philippines hotspot lie immediately to the northeast. The Nicobar Islands, which are jurisdictionally controlled by India, form part of this hotspot (Davis et al. 1995).

Sundaland has a total of 378 mammal species and contains the highest number of endemic species (173 species and 17 genera). Many of the mammals occurring in Sundaland are globally recognized flagship species, foremost of which are the two species of orang-utan, the Bornean *Pongo pygmaeus* and the Sumatran *Pongo abelii*, both seriously threatened by habitat loss. Other famous mammal flagships include the proboscis monkey *Nasalis larvatus*, also endemic to the island of Borneo, the Mentawai gibbon *Hylobates klossii* and the Sumatran rhinoceros *Dicerorhinus sumatrensis* and Javan rhinoceros *Rhinoceros sondaicus*. A total of 771 bird species are thought to occur regularly in Sundaland; of these, 146 are endemic.

2.18 Endemic Birds of Sundaland Hotspot (Nicobar Islands)

Nine species of birds are endemic to Sundaland biodiversity hotspots (Table 2.12).

Table 2.12 Endemic birds of Sundaland hotspot (Nicobar Islands)

Sl. no.	Common name	Scientific name	IUCN status
1.	Nicobar scops-owl	*Otus alius*	DD
2.	Nicobar imperial pigeon	*Ducula nicobarica*	LC
3.	Central Nicobar serpent eagle	*Spilornis minimus*	NR
4.	Great Nicobar serpent eagle	*Spilornis klossi*	NT
5.	Nicobar parakeet,	*Psittacula caniceps*	NT
6.	Nicobar bulbul	*Hypsipetes nicobariensis*	NT
7.	Nicobar sparrowhawk	*Accipiter butleri*	VU
8.	Nicobar megapode	*Megapodius nicobariensis*	VU
9.	Nicobar jungle-flycatcher	*Cyornis nicobaricus*	VU

2.19 Endemic and Threatened Species of Mammals of Sundaland Hotspot

Twenty nine species of endemic and threatened mammals were reported from this hotspot (Table 2.13). Among the recorded species, Nicobar spiny shrew is listed as critically endangered. The following species are listed as endangered category of IUCN Red List: Nicobar tree shrew, blue whale, Nicobar rat and Miller's Nicobar rat.

Table 2.13 Endemic and threatened species of mammals of Sundaland hotspot

Sl. no.	Common name	Species name	IUCN status
1.	Nicobar spiny shrew	*Crocidura nicobarica* (Miller, 1902)	CR
2.	Nicobar tree shrew	*Tupaia nicobarica* (Zelebor, 1869)	EN
3.	Blue whale	*Balaenoptera musculus* (Linnaeus, 1758)	EN
4.	Nicobar rat	*Rattus burrus* (Miller, 1902)	EN
5.		*Rattus burrescens* (Miller)	EN
6.	Miller's Nicobar rat	*Rattus burrus* (Miller, 1902)	EN
7.	Indian flying fox	*Pteropus giganteus* (Brunnich, 1782)	LC
8.	Black-bearded tomb bat	*Taphozous melanopogon* (Temminck, 1841)	LC
9.	Nicobar leaf-nosed bat	*Hipposideros nicobarulae* (Miller, 1902)	LC
10.	Little Nicobar leaf-nosed	*Hipposideros cognatus cognatus* (Anderson, 1881)	LC
11.	Bent-winged bat	*Miniopterus pusillus* (Dobson, 1876)	LC
12.	Indian pipistrelle	*Pipistrellus coromandra* (Gray, 1838)	LC
13.	Lesser yellow bat	*Scotophilus kuhlii* (Leech, 1821)	LC
14.	Blyth's pouch-bearing bat	*Taphozous saccolaimus crassus* (Blyth, 1844)	LC
15.	Common dolphin	*Delphinus delphis* (Linnaeus, 1758)	LC
16.	Nicobar wild pig	*Sus scrofa nicobarica* (Miller, 1902)	LC
17.	Cave nectar bat	*Eonycteris spelaea* (Dobson, 1871)	LC
18.	Nicobar flying fox	*Pteropus nicobaricus* (Zelebor, 1869)	VU
19.	Nicobar flying fox	*Pteropus nicobaricus* (Fitzinger, 1861)	VU
20.	Nicobar crab-eating macaque	*Macaca fascicularis umbrosa* (Miller, 1902)	VU
21.	Dugong	*Dugong dugon* (Muller, 1776)	VU
22.	Sperm whale	*Physeter catodon* (Linnaeus, 1758)	VU
23.	Zelebor's Nicobar rat	*Rattus palmarum* (Zelebor, 1869)	VU
24.	False killer whale	*Pseudorca crassidens* (Owen, 1846)	DD
25.	Nicobar tree shrew	*Tupaia nicobarica surda* (Miller, 1902)	–
26.		*Rattus pulliventer* (Miller, 1902)	–
27.	Fulvus leaf-nosed bat	*Hipposideros diadema nicobarensis* (Dobson, 1871)	–
28.	House rat	*Rattus rattus alexandrines* (Geoffroy)	–
29.		*Rattus rattus holchu* (Chaturvedi)	–

2.20 Hotspot: Himalaya

The Eastern Himalayas is the region encompassing Bhutan, northeastern India and southern, central and eastern Nepal. The abrupt rise of the Himalayan mountains from less than 500 m to more than 8000 m results in a diversity of ecosystems that range from alluvial grasslands and subtropical broadleaf forests along the foothills to temperate broadleaf forests in the mid hills, mixed conifer and conifer forests in the higher hills and alpine meadows above the tree line. Two hundred and sixty nine species of freshwater fishes were reported, and among these, 33 are endemic. One hundred and five species of amphibians are known to occur in the hotspot, of which 42 are endemic. One hundred and seventy six species have been documented, of which 48 are endemic. About 977 birds have been recorded in the hotspot, but only 15 are endemic. Birdlife International has identified four Endemic Bird Areas in this hotspot.

A total of 175 mammal species and more than 500 bird species are known from this hotspot (WWF and ICIMOD 2001). Of the recorded mammal species, 45 species were listed under various threatened (14 endangered, 29 vulnerable and 2 critically rare) (CEPF 2005; Chettri 2010) (Tables 2.14 and 2.15).

Table 2.14 Threatened mammals recorded from the Himalayan biodiversity hotspot

Common name	Scientific name	IUCN status
Anthony's pipistrelle	*Pipistrellus anthonyi*	CR
Argali	*Ovis ammon*	NT
Asian elephant	*Elephas maximus*	EN
Asiatic black bear	*Ursus thibetanus*	VU
Assamese macaque	*Macaca assamensis*	VU
Back-striped weasel	*Mustela strigidorsa*	VU
Capped leaf monkey	*Trachypithecus pileatus*	VU
Cheetah	*Acinonyx jubatus*	VU
Clouded leopard	*Neofelis nebulosa*	VU
Eld's deer/thamin	*Rucervus eldii*	VU
François leaf monkey	*Trachypithecus francoisi*	EN
Ganges river dolphin	*Platanista gangetica*	EN
Gaur	*Bos frontalis*	VU
Gee golden langur	*Trachypithecus geei*	EN
Golden snub-nosed monkey	*Rhinopithecus roxellana*	VU
Himalayan tahr	*Hemitragus jemlahicus*	VU
Hispid hare	*Caprolagus hispidus*	EN
Hoolock gibbon	*Hoolock hoolock*	EN
Irrawaddy squirrel	*Callosciurus pygerythrus*	VU
Joffre's pipistrelle	*Pipistrellus joffrei*	CR
Kozlov's shrew	*Sorex kozlovi*	CR
Markhor	*Capra falconeri*	EN
Namdapha flying squirrel	*Biswamoyopterus biswasi*	CR
Parti-coloured flying squirrel	*Hylopetes alboniger*	EN

(continued)

Table 2.14 (continued)

Common name	Scientific name	IUCN status
Pig-tailed macaque	*Macaca nemestrina*	VU
Pygmy hog	*Sus salvanius*	CR
Red goral	*Naemorhedus baileyi*	VU
Red panda	*Ailurus fulgens*	EN
Serow	*Capricornis sumatraensis*	VU
Smooth-coated otter	*Lutrogale perspicillata*	VU
Snow leopard	*Uncia uncia*	EN
Stump-tailed macaque	*Macaca arctoides*	VU
Swamp deer	*Rucervus duvaucelii*	VU
Takin	*Budorcas taxicolor*	VU
Tiger	*Panthera tigris*	EN

Source: CEPF (2005, 2007) and Chettri (2010)

Table 2.15 Threatened birds of Himalayan biodiversity hotspot

Common name	Scientific name	IUCN status
Nuthatch	*Sitta formosa*	VU
Black-breasted parrotbill	*Paradoxornis flavirostris*	VU
Black-necked crane	*Grus nigricollis*	VU
Blyth's tragopan	*Tragopan blythii*	VU
Chestnut-breasted partridge	*Arborophila mandellii*	VU
Greater spotted eagle	*Aquila clanga*	VU
Grey-crowned prinia	*Prinia cinereocapilla*	VU
Hodgson's bushchat	*Saxicola insignis*	VU
Lesser kestrel	*Falco naumanni*	VU
Palla's fish eagle	*Haliaeetus leucoryphus*	VU
Rufous-necked hornbill	*Aceros nipalensis*	VU
Slender-billed babbler	*Turdoides longirostris*	VU
Wood snipe	*Gallinago nemoricola*	VU
Blyth's kingfisher	*Alcedo hercules*	NT
Cinereous vulture	*Aegypius monachus*	NT
Firethroat	*Luscinia pectardens*	NT
Giant babax	*Babax waddelli*	NT
Great hornbill	*Buceros bicornis*	NT
Rufous-throated wren-babbler	*Spelaeornis caudatus*	NT
Rusty-bellied shortwing	*Brachypteryx hyperythra*	NT
Satyr tragopan	*Tragopan satyra*	NT
Ward's trogon	*Harpactes wardi*	NT
Yellow-rumped Honeyguide	*Indicator xanthonotus*	NT

Out of 35 biodiversity hotspots, 4 hotspots have been identified in Indian subcontinent, namely, the Himalayan hotspot, the Indo-Burma hotspot, the Western Ghats-Sri Lanka hotspot and the Sundaland hotspot. In this paper, an attempt has been made to present the endemic, endangered and threatened fauna based on the field studies and available information. According to Myers (1988), the hotspots are referred to as the areas with high concentrations of endemic species and with high habitat loss. The hotspot approach can be applied at any geographical scale and both in terrestrial and marine environments. However, hotspots represent conservation priorities in terrestrial ecosystems but remain largely unexplored in marine habitats where the amount of data is still poor (Mittermeier et al. 2011). India is one among the 17 mega biodiversity countries of world. Mittermeier (1988) and Mittermeier and Werner (1990) recognized mega diversity countries with most of them in tropics. According to Gentry (1986), tropical forests deserved more attention than temperate ecosystems; this is because of their greater species richness and also the greater concentration of high endemism. India has very rich biological resource due to various habitats and ecosystem (Haleem 2014). A total of 5416 mammals belonging to 154 families, under 29 orders, were recorded from the world. Of these, 426 species were reported and belong to 52 families and 14 orders in India. Among these, 394 species belong to terrestrial ecosystem from 43 families, and 12 orders, and the remaining 31 species are aquatic mammals belonging to 9 families and 2 orders (Srinivasulu and Srinivasulu 2012). Due to various anthropogenic pressures, several species of mammals face severe threats. Four species of mammals were already extinct in the country, namely, *Acinonyx jubatus*, *Bos javanicus*, *Dicerorhinus sumatrensis* and *Rhinoceros sondaicus*. Of the 426 species of mammals, 43 are considered to be endemic to the Indian political boundary (Sharma 2015).

Among the Indian biodiversity hotspots, the Himalayas contributes maximum in maintaining ecosystem. The resources were used in different ways as it provides fodder, fuel wood, timber and leaf litter for manuring crop fields, construction, industrial raw material and several non-timber forest produce. Indo-Burma hotspot has been facing a serious problem of biodiversity declining over the years because of the resource exploitation and habitat loss. It is one of the places where people started agriculture first. It has a long history of using fire to clear land for agricultural purposes and other needs (Solheim 1972; Diamond 1997). The Indian hotspots are very rich in faunal diversity with special reference to endemic, endangered and threatened species. The forests in the hotspots are important natural resource of the country. They play a very vital role for the sustenance of the population and are the storehouse of the biodiversity as well as play a vital role in regulating climate cycle. Biodiversity is very important for the wellbeing of particular area. Survival of many people is totally dependent on bio-resources. Biodiversity has been affected due to large scale construction, over-exploitation and disasters. For the conservation of biodiversity world, over many initiatives have been taken by the government and tried to inform people about its usefulness. The Indian government has also taken several initiatives to conserve the biodiversity by developing protected area network, national parks, wildlife sanctuaries and zoos (Negi 2002). The extreme anthropogenic pressures, causing habitat modification and destructions in recent times, have been resulting in loss of biodiversity affecting fragile ecosystem. The presence of unique and endangered in the Indian hotspots is encouraging and need to conserve. This paper is providing vital information and a baseline data on the endemic and threatened species for the future research and conservation planning which it requires at the earliest as it represents a fragile and sensitive ecosystem where increasing anthropogenic activities are beginning to show its baleful effect on the total biodiversity.

In terms of research gaps, policy-based sound scientific analysis needs to be carried out and strengthened. Addressing the challenges imposed on biodiversity by climate change and other factors. Complete inventories of species and their population within the hotspots are scarce and very limited, which is in urgent need of documentation. Assessment of invasive and exotic species, analysis of population trends of flagship and threatened species, strengthening of the effectiveness and extent of coverage of protected areas and fire management strategies are needed urgently. The importance and need for establishing long-term, consistent monitoring of climate change and its impact on biodiversity. Permanent plots and/or units need to be established on an altitudinal transect spanning the tropics to the alpine regions in order to monitor diverse ecosystems.

References

Aengals R, Sathish Kumar VM, Palot MJ (2011). Updated checklist of Indian reptiles. Zoological Survey of India, Kolkata, 24 p

Alfred JRB, Sinha NK, Chakraborty S (2002) Checklist of mammals of India. Rec Zoo Surv India Occ Paper 199:1–289

Anon (2008) Checklist of Indian mammals revised and updated. Zoos' Print 23(8): 11 p

Birdlife International (2003) Saving Asia's threatened birds: a guide for government and civil society. BirdLife International, Cambridge

BirdLife International (2017) Endemic bird areas factsheet: Assam plains. Downloaded from http://www.birdlife.org on 05/06/2017

Birdlife International CEPF (2005) Ecosystem profile: Indo-Burman hotspot, eastern Himalayan region. WWF, US-Asian Programme/CEPF, Kathmandu

CEPF (2007) Ecosystem profile: Indo-Burma hotspot, Indo- China region. Critical Ecosystem Partnership Fund, Cambridge

Chettri N, Sharma E, Shakya B, Thapa R, Bajracharya B, Uddin K, Oli KP, Choudhury D (2010) Biodiversity in the Eastern Himalayas: Status, trends and vulnerability to climate change. Climate change impact and vulnerability in the Eastern Himalayas – Technical report 2. ICIMOD, Kathmandu

Datta A, Pansa J, Madhusudan MD, Mishra C (2003) Discovery of the leaf deer *Muntiacus putaoensis* in Arunachal Pradesh: an addition to the large mammals of India. Curr Sci 84:454–458

Davis SD, Heywood VH, Hamilton AC (1995) Centres of plant diversity. A guide and strategy for their conservation. Volume 3. The Americas. The World Wildlife Fund (WWF) and IUCN – The World Conservation Union. IUCN Publications Unit, Cambridge, UK

De Candolle AP (1855) Geographic botanique raisonne, vol 2. V Masson, Paris

Diamond J (1997) Guns, germs and steel: a short history of everybody for the last 13,000 years

Dinesh KP, Radhakrishnan C, Channakeshavamurthy BH, Kulkarni NU (2015) Checklist of Amphibia of India, updated till January 2015 available at http://mhadeiresearchcenter.org/resources (online only)

Gentry AH (1986) Endemism in tropical versus temperate plant communities. In: Soule ME (ed) Conservation biology: science of scarcity and diversity. Sinauer Associates, Sunderland, pp 153–181

Haleem A, Ilyas O, Syed Z, Arya SK, Imam E (2014) Distribution, status and aspects of ecology of mammalian species in Kedarnath Wildlife Sanctuary, Uttarakhand Himalayas, India. J Mater Environ Sci 5(3):683–692

Karmakar S, Bhattacharya T, Karmakar S (2010) The status of endemic birds in three Indian hotspots: a review of available data. Sci Cult (November–December): 524–528

Mishra C, Datta A, Madhusudan MD (2004) The high altitude wildlife of western Arunachal Pradesh: a survey report, CERC Technical Report No. 8, Nature Conservation Foundation, International Snow Leopard Trust and Wildlife Conservation Society (India Program), Mysore

Mittermeier RA (1988) Primate diversity and the tropical forest: Case studies from Brazil and Madagascar, and the importance of the megadiversity countries. In: WILSON EO (ed) Biodiversity. National Academy Press, Washington, DC, pp 145–154

Mittermeir R, Werner T (1990) Wealth of plants and animals unites "megadiversity" countries. Tropicus 4(1):1, 4–5

Mittermeier CG, Turner WR, Larsen FW, Brooks TM, Gascon C (2011) Global biodiversity conservation: the critical role of hotspots. In: Zachos FE, Habel JC (eds) Biodiversity hotspots: Distribution and protection of priority conservation areas. Springer, Berlin, pp 3–22

Mittermeier RA, Robles-Gil P, Hoffmann M, Pilgrim J, Brooks T, Mittermeier GC, Lamoreux J, Da Fonseca GAB (2005) Hotspots revisited: earth's biologically richest and most endangered terrestrial ecoregions. Conservation International

Mohnot SM (1980) Intergroup infant kidnapping in Hanuman langur. Folia Primatol 34:259–277

Myers N (1988) Threatened biotas: "Hot spots" in tropical forests. Environ 8:1–20

Myers N, Mittermeier RA, Mittermeier CG, da Fonseca GA, Kent J (2000) Biodiversity hotspots for conservation priorities. Nature 403:853–858

Nayar MP (1996) Hotspots of endemic plants of India, Nepal and Bhutan. SB Press, Trivandrum

Negi SS (2002) Handbook of national parks, sanctuaries, and biosphere reserves in India. Indus Publishing, New Delhi. 94 p

Pande H, Arora S (2014) India's fifth national report to the convention on biological diversity. Ministry of environment and forests, Government of India, New Delhi, 142 p

Pascal JP (1988) Wet evergreen forests of the western Ghats on India. Ecology, structure, floristic composition and succession. French Institute, Pondicherry, 345 p

Radhakrishnan C, Rajmohana K (2012) Fauna of ecosystem of India – western Ghats: 1–14. Zoological Survey of India, Kolkata

Roonwal ML, Mohnot SM (1977) Primates of South Asia: ecology, sociobiology, and behavior. Harvard University Press, Cambridge, MA, 421 p

Sen N (2003) Fish fauna of north east India with special reference to endemic and threatened species. Rec Zool Surv India 101(Part 3–4):81–99

Sharma G, Kamalakannan M, Venkataraman K (2013) A checklist of mammals with their distribution and conservation status. Zoological Survey of India, Kolkata. 121 p

Sharma G, Kamalakannan M, Venkataraman K (2015) A checklist of mammals of India with their distribution and conservation status. Zool Surv. India, 2015

Solheim WG (1972) An earlier agricultural revolution. Sci Am 266(4):34–41

Srinivasulu C, Srinivasulu B (2012) South Asian mammals. In: South Asian mammals: their diversity, distribution, and status. Springer, New York, pp 9–98

Stattersfield AJ, Crosby MJ, Long AJ, Wege DC (1998) Endemic bird areas of the world: priorities for biodiversity conservation. BirdLife International, Cambridge

Tordoff AW, Baltzer MC, Fellowes JR, Pilgrim JD, Langhammer PF (2012) Key biodiversity areas in the Indo-Burma hotspot: process, progress and future directions. J Threatened Taxa 4(8):2779–2787

UNDP (1998) Eco-regional co-operation for biodiversity conservation in the Himalayas. Proceedings of a regional meeting organized by UNDP in co-operation with WWF and ICIMOD

van Steenis CGGJ (1950) The delimitation of Malesia and its main plant geographical divisions. Flora Malesiana I(1):lxx–lxxv

Whitmore TC (ed) (1984) Tropical rain forests of the far East, 2nd edn. Oxford University Press, Oxford

WWF; ICIMOD (2001) Eco-region-based conservation in the eastern Himalaya: identifying important areas for biodiversity conservation. WWF-Nepal, Kathmandu

Diversity and Conservation of Chiropteran Fauna

3

Vadamalai Elangovan, Virendra Mathur, Mukesh Kumar, and Yuvana Satya Priya

Abstract

Bats are the second largest order of mammals, accounting for about 1240 species. Bats utilize almost all the niches and make up a quarter of all known mammals on earth. India is known for its rich biodiversity and its natural history abound, but the bats of India are hardly ever mentioned, and very little is known about them. In India, about 119 species of bats incorporated in nine families, namely, Pteropodidae, Emballonuridae, Megadermatidae, Rhinolophidae, Hipposideridae, Vespertilionidae, Miniopteridae, Rhinopomatidae, and Molossidae. The present study was conducted at different districts of Uttar Pradesh. The present study was aimed to assess the diversity and conservation status of chiropteran fauna in Uttar Pradesh. Visual observations were conducted in old abandoned building, caves, crevices, historical monuments, and forest areas to assess the distribution of bats. A total of 15 species of bats were identified from the study area. Out of which, 3 species belong to suborder Megachiroptera, namely, *Rousettus leschenaulti*, *Cynopterus sphinx*, and *Pteropus giganteus*, and the remaining 12 species belong to the suborder Microchiroptera, namely, *Rhinopoma microphyllum*, *R. hardwickii*, *Hipposideros fulvus*, *H. lankadiva*, *Megaderma lyra*, *Pipistrellus coromandra*, *P. dormeri*, *P. tenuis*, *P. ceylonicus*, *Scotophilus heathii*, *S. kuhlii*, and *Taphozous nudiventris*. The colonies of *P. giganteus* were observed on tall trees such as *Mangifera indica*, *Ficus religiosa*, *F. glomerata*, *F. benghalensis*, *Eucalyptus* sp., *Azadirachta indica*, *Dalbergia sissoo*, *Madhuca indica*, and *Bambusa spinosa*. The

V. Elangovan (✉) · V. Mathur · M. Kumar
Department of Applied Animal Sciences, School for Biosciences and Biotechnology, Babasaheb Bhimrao Ambedkar Central University, Lucknow, Uttar Pradesh, India
e-mail: elango70@yahoo.com

Y. S. Priya
Residential Coaching Academy, Babasaheb Bhimrao Ambedkar Central University, Lucknow, Uttar Pradesh, India

C. Sivaperuman, K. Venkataraman (eds.), *Indian Hotspots*,
https://doi.org/10.1007/978-981-10-6605-4_3

medium-sized fruit bat *R. leschenaulti* occupied the abandoned buildings or old monuments such as Chunar Fort at Mirzapur and Bari Mosque at Jaunpur. The short-nosed fruit bat, *C. sphinx*, preferably occupied tree canopy, palm fronts, tree holes, and flower cluster; however, few colonies of *C. sphinx* were observed at abandoned buildings or monuments. The colonies of microchiropteran bats were observed at tree cleft, abandoned buildings, caves, and crevices throughout the study area. The results of field survey showed that *H. lankadiva* and *P. ceylonicus* were not earlier reported in Uttar Pradesh. The IUCN red list of threatened species 2011 categorized all 15 species of bats as least concern (LC version – 3.1). Scientific research coupled with education and awareness is a key to success for any conservation program. During field surveys, a number of discussions were held with students, teachers, forest officials, and public to increase the level of understanding on conservation of bats. The current study reveals the highest distribution of bats in eastern Uttar Pradesh due to the presence of a large number of old monuments, palaces, caves, deep well, and forests which harbor bats. These permanent structures give stable roosting sites to the bats. In general, there was no major threat to the bats in the study area, except sporadic observations at times. Another support is that a maximum of bat colonies are located in old monuments which are governed by the Archaeological Survey of India, while few more colonies are located in caves. Thus, the state Uttar Pradesh provides a range of suitable habitats for the distribution of both frugivorous and insectivorous bats.

Keywords
Chiroptera · Conservation · Distribution · Taxonomy

3.1 Introduction

Bats are the second largest order of mammals and second most species-rich order of mammals in the world, only surpassed by the rodents. In several localities, bat species can represent more than 50% of the local mammalian diversity (Wilson 1983), and this number is likely to increase as more bat surveys are being conducted in many parts of the world. The general pattern of worldwide bat distributions based on latitudinal gradients is similar to that of other mammals (Pagel et al. 1991), with bat communities with the highest diversity concentrated in the tropical regions (Findley 1976; Patterson et al. 2003). Bats constitute most diverse groups of mammals in the tropical regions, with only the rodents approaching them in number (Handley 1966; Nowak 1994; Emmons 1997). Bats are ecologically important fauna of our environment because many of the plant species depend upon bats for their pollination and seed dispersal.

The order Chiroptera is divided into two major suborders, Megachiroptera and Microchiroptera. In the Indian subcontinent, about 119 species of bats are incorporated into nine families, namely, Pteropodidae, Rhinopomatidae, Emballonuridae, Megadermatidae, Rhinolophidae, Hipposideridae, Molossidae, Vespertilionidae, and Miniopteridae (Bates and Harrisson 1997). The suborder Megachiroptera consists only one family Pteropodidae which restrict to the Old World tropics of Africa and Asia. The family Pteropodidae is one of the important pollinators and seed dispersers for a number of ecologically and economically important plants (Marshal 1985; Cox et al. 1991; Rainey et al. 1995). They are the only pollinator able to carry large-seeded fruits. Thus, they play a key role in structuring the forest community (Rainey et al. 1995), but the Indian legislation and policies have given poor recognition, and fruit bats are kept under Schedule V (along with vermin) of the Indian Wildlife Protection Act 1972. India being a tropical country has a rich diversity of chiropteran fauna. The microchiropteran or microbats are important for the agro-ecosystem as they consume nocturnal insect pest relatively in large volume up to 100% of their body weight (Davison and Zubaid 1992; Eckrich and Neuweuilar 1988). They play a major role in the regulation of insect population the landscape. Among the world's mammals, bats make up 25% of the total number. Bats are considered as the important regulator of the complex ecological processes through interaction such as seed dispersal, pollination, and insect regulation (Findley and Black 1983; Fleming 1986; Findley 1993). This study was aimed to investigate the status, distribution, and conservation of chiropteran fauna in eastern Uttar Pradesh.

3.1.1 Study Area

The state Uttar Pradesh has a geographical area of 29.44 million hectares which is about 9% of the land area of the country. Uttar Pradesh covers a large part of highly fertile and densely populated Upper Gangetic Plain. It lies between lat. 23° 52′ and 31° 28′ N and long. 77° 5′ and 84° 38′ E. The state Uttar Pradesh is known for its variety of natural resources like reserve forests, national parks, sanctuaries, etc. The recorded forest area of the state is about 5.17 million hectares which constitutes 17.55% of its geographical area. By legal status, reserve forest constitutes 70.51%, protected forest 2.90%, and unblessed forest 26.59%. The state has quite a good number of palaces, ancient temples, and historical monuments. Historical monuments and temples tend to be ideal roost sites for bats. Roosts are vital for bat survival and provide site for mating, rearing of young, consumption of prey, protection from predators and shelter from adverse weather conditions. This study was conducted at different districts of Uttar Pradesh (Fig. 3.1).

Fig. 3.1 The map of Uttar Pradesh and the study area is colored

3.2 Methods

Field surveys were carried out at different districts of eastern Uttar Pradesh through visual observations and mist netting/hoop netting. The study was conducted at Barabanki old temple, Banki village, Budhain Purva, Basauri, and Ram Sanehi Ghat in Barabanki district (26° 56′ N, 81°. 13′ E); Jais, Rani Harbans Ganj, Gangaganj, Harchandpur, and Balban Singh ka Purva in Raebareli district (26°.14′ N, 81°.16′ E); Diyara Fort (King Rudra Pratap Shahi), Kadipur, and Amethi Fort (King Ranvir Singh) in Sultanpur district (26°.16′ N, 81°.16′ E); Bara Imambara, Mohanlalganj, Sisandy house, Residency, Roomi Gate, and Telibagh in Lucknow district (26°.55′ N, 80°.59′ E); Makbara (Bahu Begum Shahiba), Begamganj, Gulab Bari, Rushi Temple, and Roosi temple in Faizabad (26°.47′ N, 81°.12′ E); Jaunpur Fort, Bari Masjid, and Atala Masjid in Jaunpur (25°. 46′ N, 82°. 44′ E); Allipur, Rafi Ahmad Intermediate College, Raja Rookmangal Singh Inter College, Kashipur,

Masit, and Kaimau in Hardoi (27°.23′ N, 80°.10′ N); Kunda, Pratapgarh, and railway station in Pratapgarh (25°.34′ N, 81°.59′ E); Sangam Fort, Jhushi Fort, and Khushroo Bagh in Allahabad (25°.28′ N, 81°.54′ E); hilly areas, Ram Ghat temple, Ashok Darwaja, and Hanuman Dhara Caves in Chitrakoot (25°.28′ N, 81°.54′ E); Mirzapur Fort, Chunar Fort, Durgaji Cave temple Chunar, Vindhyachal, hill areas, and Kali Khoh Temple in Mirzapur (25°.10′ N, 82°.37′ E); and Ramnagar Fort, Sarnath, Chaukhandi stoop, Paal kothi near Ganga River, Dr. Sampurnanand Sanskrit University, Queen Mary Inter College, and Banaras Hindu University campus in Varanasi (25°.20′ N, 83°.00 E).

The periodical visits were made through roost search and visual observations at roost sites twice in a month. Bats roosts were located based on the information given by local residents. Field surveys were conducted at historical monuments, tunnels, caves, crevices, old temples/buildings, wildlife sanctuaries, and forest areas. Bats were captured using nylon mist nets (9.0 m length, 2.0 m width, and 38.0 mm mesh size – Avinet, Dryden, USA). Mist nets were erected at 1800 h and closed at 0500 h. Individuals were kept inside the bat cage until morphological measurements were taken, and thereafter they were released at the site of capture. The morphological measurements such as forearm length, head and body length, and tail length were measured using digital venire calipers (Mitutoyo, Japan), and body mass was measured to the nearest 0.5 g using 50.0 g, 100 g, and 200 g spring balances. In addition, morphological measurements such as wingspan (cm); length of metacarpals 2nd, 3rd, 4th, and 5th (mm); length of tibia (mm); length of toe (mm); length of thumb (mm); and length of lower and upper jaw (mm) were recorded. Data were collected on distribution and diversity of bats in the study area. Population counts were made through photography with the help of digital camera (Kodak C 173). Dead bats were collected from the roosting sites, and bones were preserved for osteological studies.

3.3 Results

A total of 15 species of bats belong to six families observed in the study area. The distribution pattern of bats in the study area is given in Table 3.1.

3.3.1 Pteropodidae

The pteropodid bats can be easily identified by their simple external ear without tragus, and the edge of the pinna forms an unbroken ring around the ear canal. No nose leaf is present. The eyes are large with well-developed vision in all megachiropterans. The second finger has three bony phalanges, the last of which is very small or rudimentary and usually bears a small claw. The tail is usually either very short or absent. The shoulder joint is simple. The head looks like fox- or doglike appearance; hence, the common name "flying fox" is due to their fox-like faces. The canines are prominent, and the molar teeth are highly modified for fruit eating.

Table 3.1 Distribution of bats in Uttar Pradesh

Family	Species	Lucknow	Hardoi	Barabanki	Raebareli	Sultanpur	Faizabad	Ambedkar Nagar	Ballia	Mau	Mirzapur	Gorakhpur	Gonda	Allahabad	Pratapgarh	Varanasi	Jaunpur	Chandauli	Bahraich	Basti	Balrampur	S. Kabir Nagar	Deoria	Azamgarh	Chitrakoot
Pteropodidae	*R. leschenaulti*	+	+	+	–	–		+	+	+	+	–	+	–	+	+	+	–	–	–	–	–	–	–	+
	C. sphinx	+	+	+	+	+	+	+	+	+	+	+	+	+	+	+	+	+	+	+	+	+	+	+	+
	P. giganteus	+	+	+	+	+	+	+	+	+	+	+	+	+	+	+	+	+	+	+	+	+	+	+	+
Rhinopomatidae	*R. microphyllum*	–	–	–	–	–	–	–	–	–	+	–	–	–	–	+	–	+	–	–	–	–	–	–	+
	R. hardwickii	+	–	–	–	–	–	–	–	–	–	–	–	+	–	+	+	–	–	–	–	–	–	–	+
Hipposideridae	*H. lankadiva*	–	–	–	–	–	–	–	–	–	–	–	–	–	–	–	–	–	–	–	–	–	–	–	+
	H. fulvus	–	–	–	–	+	–	–	–	–	–	–	–	–	–	–	–	–	–	–	–	–	–	–	+
Megadermatidae	*M. lyra*	+	+	–	–	–	–	–	–	–	–	–	–	–	–	–	–	–	–	–	–	–	–	–	–
Vespertilionidae	*P. coromandra*	+	+	+	+	+	+	+	+	+	+	–	–	+	–	+	–	+	–	+	–	+	–	–	–
	P. dormeri	–	–	–	–	–	+	–	–	–	–	–	–	–	–	–	–	–	–	–	–	–	–	–	–
	P. tenues	+	+	+	+	–	–	–	–	–	–	–	–	–	–	–	–	–	–	–	–	–	–	–	–
	S. heathii	+	+	–	–	–	–	–	–	–	–	–	–	–	+	+	–	–	–	–	–	–	–	–	–
	S. kuhlii	–	+	–	–	–	–	–	–	–	–	–	–	–	–	–	–	–	–	–	–	–	–	–	–
	P. ceylonicus	–	–	–	+	–	–	–	–	–	–	–	–	–	–	–	–	–	–	–	–	–	–	–	–
Emballonuridae	*T. nudiventris*	–	–	–	–	–	–	–	–	–	+	–	–	–	–	+	–	–	–	–	–	–	–	–	–

+ = Present; – = Absent

These bats lack acoustic orientation except rousette bats (*Rousettus*), generally roost in trees, colonial, and often show cryptic markings or bright fur colors or patterns. The Indian pteropodids are predominantly frugivorous. Flying foxes gather in large numbers to roost, and in the evening, they start emerging from roost one by one at initial period and then large numbers. The distribution of pteropodid bats in the study area is shown in Table 3.1.

3.3.2 Fulvous Fruit Bat *Rousettus leschenaulti* (Desmarest, 1820)

It is a very common species and generally known as Indian fulvous fruit bat, *Rousettus leschenaulti*. It has a widespread distribution in the study area. This bat is intermediate in size between short-nosed fruit bat, *Cynopterus sphinx*, and Indian flying foxes, *Pteropus giganteus*, with an average forearm length of 80.6 mm (75–86 mm), and the hind feet and thumb are shorter. The muzzle is short and slender, and its body color on dorsal surface dark brown and ventral surface light brown. The pelage is soft fine and silky (Fig. 3.2). A small group of four individuals were observed in a building roost at Barabanki. A colony consists 10,000–10,500 individuals of *R. leschenaulti* observed in a dome of Bari Masjid (Mosque) located at Jaunpur, Uttar Pradesh. The mosque was constructed by the late Firuz Shah Tughlak during the thirteenth century. According to the local residents, the colony occupies the mosque for the last 20 years. Another colony of *R. leschenaulti* consists of 800–900 individuals observed in the tunnel of a deep well at Chunar Fort located 35 km east of Mirzapur.

A total of seven individuals (four males and three females) caught from different locations, namely, Banki, Jais in Raebareli, Bari Masjid in Jaunpur, and Gopal Khera in Lucknow to study the morphological measurements and the morphological measurements, are given in Table 3.2. There was no major threat to this species in the study area. However, little disturbance was observed due to human interference to the roost sites, since the roost sites are located in national monuments which

Fig. 3.2 Indian fulvous fruit bat *Rousettus leschenaulti* with a pup

Table 3.2 Morphological measurements of *Rousettus leschenaulti*

	Male ($n = 4$)		Female ($n = 3$)	
Morphological parameters	Mean	SD	Mean	SD
Body wt (g)	94.3	9.1	88.4	30.4
FAL (mm)	54.0	19.1	62.2	23.2
Head length	37.5	1.5	35.4	12.9
HB length (mm)	115.1	2.9	103.4	37.7
Ear length	27.2	6.1	21.7	7.5
Thumb length(mm)	34.4	10.7	30.2	9.2
Tibia length (mm)	22.9	7.0	29.3	12.9
Toe length (mm)	18.0	3.5	16.6	5.2
Tail length (mm)	16.7	0.6	14.6	5.3
WSP length (cm)	51.4	0.6	51.7	0.9
MET II (mm)	32.3	1.8	30.8	11.1
MET III (mm)	52.9	2.3	37.9	17.0
MET IV (mm)	49.7	7.6	46.8	15.3
MET V (mm)	29.1	10.2	37.9	17.0
Lower jaw length (mm)	21.5	1.9	19.6	6.7
Upper jaw length (mm)	20.4	0.8	17.4	6.3

are being visited by tourists. In addition, the roost sites are disturbed by the developmental activities of monuments. The IUCN red list of threatened species 2011 categorized *R. leschenaulti* as least concern, i.e., LC version – 3.1 (Table 3.17).

3.3.3 Short-Nosed Fruit Bat *Cynopterus sphinx* (Vahl, 1797)

The short-nosed fruit, *Cynopterus sphinx*, is commonly distributed throughout the study area. It can be easily identified by its dog-shaped head, divergent nostril, large prominent eyes, and short ears with white margin as unique morphological characters. The muzzle is short, broad, and covered with the hairs as far as nostril which project well forward. In males, the chin, anterior part of shoulder, sides of the chest, belly, and thighs are characteristically orange tinted.

In females, the color is usually towny brown; the rump is gray brown with paler gray belly. The forehead and the neck are darker and rich russet brown; posteriorly the back is gray brown (Fig. 3.3). A colony consists 250–300 individuals of *C. sphinx* observed in 12 harems at different locations in the study area. In addition, a huge colony of *C. sphinx* consists 55–60 individuals observed in the dome of Khushroo Bagh (a historical monument) located in Allahabad. According to local residents, bats are living in the monuments for the last 12 years. A total of 12 individuals (6 males and 6 females) were captured using mist nets, and they were released after taking the morphological measurement (Table 3.3).

Fig. 3.3 Short-nosed fruit bat *Cynopterus sphinx* (male)

Table 3.3 Morphological measurements of *Cynopterus sphinx*

	Male (n = 6)		Female (n = 6)	
Morphological parameters	Mean	SD	Mean	SD
Body wt (g)	51.1	6.1	59.1	3.1
FAL (mm)	71.3	1.9	72.4	1.6
Head length	41.1	1.1	41.9	0.9
HB length (mm)	106.5	5.5	108.7	2.3
Ear length	21.1	1.7	20.7	1.3
Thumb length (mm)	18.1	0.8	17.5	0.7
Tibia length (mm)	31.0	1.2	31.2	0.9
Toe length (mm)	8.3	0.6	8.0	0.3
Tail length (mm)	12.1	0.4	12.1	0.8
WSP length (cm)	43.5	0.7	43.5	0.6
MET II (mm)	42.6	1.0	43.2	0.5
MET III (mm)	46.0	1.0	45.8	1.1
MET IV (mm)	45.1	0.8	45.1	0.5
MET V (mm)	46.6	0.6	45.9	0.3
Lower jaw length (mm)	11.3	0.6	11.7	0.9
Upper jaw length (mm)	13.0	0.5	12.8	0.7

There was no major threat observed to this species in the study area. However, in some parts of the study area, it was observed that people are hunting them with a myth that the flesh of *C. sphinx* has some medicinal values. Though the fruit bats are playing a key role in pollination and seed dispersal, the farmers of guava and mango orchards at Malihabad consider them as pest as the bats are damaging their crops and reducing the yield. The IUCN red list of threatened species 2011 categorized this species as least concern (LC version – 3.1).

Table 3.4 Morphological measurements of *Pteropus giganteus*

Morphological parameters	Male	Male
Body weight (g)	600.0	400.0
Head and body length (mm)	220	155.0
Tail length (mm)	0	0
Toe (mm)	22.9	15.2
Length of tibia (mm)	74.7	63.3
Forearm length (mm)	155.1	137.5
Ear length (mm)	36.2	26.3
Ear width (mm)	18.7	14.1
Wing span (cm)	104.0	93.0
Length of thumb (mm)	43.8	33.9
2 MT (mm)	79.1	71.8
3 MT (mm)	108.2	93.1
4 MT (mm)	103.9	90.6
5 MT (mm)	114.1	96.5

Fig. 3.4 Indian flying fox *Pteropus giganteus* with a pup

3.3.4 Indian Flying Fox *Pteropus giganteus* (Brunnich, 1782)

Indian flying fox, *Pteropus giganteus*, has widespread distribution in eastern Uttar Pradesh. A total of about 15,906 individuals were observed from 35 colonies of *P. giganteus* roosting in different locations of the study area. The colonies of *P. giganteus* were observed in at Mohanlalganj, Hullaskhera, Masit, Nakarsen, and Hardoi. It is the largest fruit bat in India with the forearm length of 152–183 mm. The morphological measurements of two bats which died due to electrical shock were collected and given in Table 3.4.

Pteropus giganteus has long snout, well-developed nostril, and long pointed black ears. The pelage is chestnut brown on the crown of the head and relatively darker around the eyes (Fig. 3.4). There was no major threat observed to *P. giganteus* in the study area. In some part of study area, this species found locally threatened by cutting down of roost trees because of road expansion or other domestic purposes. The species is rarely hunted by people for bush meat. The IUCN red list of threatened species 2011 categorized this species as least concern (LC version – 3.1).

3.3.5 Rhinopomatidae

The family Rhinopomatidae is traditionally considered to be one of the most ancient chiropteran clades. It is a small family of insect-eating bats of primitive structure found in the arid and semiarid region. The tail is very long and slender with the longest part projecting free from the membrane. The family Rhinopomatidae consists of three known species, namely, *Rhinopoma hardwickii*, *R. microphyllum*, and *R. muscatellum*. Two species of rhinopomatid bats, namely, *R. microphyllum* and *R. hardwickii*, were observed at specific locations in the study area.

3.3.6 Greater Mouse-Tailed Bat *Rhinopoma microphyllum* (Brunnich, 1782)

The greater mouse-tailed bat, *Rhinopoma microphyllum*, is widespread and relatively common species. *Rhinopoma microphyllum* roosts in historical monuments in the study area. The distribution of greater mouse-tailed bat was very limited in eastern Uttar Pradesh and observed only three colonies in the study area. It is the largest of three species of *Rhinopoma* in the Indian subcontinent with the forearm length 68.54–74.6 mm with short tail. The face, ears, and connecting membrane on the forehead are found naked. The ears are well developed with bluntly sickle-shaped tragus. The pelage is short and fine, and its body color is gray brown on the dorsal surface and paler in ventral side (Fig. 3.5). A colony of greater mouse-tailed bat consists 11,000–11,500 individuals observed in Chunar Fort located 30 km away from east of Mirzapur. The second colony of *R. microphyllum* was observed in an old monument, namely, Pal Kashi Naresh Fort located at Varanasi. The colony consists of 2500–3000 individuals of *R. microphyllum*.

In addition, a colony consists 1800–2000 individuals of *R. microphyllum* observed in a cave located at Chitrakoot. According to the local residents, *R. microphyllum* lives in the cave for the last 25 years. Human disturbance was observed at roost sites during the study period; however, hunting was not observed. The IUCN red list of threatened species 2011 categorized this species as least concern (LC version – 3.1). A total of seven adults (three males and four females) were captured using mist nets, morphological measurements were taken, and the bats were released at the site of capture. A list of morphological measurements is given in Table 3.5.

Fig. 3.5 Greater mouse-tailed bat, *Rhinopoma microphyllum*

Table 3.5 Morphological measurements of *Rhinopoma microphyllum*

	Male ($n = 3$)		Female ($n = 4$)	
Morphological parameters	Mean	SD	Mean	SD
Body wt (g)	22.8	0.8	22.1	1.9
FAL (mm)	67.0	2.9	67.7	1.5
Head length	26.6	0.5	26.3	0.7
HB length (mm)	77.0	3.0	77.1	0.6
Ear length	18.5	0.8	19.6	0.5
Thumb length (mm)	14.6	0.5	15.2	0.3
Tibia length (mm)	28.7	0.4	29.4	0.5
Toe length (mm)	13.1	0.3	14.4	0.6
Tail length (mm)	54.7	2.3	57.5	0.5
WSP length (cm)	36.6	0.3	36.0	0.8
MET II (mm)	50.9	0.9	52.2	0.6
MET III (mm)	51.7	0.5	53.3	0.4
MET IV (mm)	42.8	0.7	43.5	0.4
MET V (mm)	46.9	0.2	47.3	0.3
Lower jaw length (mm)	11.1	0.2	12.0	0.6
Upper jaw length (mm)	18.0	0.2	18.6	0.3

3.3.7 Lesser Mouse-Tailed Bat *Rhinopoma hardwickii* (Gray, 1831)

The lesser mouse-tailed bat, *Rhinopoma hardwickii*, has a wide distribution in Uttar Pradesh. It is a small bat with slender long free tail. The face is glandular, and the ears are connected across the forehead. The eyes and tragus are well developed. The muzzle bears small trigonal nose leaf terminally (Fig. 3.6). The roosts of *R. hardwickii* were observed in caves and roofs of deserted buildings. The colony size varies from 250 to 1000. The distribution of *R. hardwickii* was observed in the cave (Jhushi Fort, Ulta Kila), historical monument (Khushroo Bagh), and Atala Mosque, Jaunpur.

Fig. 3.6 Lesser mouse-tailed bat *Rhinopoma hardwickii*

Table 3.6 Morphological measurements of *Rhinopoma hardwickii*

Morphological parameters	Male (n = 3)		Female (n = 3)	
	Mean	SD	Mean	SD
Body wt (g)	20.0	1.1	21.3	0.5
FAL (mm)	62.3	0.6	62.7	0.8
Head length	23.8	0.2	24.1	0.2
HB length (mm)	71.6	0.3	72.3	0.6
Ear length	17.8	0.2	17.9	0.1
Thumb length (mm)	13.7	0.2	14.0	0.2
Tibia length (mm)	31.3	0.6	31.8	0.4
Toe length (mm)	13.8	0.2	13.6	0.2
Tail length (mm)	87.1	0.8	87.0	0.3
WSP length (cm)	34.4	0.1	34.9	0.1
MET II (mm)	46.5	0.4	46.5	0.3
MET III (mm)	46.7	0.3	46.6	0.3
MET IV (mm)	39.4	0.6	39.7	0.1
MET V (mm)	45.4	0.3	45.7	0.3
Lower jaw length (mm)	6.9	0.1	6.8	0.2
Upper jaw length (mm)	6.8	0.2	7.4	0.3

The IUCN red list of threatened species 2011 categorized this species as least concern (LC version – 3.1). A total of six adults (three males and three females) were captured using mist nets, morphological measurements were taken, and the bats were released at the site of capture. The morphological details are given in Table 3.6.

3.3.8 Hipposideridae

In general, most hipposiderids have brown or reddish-brown shade, but fur color varies intraspecifically. Like their close relatives, the rhinolophids, members of

Hipposideridae, possess an ornate nose leaf and broad mobile ears. The nose leaf is basically a horseshoe shape. Behind the anterior leaf, there is an intermediate swollen area which sometimes has a small central projection. The intermediate leaf forms a base for a thinner, more elaborate, and erect posterior element. This posterior leaf is not pointed, as in rhinolophids, but usually rounded or flat across the top. In addition, the face of the posterior leaf may have several thin-walled compartments. The complexity of the nose leaf may be further enhanced by secondary foliations of skin from under the edges of the horseshoe. Hipposiderids fly 1–2 m above the ground, frequently avoid bushes and use their short broad wings for slow and maneuverable flight. Hipposiderids hang free by their toes and usually roosts in cave ceiling. The family has found scanty distribution in the study area.

3.3.9 Indian Leaf-Nosed Bat *Hipposideros lankadiva* (Kelaart, 1850)

It is commonly known as the Kelaart's leaf-nosed bat. This large *Hipposideros* has an average forearm length of 83.5 mm (75.0–99.0 mm). Its nose leaf usually has four supplementary leaflets bordering the horseshoe. The intermediate leaf is expanded, its central part is inflated and swollen, and its upper surface is evenly concave. The posterior leaf is broad, and the pelage color ranges from pale cream to brown (Fig. 3.7).

A colony of *H. lankadiva* consists 1000–1100 individuals observed in a cave at Chitrakoot. No major threat was observed to this species during study period. However, *H. lankadiva* faces disturbance at times due to pilgrims who are visiting this religious place. Morphological measurements were taken from six adult bats (three males and three females), and the details are given in Table 3.7.

Fig. 3.7 Indian leaf-nosed bat *Hipposideros lankadiva* (male)

Table 3.7 Morphological measurements of *Hipposideros lankadiva*

	Male (n = 3)		Female (n = 3)	
Morphological parameters	Mean	SD	Mean	SD
Body wt (g)	42.0	1.0	41.7	2.1
FAL (mm)	84.3	0.4	84.9	1.0
Head length	33.0	0.2	32.5	0.4
HB length (mm)	91.4	0.7	92.3	0.6
Ear length	22.6	0.4	22.6	0.4
Thumb length (mm)	13.3	0.2	13.6	0.3
Tibia length (mm)	33.5	0.3	34.0	0.1
Toe length (mm)	12.5	0.2	12.5	0.1
Tail length (mm)	42.9	0.6	44.3	0.7
WSP length (cm)	49.6	0.4	49.5	0.5
MET II (mm)	62.7	0.3	62.7	0.3
MET III (mm)	62.4	0.2	62.6	0.3
MET IV (mm)	59.3	0.3	59.4	0.2
MET V (mm)	56.4	0.1	56.8	0.3
Lower jaw length (mm)	13.6	0.3	13.7	0.3
Upper jaw length (mm)	12.7	0.2	12.6	0.3

Fig. 3.8 Fulvus leaf-nosed bat *Hipposideros fulvus* (female)

3.3.10 Fulvus Leaf-Nosed Bat *Hipposideros fulvus* (Gray, 1838)

This is a small-sized leaf-nosed bat and commonly known as fulvous leaf-nosed bat with characteristically very large ear, and the tips are broadly rounded off. The feet are small, and the length of the nose leaf is about 5.0 mm. The pelage is pale gray (Fig. 3.8).

Table 3.8 Morphological measurements of *Hipposideros fulvus*

	Male (n = 3)		Female (n = 3)	
Morphological parameters	Mean	SD	Mean	SD
Body wt (g)	11.5	0.5	12.6	0.4
FAL (mm)	40.8	0.5	41.3	0.5
Head length	20.6	0.4	20.5	0.5
HB length (mm)	48.6	0.5	49.1	0.2
Ear length	20.0	0.4	21.0	0.6
Thumb length (mm)	11.0	0.2	10.5	0.7
Tibia length (mm)	18.0	0.3	18.0	0.5
Toe length (mm)	7.5	0.3	8.0	0.2
Tail length (mm)	27.0	0.3	28.0	0.6
WSP length (cm)	25.8	0.4	25.9	0.9
MET II (mm)	33.8	0.5	34.2	0.3
MET III (mm)	23.7	0.2	24.9	0.7
MET IV (mm)	31.4	0.5	31.7	0.5
MET V (mm)	30.6	0.2	31.3	0.4
Lower jaw length (mm)	8.7	0.4	8.8	0.1
Upper jaw length (mm)	10.0	0.2	9.9	0.7

It has limited distribution in the eastern Uttar Pradesh. A colony of *H. fulvus* consists 300–350 individuals observed in an abandoned palace (King Rudra Pratap Shahi Diyara) at Sultanpur. The distribution of *H. fulvus* was also observed in Gupt Godavari cave at Chitrakoot. No major threat was observed to *H. fulvus* in the study area. However, *H. fulvus* faces roost site destruction due to developmental activities in the monument and Gupt Godavari religious site. Morphological measurements of six adults (three males and three females) were taken and presented in Table 3.8.

3.3.11 Megadermatidae

The family Megadermatidae comprises four genera and five species; however, only one species *Megaderma lyra* was observed in the study area. Externally the members of the family can be recognized by the very large ears which are joined at their bases across the head and have very large and bifurcate tragus. The tail is very short or absent, and a large well-developed nose leaf is present. This species may roost singly or in small groups.

3.3.12 Indian False Vampire Bat *Megaderma lyra* (Geoffroy, 1810)

Megaderma lyra is commonly known as Indian greater false vampire bat. It has an average forearm length of 66.5 mm (56–71.5 mm). The head is characterized by its large oval ears which have fringe of white hairs on their inner margins. The ears are joined medially between one third and half of their length. Each ear has bifid tragus, the posterior process of which is taller. The face of *M. lyra* is hairy on the forehead

Fig. 3.9 Indian false vampire bat *Megaderma lyra* (male)

Table 3.9 Morphological measurements of *Megaderma lyra*

Morphological parameters	Male (n = 3)		Female (n = 3)	
	Mean	SD	Mean	SD
Body wt (g)	21.3	1.2	22.7	1.5
FAL (mm)	65.5	0.6	67.2	0.8
Head length	29.5	0.5	29.6	0.3
HB length (mm)	88.9	0.3	89.8	0.1
Ear length	40.4	0.6	40.5	0.5
Thumb length (mm)	15.8	0.4	16.2	0.3
Tibia length (mm)	38.0	0.2	38.0	0.2
Toe length (mm)	17.8	0.4	18.0	0.3
Tail length (mm)	0.0	0.0	0.0	0.0
WSP length (cm)	43.62	0.21	43.22	0.6
MET II (mm)	60.9	0.8	61.3	0.6
MET III (mm)	52.7	0.4	52.7	0.4
MET IV (mm)	56.9	0.2	57.0	0.2
MET V (mm)	55.5	0.5	55.6	0.5
Lower jaw length (mm)	15.0	0.1	15.6	0.2
Upper jaw length (mm)	14.9	0.3	14.5	0.4

and upper cheeks and naked at the snout. The nose leaf is erect, straight, and about 9 mm in height, and it has a longitudinal ridge and a simple rounded horizontal base. The pelage is fine soft and moderately long. The color of the body is grayish brown dorsally and paler at ventral side. The membrane of the ear is grayish black and semitranslucent, and the wings are broad (Fig. 3.9).

A colony of *M. lyra* consists about 550 individuals observed roosting beneath the staircase of a building (Rafi Ahmad Inter College) at Hardoi. No major threat was observed to this species in the study area. However, the bats used to get little disturbance due to human interference to the roost sites. A total of six adult bats (three males and three females) were captured; morphological measurements were taken and presented in Table 3.9.

3.3.13 Vespertilionidae

Vespertilionidae is the largest family of Indian bats and one of the most widely dispersed group of mammals in the world. As might be expected, this large and adaptable group of bats displays a correspondingly diverse range of morphological variation. A number of features distinguish the group as a whole from other bats. This group lacks nose leafs and has simple, unmodified lips and nostrils; they are commonly called the "plain-faced" bats. The tragus is usually well developed, and the tail is not free from the uropatagium. The facial region may have a variety of swollen glands and related structures. The eyes are usually small and the ears of vespertilionids are generally separate, small, and simple in structure. The tragus is usually a simple tongue-shaped structure. Occasionally, the lower margin of the ear is attached on the side of the head just behind the corner of the mouth. Most vespertilionids are brown, gray, or blackish brown in color. Belly fur is generally lighter than back fur. Internally, vespertilionids are distinguished by the highly developed double articulation between the scapula and humerus, the very rudimentary ulna, the essentially unmodified shoulder girdle and pelvis, and the conspicuous anterior emargination of their bony palate. There is a general trend throughout the family whereby the jaws are shortened to increase the effectiveness of the chewing muscles. The ulna is usually fused with the radius at its head, and the shaft is reduced to a fine ossified fibrous strand. The third finger bears three phalanges, of which the distal one is cartilaginous throughout except at the extreme base where a distinct joint is formed with the middle phalanx.

3.3.14 Little Indian Bat *Pipistrellus coromandra* (Gray, 1838)

The little Indian bat, *Pipistrellus coromandra*, is average larger, but there are significant overlaps in all external measurement. Pelage color is generally uniform brown on the dorsal surface ranging from dark chestnut to dark clove brown. The ventral surface is conspicuously pale brown. The ears and its membrane are mid to dark brown and essentially naked although there are some hairs on the interfemoral membrane adjacent to the body and tail above and below (Fig. 3.10).

Pipistrellus coromandra colonies were observed in crevices of Rushi temple, Ayodhya, in tree cavities at Hardoi Railway Station, and in wall crevices at Allipur; Telibagh; Banki (Barabanki); Jais, Raebareli; and B.B. Ambedkar University hostel building, Lucknow. No major threat was observed to this species and its habitat. Morphological measurements were collected from 12 adult bats (6 males and 6 females), and the details are given in Table 3.10.

3.3.15 Dormer's Bat *Pipistrellus dormeri* (Dobson, 1875)

It is commonly known as Dormer's bat. It is a medium-sized *Pipistrellus*, with an average forearm length of 34–30 mm. The tail is considerably shorter than the head

Fig. 3.10 Little Indian bat *Pipistrellus coromandra*

Table 3.10 Morphological measurements of *Pipistrellus coromandra*

	Male ($n = 6$)		Female ($n = 6$)	
Morphological parameters	Mean	SD	Mean	SD
Body wt (g)	10.2	0.7	10.8	0.7
FAL (mm)	28.0	0.3	28.2	0.6
Head length	14.0	0.3	14.1	0.6
HB length (mm)	42.8	1.0	44.5	0.5
Ear length	8.8	0.2	9.3	0.4
Thumb length (mm)	6.3	0.3	7.1	0.4
Tibia length (mm)	10.9	0.4	11.1	0.5
Toe length (mm)	5.6	0.1	5.8	0.6
Tail length (mm)	28.0	0.6	29.2	0.9
WSP length (cm)	21.0	0.6	21.1	0.5
MET II (mm)	26.5	0.4	27.4	0.4
MET III (mm)	27.5	0.3	27.9	0.2
MET IV (mm)	26.7	0.3	26.4	0.2
MET V (mm)	25.7	0.2	25.5	0.3
Lower jaw length (mm)	4.8	0.3	4.6	0.2
Upper jaw length (mm)	5.7	0.2	6.3	0.4

and body. The pelage on the dorsal surface is grayish brown with some of the tip almost silver in color, while the roots are dark brown/black; overall there is a slight glossy sheen. The ventral surface is contrastingly paler, with all hair tips white or pale white, and the roots are very dark. The ears, naked areas of the face, and the membrane are brown. In some specimen, the veins in the interfemoral membrane are conspicuously marked in white (Fig. 3.11).

It has a limited distribution in the study area and found a single colony of *P. dormeri* was observed in the crevice of Makbara of Bahu Begum Shahiba, Faizabad. Overall there was no major threat observed to *P. dormeri* in the study area. Morphological measurements of six bats (three males and three females) were taken, and details are presented in Table 3.11.

Fig. 3.11 Dormer's bat *Pipistrellus dormer*

Table 3.11 Morphological measurements of *Pipistrellus dormeri*

	Male (n = 3)		Female (n = 3)	
Morphological parameters	Mean	SD	Mean	SD
Body wt (g)	11.0	1.0	10.3	1.5
FAL (mm)	33.0	3.5	36.1	0.2
Head length	17.3	2.2	18.6	0.2
HB length (mm)	35.1	23.5	49.1	0.7
Ear length	9.8	0.3	12.0	0.6
Thumb length (mm)	6.8	0.5	7.1	0.3
Tibia length (mm)	12.7	0.9	13.8	0.2
Toe length (mm)	6.1	0.2	6.1	0.4
Tail length (mm)	30.1	0.8	31.3	0.6
WSP length (cm)	22.2	0.7	23.3	0.6
MET II (mm)	30.6	3.5	32.3	0.5
MET III (mm)	31.8	3.2	34.3	0.4
MET IV (mm)	30.9	3.4	33.1	0.1
MET V (mm)	29.6	3.3	32.0	0.6
Lower jaw length (mm)	6.7	1.9	7.9	0.4
Upper jaw length (mm)	8.1	1.3	8.8	0.15

3.3.16 Indian Pygmy Bat *Pipistrellus tenuis* (Temminek, 1840)

It is a small *Pipistrellus* and commonly known as Indian pygmy bat. The dorsal pelage is uniform, brown varying in tone from mid brown to deep brown, the ventral surface is paler, and hair tips are buffy brown. The ear membrane is dark and essentially naked (Fig. 3.12). Colonies of *P. tenuis* were observed roosting in wall crevices of old buildings at Telibagh, Lucknow. No major threat was observed to this species in the study area. Morphological measurements were collected from six bats (three males and three females), and details are given in Table 3.12.

Fig. 3.12 Indian pygmy bat *Pipistrellus tenuis*

Table 3.12 Morphological measurement of *Pipistrellus tenuis*

	Male (n = 3)		Female (n = 3)	
Morphological parameters	Mean	SD	Mean	SD
Body wt (g)	11.0	1.0	11.0	2.0
FAL (mm)	31.2	4.0	28.7	0.8
Head length	13.9	4.2	13.0	0.2
HB length (mm)	41.0	7.6	38.1	0.9
Ear length	9.9	1.1	10.3	1.3
Thumb length (mm)	7.3	0.3	7.1	0.8
Tibia length (mm)	13.6	0.5	14.6	0.5
Toe length (mm)	6.5	0.4	6.1	0.3
Tail length (mm)	28.1	2.5	28.4	1.4
WSP length (cm)	22.8	1.3	20.0	0.6
MET II (mm)	26.2	5.4	24.0	0.3
MET III (mm)	28.4	4.8	26.4	1.7
MET IV (mm)	29.5	2.9	28.1	0.4
MET V (mm)	28.5	2.8	27.1	1.8
Lower jaw length (mm)	6.6	2.0	6.0	0.2
Upper jaw length (mm)	7.4	1.9	7.3	0.6

3.3.17 Kelaart's Pipistrelle *Pipistrellus ceylonicus* (Kelaart, 1852)

Kelaart pipistrelle is relatively large with an average forearm length of 33–60 mm. The ears, naked area of face, wings, and interfemoral membrane are uniform dark brown in color. There are some hairs found on the interfemoral membrane above and below, adjacent to the body tail and femora. The dorsal pelage is variable color ranging from grayish brown to chestnut, reddish, or golden brown. The ventral surface has dark hair bases and pale gray tips (Fig. 3.13).

Colonies of *P. ceylonicus* were observed in roof crevices of abandoned building at Kasharawan, Raebareli, and in wall crevices of abandoned building at Jais, Raebareli. There was no major threat observed to *P. ceylonicus*. Morphological measurements of six adult bats (three males and three females) were taken and presented in Table 3.13.

Fig. 3.13 Kelaart's pipistrelle *Pipistrellus ceylonicus*

Table 3.13 Morphological measurements of *Pipistrellus ceylonicus*

	Male (*n* = 3)		Female (*n* = 3)	
Morphological parameters	Mean	SD	Mean	SD
Body wt (g)	9.0	2.6	10.0	1.0
FAL (mm)	35.6	4.8	38.7	1.1
Head length	15.0	1.6	16.3	0.7
HB length (mm)	49.8	6.5	54.0	1.0
Ear length	12.6	0.7	10.7	0.7
Thumb length (mm)	5.6	1.3	5.2	0.6
Tibia length (mm)	13.8	0.7	13.6	0.8
Toe length (mm)	5.7	1.7	5.4	0.4
Tail length (mm)	31.5	1.5	34.9	0.4
WSP length (cm)	21.3	0.5	21.0	0.1
MET II (mm)	32.4	3.6	35.1	0.6
MET III (mm)	33.9	3.0	36.2	0.6
MET IV (mm)	34.5	3.5	36.1	0.8
MET V (mm)	33.5	3.1	35.2	0.6
Lower jaw length (mm)	5.2	0.4	5.7	0.3
Upper jaw length (mm)	6.0	0.7	6.6	0.4

3.3.18 Greater Asiatic Yellow House Bat *Scotophilus heathii* (Horsefield, 1831)

Asiatic greater yellow house bat is widely distributed in the study area. It is a robust insectivorous bat. The tail is long with only the terminal 2.0–3.0 mm projecting free from the interfemoral membrane. The muzzle is broad and blunt; it is swollen on the sides, dark in color, and naked. The nostrils are simple in form, round, and slightly outward facing. The ears are small in relation to the size of the head; they are naked and have a number of transverse ridge. The pelage is short and fine, it is longer on the nap of the neck and throat. The head and back have pale buffy brown hairs. The throat, chest, and belly are pale yellow buff. In some individuals, the back is chestnut brown with reddish or golden yellow belly. The interfemoral membrane and

Fig. 3.14 Greater Asiatic house bat *Scotophilus heathii*

wings are uniformly dark brown and essentially naked except some hairs adjacent to the body and forearm on the ventral surface of each wing (Fig. 3.14).

Colonies of *S. heathii* were found roosting in tree holes, wall crevices, and door crevices at Hardoi, Allipur, Pratapgarh, Kunda, Raebareli, and Sisandy house in Lucknow. Destruction of roost site was observed during the study period. Morphological measurements of eight adult bats (four males and four females) were taken, and the details are given in Table 3.14.

3.3.19 Lesser Asiatic Yellow House Bat *Scotophilus kuhlii* (Leach, 1821)

The lesser yellow house bat is found in limited distribution in the study area. Dorsally, the pelage is soft and olive brown in color and ventrally creamish. The muzzle is broad and blunt. Ears are small compared to the head, and the tragus is about half the size of the ear and is crescent shaped. The tragus is separated from the pinna by a distinct notch. The pelage is chestnut brown but usually without the characteristic yellowish (Fig. 3.15).

Colonies of *S. kuhlii* were found roosting inside the tree hole of Banyan tree at Mamman purva and wall crevices at Hardoi. There was no major threat observed to this species in the study area. Morphological measurement of *S. kuhlii* taken from six individuals (three males and three females), and details are given in Table 3.15.

Table 3.14 Morphological measurement of *Scotophilus heathii*

	Male ($n = 4$)		Female ($n = 4$)	
Morphological parameters	Mean	SD	Mean	SD
Body wt (g)	31.3	2.2	31.5	1.3
FAL (mm)	60.1	3.1	58.4	1.3
Head length	20.2	0.9	19.3	1.2
HB length (mm)	87.4	6.0	90.1	1.8
Ear length	12.6	2.0	13.3	1.0
Thumb length (mm)	12.4	4.4	11.3	1.0
Tibia length (mm)	22.5	1.3	24.9	0.9
Toe length (mm)	8.9	1.4	10.6	0.9
Tail length (mm)	51.3	2.4	54.4	3.5
WSP length (cm)	42.5	0.9	41.0	0.4
MET II (mm)	55.8	1.3	55.6	0.6
MET III (mm)	56.4	2.0	58.4	0.4
MET IV (mm)	55.9	2.9	57.0	0.6
MET V (mm)	53.5	1.9	53.2	1.1
Lower jaw length (mm)	10.1	1.7	10.8	0.5
Upper jaw length (mm)	11.7	2.1	12.6	0.47

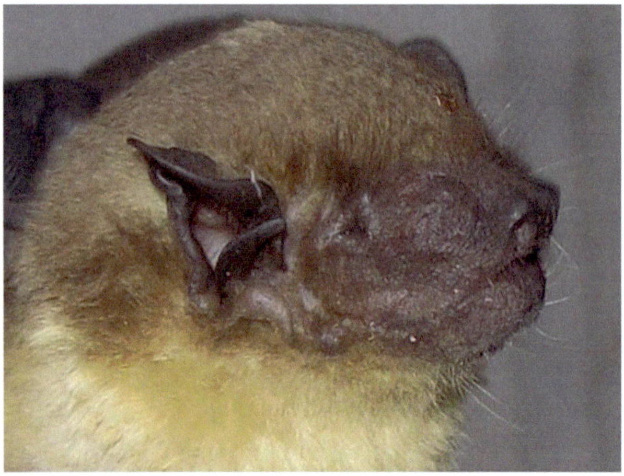

Fig. 3.15 Lesser Asiatic house bat *Scotophilus kuhlii*

3.3.20 Emballonuridae

The family Emballonuridae consists of small to reasonably large microchiropteran bats with a forearm ranging in length from 35 to 95 mm. It has scanty distribution in the study area.

Table 3.15 Morphological measurement of *Scotophilus kuhlii*

	Male ($n = 3$)		Female ($n = 3$)	
Morphological parameters	Mean	SD	Mean	SD
Body wt (g)	30.3	2.1	31.3	1.5
FAL (mm)	61.5	0.7	63.6	2.0
Head length	20.2	0.5	19.9	0.9
HB length (mm)	79.2	0.7	77.2	2.1
Ear length	8.5	0.5	8.6	1.1
Thumb length (mm)	20.3	0.6	20.5	0.5
Tibia length (mm)	21.5	0.3	21.5	0.5
Toe length (mm)	7.0	0.6	7.1	0.1
Tail length (mm)	49.0	0.3	49.1	0.3
WSP length (cm)	41.9	0.4	42.7	0.1
MET II (mm)	46.2	10.2	51.9	0.7
MET III (mm)	47.1	9.1	53.5	0.4
MET IV (mm)	46.7	9.5	52.7	0.3
MET V (mm)	45.5	8.3	49.9	0.3
Lower jaw length (mm)	6.9	1.4	6.9	0.4
Upper jaw length (mm)	7.9	1.3	8.7	0.37

3.3.21 Naked-Rumped Tomb Bat *Taphozous nudiventris* (Cretzschmar, 1830)

It is a medium-sized species of family Emballonuridae and commonly known as naked-rumped tomb bat or sheath-tailed bat. It has limited distribution in the study area. It is characterized by the naked rump which may contain copious fat reserve, especially in the post-monsoon season. The head is flattened in appearance while the jaw extremely powerful. The ears are long and rather narrow, semitranslucent, and widely separated from each other and the tip bluntly rounded off. The pelage is short, fine, and dense its body dark brown on the dorsal surface and paler brown in the ventral surface (Fig. 3.16).

Colonies of *T. nudiventris* were observed in Chunar Fort at Mirzapur district and Kashi Naresh Pal kothi, Munshi Ghat, Varanasi. The roosting sites of *T. nudiventris* were ruined due to renovation of old buildings. The IUCN red list of threatened species 2011 categorized this species as least concern (LC version – 3.1). A total of six adult bats (three males and three females) were captured for morphometry. The morphological measurements are given in Table 3.16.

Fig. 3.16 Naked-rumped tomb bat *Taphozous nudiventris*

Table 3.16 Morphological measurements of *Taphozous nudiventris*

	Male (n = 3)		Female (n = 3)	
Morphological parameters	Mean	SD	Mean	SD
Body wt (g)	42.7	7.6	54.0	7.9
FAL (mm)	71.5	10.0	77.9	0.8
Head length	28.1	6.4	32.9	0.9
HB length (mm)	97.0	4.0	100.9	0.8
Ear length	21.4	6.4	24.7	0.6
Thumb length (mm)	14.1	2.8	16.2	0.6
Tibia length (mm)	29.8	4.3	32.2	0.5
Toe length (mm)	14.2	3.1	15.9	0.6
Tail length (mm)	40.4	14.6	32.7	1.0
WSP length (cm)	44.3	2.7	45.7	0.9
MET II (mm)	61.8	4.4	65.0	0.5
MET III (mm)	65.5	6.9	68.8	1.0
MET IV (mm)	53.2	4.2	52.0	1.0
MET V (mm)	51.2	3.1	50.4	0.4
Lower jaw length (mm)	13.6	1.7	14.2	0.4
Upper jaw length (mm)	16.7	3.5	18.6	0.47

Table 3.17 Present status of bats of eastern Uttar Pradesh

Name of the species	Earlier status/past records		Present distribution	Remarks
	Status/location	Reference/reported by		
R. leschenaulti	Tanakpur, Chunar, Dehradun, Pauri, Almora, Nainital, Pithoragarh, and Chamoli	Wroughton (1914) and Bhat (1974)	Lucknow, Barabanki, Jaunpur, and Mirzapur	IUCN red list 2011 categorized as least concern
C. sphinx	Luck now, Pilibhit, Varanasi, Mirzapur, Dehradun, Nainital, Almora, Pithoragarh Pauri, and Chamoli	Wroughton (1914), Bhat (1974), and Khajuria (1953)	Widely distributed in study area	IUCN red list 2011 categorized as least concern
P. giganteus	Lucknow, Pilibhit, Varanasi, Mirzapur, and Farrukhabad	Wroughton (1914), Sinha (1980), and Bhatnagar and Srivastava (1974)	Widely distributed in study area	IUCN red list 2011 categorized as least concern
R. microphyllum	Agara, Fatehpur Sikri	Brosset (1962) and Sinha, (1980)	Varanasi, Chunar, and Chitrakoot	IUCN red list 2011 categorized as least concern
R. hardwickii	Fatehpur Sikri, Pratapgarh, Allahabad, and Agra	Brosset (1962) and Khajuria (1953)	Allahabad and Jaunpur	IUCN red list 2011 categorized as least concern
H. fulvus	Varanasi	Khajuria (1980)	Sultanpur and Chitrakoot	IUCN red list 2011 categorized as least concern
H. lankadiva	No record	No record	Chitrakoot	IUCN red list 2011 categorized as least concern
M. lyra	Rani Bagh, Nisarga, Agara, Gazipur, and Lucknow	Wroughton (1914) and Sinha (1980)	Hardoi and Unnao	IUCN red list 2011 categorized as least concern
P. coromandra	Dhakhuri, Lawarkhet, Mirzapur, Ramnagar, and Pilibhit	Wroughton (1914)	Lucknow, Hardoi, Barabanki, Faizabad, and Raebareli	IUCN red list 2011 categorized as least concern
P. dormeri	Khamaria	No record	Faizabad	IUCN red list 2011 categorized as least concern
P. tenuis	Sitabani, Ramnagar, Delajerna, Pilibhit, Haldwani, Varanasi, and Kaladhungi	Wroughton (1914), Bhat (1974), and Pathak and Sharma (1969)	Lucknow and Hardoi	IUCN red list 2011 categorized as least concern
P. ceylonicus	No record	No record	Observed at Bachhrawan and Raebareli	IUCN red list 2011 categorized as least concern

(continued)

Table 3.17 (continued)

Name of the species	Earlier status/past records		Present distribution	Remarks
	Status/location	Reference/reported by		
S. heathii	Haldwani, Ramnagar, Allahabad, Meerut, Bareilly, Pilibhit, and Mirzapur	Bhat (1974), Bhatanagar and Srivastava (1974), and Gandhi (1986)	Pratapgarh, Kunda, Jaunpur, Ayodhya, Hardoi, and Raebareli	IUCN red list 2011 categorized as least concern
S. kuhlii	Pilibhit and Ramnagar	Wroughton (1914)	Hardoi	IUCN red list 2011 categorized as least concern
T. nudiventris	Chunar, Fatehpur Sikri, and Agra	Brosset (1962)	Varanasi and Mirzapur	IUCN red list 2011 categorized as least concern

3.4 Discussion

The results revealed the distribution of 15 species of bats out of 119 known species of bats of Indian subcontinent. The observed 15 species of bats were distributed at different locations of study sites, namely, Sultanpur, Lucknow, Varanasi, Mirzapur, Chitrakoot, Allahabad, Pratapgarh, Hardoi, Jaunpur, Faizabad, and Raebareli. Out of 15 species, 3 species belong to Megachiroptera, namely, *R. leschenaulti*, *C. sphinx*, and *P. giganteus.* Inconsistent with the wide distribution of *R. leschenaulti* in India, the study area also had a wide distribution with good population of *R. leschenaulti*. The *Rousettus leschenaulti* was observed in permanent building roost or tunnels at deep well in the study area. *Rousettus leschenaulti* was already reported in the districts of Chunar and Pithoragarh, Uttar Pradesh (Bhat 1974).

Further studies are needed to understand abundance, reproduction, and population ecology of this species. *Cynopterus sphinx* is a common species which distributed throughout India. The study area also provides suitable roosting habitats for a wide distribution of *C. sphinx* in Uttar Pradesh. The distribution of *C. sphinx* was already reported in the districts of Lucknow, Varanasi, and Pilibhit (Wroughton 1914; Khajuria 1953). The current study revealed the distribution of *C. sphinx* in all 23 districts of eastern Uttar Pradesh. The colonies of *C. sphinx* were observed roosting in buildings as well as tree roost in the study area, while a large number of studies report the usage of tree roosts. It reveals that the population of *C. sphinx* is more adaptable and stable than *R. leschenaulti*. Molur et al. (2002) reported that *C. sphinx* is considered to be more adaptable than *C. brachyotis*. Like most other fruit bats in India, *C. sphinx* is considered as vermin under Schedule V of the Indian Wildlife Protection Act 1972. Though the Indian flying fox, *Pteropus giganteus*, is widely distributed throughout India, the current study reveals the occurrence of very high population of *P. giganteus* in Uttar Pradesh. The distribution of *P. giganteus* was already reported in Pilibhit (Wroughton 1914), Lucknow and Varanasi (Sinha 1980), and Allahabad (Bhatnagar and Srivastava 1974).

Two species of the family Rhinopomatidae were observed in the study area. The distribution of *R. microphyllum* ranges from Mauritania, Senegal, Nigeria, and Cameroon to Egypt, Arabia, Iran, Afghanistan, Pakistan, India, and Sumatra (Schliltter and Qumsiyeh 1996). In India, *R. microphyllum* has a widespread distribution. The distribution of *R. hardwickii* ranges from Niger, Morocco, Mauritania, East Africa, Arabia, Iran, Afghanistan India, to Myanmar (Koopman 1993). The distribution of *R. hardwickii* was common in the study area. Hipposiderids are found throughout the tropical areas of the Old World from Africa, Madagascar through India to southeastern Asia, the Philippines, New Guinea, Australia, New Caledonia, and the New Hebrides (Vanuatu). *Hipposideros fulvus* is distributed from Afghanistan to India and Sri Lanka. The distribution of *H. fulvus* in Uttar Pradesh was reported at Varanasi (Khajuria 1980). The current study reveals the distribution of *H. fulvus* at more sites in Uttar Pradesh. The Indian false vampire bat, *M. lyra*, has a distribution range from Afghanistan to Southern China and South to Pakistan, Sri Lanka, Malaysia, and India (Bates and Harrison 1997). The distribution of *M. lyra* in Uttar Pradesh is reported at Rani Bagh (Wroughton 1914), Agra, Ghazipur, and Lucknow (Sinha 1980).

A total of 59 species belong to the family Vespertilionidae widely distributed throughout India. However, four species belong to the genus *Pipistrellus* observed in the study area during the current study. *Pipistrellus coromandra* was distributed from Afghanistan to Southern China, India, Sri Lanka, Nicobar Island, Thailand, and Vietnam. In India, the distribution of *P. coromandra* was reported in Pilibhit and Ramnagar.

Pipistrellus tenuis was distributed from Afghanistan, Pakistan, India, and Sri Lanka to Thailand and Vietnam. In India, the distribution of *P. tenuis* was reported from Varanasi (Pathak and Sharma 1969) to Ramnagar, Pilibhit, and Sitabani (Wroughton 1914). The current study reveals the additional roost sites of *P. tenuis* in the study area. *Pipistrellus ceylonicus* was distributed from Pakistan, India, Sri Lanka, Myanmar, China, Vietnam, to northern Borneo (Bates and Harrisson 1997). The distribution of *P. ceylonicus* was not observed in Uttar Pradesh; however, the current study revealed the distribution of *P. ceylonicus* at different sites of Uttar Pradesh. *Scotophilus heathii* was distributed from Afghanistan to Southern China, Sri Lanka, Vietnam, and India. In Uttar Pradesh, the distribution of *S. heathii* was reported in Ramnagar (Bhat 1974), Allahabad (Bhatnagar and Srivastava 1974), and Meerut (Gandhi 1986). The current study reveals the distribution of *S. heathii* at Hardoi, Allipur, Pratapgarh, Kunda, Bareilly, and Sisandy house in Lucknow. The distribution of *S. kuhlii* in India is reported in Andhra Pradesh, Bihar, Gujarat, Madhya Pradesh, Maharashtra, Meghalaya, Karnataka, Kerala, Orissa, Rajasthan, Tamil Nadu, Tripura, Uttar Pradesh, and West Bengal (Bates and Harrison 1997; Molur et al. 2002). *Scotophilus kuhlii* was reported in both rural and urban landscapes and known to roost in crevices and holes in walls of huts and old buildings, caves, old temples, palm fronds, hollows in palm trees, and dried leaves on trees (Wroughton 1915; Brosset 1962; Sinha 1986). The current study also reveals that *S. kuhlii* prefers to roost in tree holes and wall crevices. In addition to the above findings, *S. kuhlii* was observed at Mamman purva and Hardoi. The current study revealed the distribution of *H. lankadiva* and *P. ceylonicus* in eastern Uttar Pradesh. Tree holes and wall crevices provide ideal roost site for *S. heathii* and *P. coromandra*.

The family Emballonuridae comprises 13 genera and about 50 species (Honacki et al. 1982). *Taphozous nudiventris* was distributed in a limited part of the study area. However, it has a widespread distribution in Africa ranging from Mauritania to Egypt in Asia (Brosset 1962). The current study reveals the highest distribution of bats in eastern Uttar Pradesh due to the presence of a large number of old monuments, palaces, caves, deep well, and forests which harbor bats. These permanent structures gave stable roosting conditions to the bats. In general, there was no major threat to the bats in the study area, except sporadic observations at times. Another support is that a maximum of bat colonies are located in old monuments which are governed by the Archaeological Survey of India, while few more colonies are located in caves. Thus, the state Uttar Pradesh provides a range of suitable habitats for the distribution of both frugivorous and insectivorous bats.

Acknowledgments The authors thank the Archaeological Survey of India for permitting us to conduct the field survey in old monuments of Uttar Pradesh. The State Forest Department is acknowledged for its permission to perform field survey in Uttar Pradesh. The financial assistance of Uttar Pradesh State Biodiversity Board, Uttar Pradesh, and the University Grants Commission, New Delhi, through research projects (No. 493/3-4-48/2013) and (No. 42-530/2013(SR)), respectively, to VE is acknowledged. VM acknowledge the university for financial support through UGC University fellowship. MK is a University Grants Commission-Rajeev Gandhi National fellowship holder.

References

Bates PJJ Harrison DL (1997) Bats of the Indian subcontinent. Harrison Zoological Museum Press, England, 258 p

Bhat HR (1974) Records and observations on bats of Himalayan region of Uttar Pradesh and West Bengal, India. J Bombay Nat Hist Soc 71(1):51–57

Bhatnagar VS, Srivastava MDL (1974) Somatic chromosomes of four common bats of Allahabad. Cytologia 39(2):327–334

Brosset A (1962) The bats of central and western India. Part I. J Bombay Nat Hist Soc 59:1–57

Cox PA, Elmqvistpierson TED, Rainey WE (1991) Flying foxes as strong interactors in South Pacific Island ecosystems: a conservation hypothesis. Conserv Biol 5:448–454

Davison GNH, Zubaid A (1992) Food habits of the lesser false vampire, *Megaderma spasma*, from Kuala Lumpur, Peninsular Malaysia fur sangetier kunde 57: 310–312

Eckrich M, Neuweiler G (1988) Food habits of the sympatric insectivorous bats *Rhinolophus rouxi* and *Hipposideros lankadiva* from Sri Lanka. J Zool (London) 215:729–737

Emmons LH (1997) Neotropical rain forest, 2nd edn. University press Chicago, Chicago

Findley JS (1976) The structure of bat communities. Am Nat 110:129–139

Findley JS (1993) Bats: a community perspective. Cambridge University Press, Cambridge

Findley JS, Black HL (1983) Morphological and dietary structuring of a Zambian insectivorous bat community. Ecology 64:625–630

Fleming TH (1986) Opportunism versus specialization: the evolution of feeding strategic in frugivorous bats. In: Estrada A, Fleming TH (eds) Frugivorous and seed dispersal. Junk, Dordrecht, pp 105–118

Gandhi S (1986) A new record of Plagiorchis vespertillionides from a bat in India. Res Bull Punjab Univ 37(1–2):147–148

Handley CO (1966) Checklist of the mammals of Panama. In: Wenzel RL, Tipton VJ (eds) Ectoparasite in Panama. Field Museum Natural History, Chicago, pp 753–795

Honacki JH, Kinman KE, Koeppl JW (eds) (1982) Mammal species of the world: a taxonomic and geographic reference. Allen Press and The Association of Systematics Collections, Lawrence, p 694

Khajuria H (1953) Taxonomic studies on some Indian Chiroptera. Rec Indian Mus 50:113–128

Khajuria H (1980) Taxonomical and ecological studies on bats of Jabalpur district Madhya Pradesh, India, part 2nd. Families Megadermatidae, Rhinolophidae Vespertilionidae. Miscellaneous publication occasional paper. Rec Zool Surv 19:1–69

Koopman KF (1993) Order Chiroptera. In: Wilson DE, Reeder DM (eds) Mammals species of the world: a taxonomic and geographic reference. Smithsonian Institution Press, Washington, DC, pp 137–241

Marshall AG (1985) Old World phytophagous bats (Megachiroptera) and their food plants: a survey. Zool J Linnean Soc 83:351–369

Molur S, Marimuthu G, Srinivasulu C, Mistry S, Hutson AM, PJJ B, Walker S, Padma KP, ARP B (2002) Status of south Asian Chiroptera: conservation assessment and management plan (C.A.M.P.) workshop report. Zoo Outreach Organization/CBSG-South Asia, Coimbatore

Nowak RM (1994) Walkers bats of the world, 5th edn. John Hopkins University Press, Baltimore, pp 1–187

Pagel MD, May RM, Collie AR (1991) Ecological aspects of the geographic distribution and diversity of mammalian species. Am Nat 137:791–815

Pathak S, Sharma T (1969) Chromosomes of five species of Indian species Vespertilionidae bats. Caryologia 22(1):35–46

Patterson BD, Willing MR, Stevens RD (2003) Trophic strategies, niche partitioning and patterns of ecological organization. In: Kunz TH, Fenton MB (eds) Bat ecology. The University of Chicago Press, Chicago, pp 536–579

Rainey WE, Pierson ED, Elmqvist T, Cox PA (1995) The role of Pteropodids in oceanic island ecosystems of the Pacific. In: Racey PA, Swift M (eds) Ecology, behaviour and evolution of bats. Symposium Zoological Society London. Oxford University Press, Oxford

Sinha YP (1980) The bats of Rajasthan: taxonomy and zoogeography. Rec Zool Surv India 76(1–4):7–63

Sinha YP (1986) The bats of Bihar: taxonomy and field ecology. Records Zool Surv India Misc Publ Occas Pap 77:1–60

Wilson DE (1983) Check list of mammals. In: Janzen DH (ed) Costa Rican natural history. University of Chicago Press, Chicago, pp 443–447

Wroughton RC (1914) Report no 15: Bombay natural history society's mammal survey of India, Kumaon. J Bombay Nat Hist Soc 22(4):282–301

Wroughton RC (1915) Report no 16: dry zone, Central Burma & Mount Popa. Bombay natural history society's mammal survey of India, Burma and Ceylon. J Bombay Nat Hist Soc 23:460–480

An Overview of the Vertebrate Diversity of Sriharikota (India's Spaceport), Southern India: Conservation Perspectives

4

J. Patrick David, Ranjit Manakadan, and S. Sivakumar

Abstract

Sriharikota is bestowed with one of the last remaining, largest and best preserved tracts of coastal tropical dry evergreen forests in India. Surveys were carried out from 2001 to 2007 to inventory the faunal species. In this chapter, we provide an overview of the status and distribution of the fauna of this region and discuss the significance of the island in the conservation scenario of the biodiversity of the Coromandel Coast. A total of 26 species of mammals, 215 species of birds, 29 species of reptiles, 12 species of amphibians and 53 species of fish were recorded. Some of the noteworthy findings were the occurrence of the grey slender loris *Loris lydekkerianus* and rusty-spotted cat in the island and a new species of yellow-green cat snake *Boiga flaviviridis* and the discovery of three unreported heronries in the island, including a large breeding colony of the threatened painted stork (*Mycteria leucocephala*). The beach on the eastern coast of the island is the nesting ground of the olive ridley turtle (*Lepidochelys olivacea*). The main factor responsible for the rich biodiversity of the island, besides due to the variety of habitat types, is the takeover of the island by the Indian Space Research Organisation (ISRO) and the shifting of the human population from the island and the pro-conservation attitude of the authorities. It is suggested that such high-security and disturbed enclaves could play a role in protecting India's wildlife. However, the development and expansion plans of the spaceport are of concern, and there needs to be a judicious mix of wildlife conservation initiatives

J. P. David (✉) · R. Manakadan
Bombay Natural History Society, Mumbai, India
e-mail: patdavid28@gmail.com

S. Sivakumar
Centre for Environment Education, Ahmedabad, India

C. Sivaperuman, K. Venkataraman (eds.), *Indian Hotspots*,
https://doi.org/10.1007/978-981-10-6605-4_4

with developmental activities in Sriharikota and for other such biodiversity-rich enclaves.

Keywords
Amphibians · Birds · Conservation · Fish · Mammals · Reptiles · Tropical dry evergreen forest

4.1 Introduction

The tropical dry evergreen forest (TDEF) occurs mainly in northern Sri Lanka, parts of the Coromandel Coast of India, the Caribbean and North and East Africa (Blasco and Legris 1973; Mani and Parthasarathy 2009; Parthasarathy et al. 2008). In India, historically, the forest extended from northern Andhra Pradesh to southern Tamil Nadu, as a belt of vegetation about 30–50 km wide. However, due to deforestation over the decades, TDEF is now found in small isolated patches along this stretch.

Sriharikota is known as the site of India's spaceport, but few know that the island has more than 90% of its area under TDEF forests, scrub forests, plantations, miscellaneous forest, sand dune vegetation, fresh and brackish streams, ponds, lakes and creeks. These diverse habitats support a variety of flora and fauna that enjoy good protection due to the high-security status of the island and the pro-conservation-oriented officials of the Indian Space Research Organisation (ISRO). The flora and fauna of Sriharikota remained largely unknown till ISRO invited the Bombay Natural History Society (BNHS) in the late 1970s to carry out a survey of the birds of the island (BNHS 1977). This was followed by an intensive study of the avifauna (Rao 1998; Samant and Rao 1996). A botanical exploration of the island was carried out by Suryanarayana et al. (1989, 1998), recording 445 species of plants. Again, on the invitation of ISRO, the vertebrate diversity of Sriharikota was documented and published (Manakadan and Sivakumar 2004a, b; Manakadan et al. 2004; Sivakumar and Manakadan 2004), during which the butterfly fauna was also documented (Sivakumar et al. 2004). This was followed by a project studying the interactions between fruits and fruit-eating animals (David et al. 2011, 2012) and the ecology of the slender loris (Manakadan et al. 2008).

4.2 Sriharikota

Sriharikota (13° 71′ N; 80° 20′ E; 181 km^2) is a spindle-shaped island situated in Nellore and Tiruvallur districts of Andhra Pradesh and Tamil Nadu. The forest here is classified as Tropical Dry Evergreen Forest (Champion and Seth 1968). The island is bordered on the east by the Bay of Bengal and on the north, south and west by Pulicat Lake. The island comprises of low ridges of sand, marine and aeolian in origin, rising 4.5–6 m above msl and sloping from west to east. The water table is at a depth of *c.* 2–5 m. The rainfall is largely from the northeast monsoon (October–December). Some rainfall is also received from the southwest monsoon

(June–September). The annual rainfall is *c.* 1200 mm. December to February is the cold season with temperatures as low as 10 °C; March to September is the hot season with temperatures soaring over 40 °C. Relative humidity is lowest during May (18%) and maximum during October (99%).

The forests in Sriharikota have a long history of systematic clear-felling for fuelwood and timber, starting with the British era. Plantations of eucalyptus (*Eucalyptus* sp.), casuarina (*Casuarina equisetifolia*) and cashew (*Anacardium occidentale*) were also grown over the years, these now cover ca. 20% of the landmass. *Acacia auriculiformis* was introduced on a small scale in the 1970s. The invasive *Prosopis juliflora* has proliferated in some areas, especially along the western edge of the island that borders Pulicat Lake and towards the southern parts of the island. Another invasive cane, *Calamus rotang*, introduced during the late nineteenth century by the British, has colonised the edges of freshwater bodies and water courses. Patches of abandoned coconut, tamarind, mango and palmyra (now overgrown with native vegetation) planted by the former settlers also occur, especially in the southern tracts of the island (Reddy 1981). There are extensive grasslands with scattered shrubs and trees in the southern part of the island and remnants of mangrove and salt marsh vegetation along the western edge of the island. Other than the terrestrial habitats, Sriharikota has both inland and coastal wetlands. The freshwater habitats are represented by ponds, lakes and streams. There are also brackish streams and lakes and creeks adjoining Pulicat Lake or the Bay of Bengal.

4.3 Vertebrate Fauna

Other than publication of the finding of the studies in the form of project reports (BNHS 1977; Samant and Rao 1998; Manakadan and Sivakumar 2004a, b; Sivakumar and Manakadan 2004, 2008; Manakadan et al. 2008), a number of papers have been also brought out (Manakadan and Sivakumar 2004c; Sivakumar and Manakadan 2005, 2007, 2010, Kannan et al. 2008; Manakadan et al. 2009a, b, 2013; David et al. 2011, 2012; Sivakumar et al. 2011). In this chapter, we provide an overview of occurrence of the vertebrate fauna in the island (for more details, refer to the cited papers and reports).

4.4 Mammals

A total of 26 species of mammals were recorded in Sriharikota (Table 4.1). Significant records are the grey slender loris *Loris lydekkerianus*, classified under Schedule I of the Indian Wildlife (Protection) Act 1972 and in the 'vulnerable' category by IUCN (2000). The population of the species in Sriharikota is estimated at 250–300 individuals. Another species of interest is the arboreal rusty-spotted cat *Prionailurus rubiginosus*. Earlier considered to be confined to southwestern India, there have been sightings from other areas, since the 1970s (Jackson 1998; Mukherjee 1998). Overall, the encounter rates of mammals were found to be higher in natural habitats than in plantations.

Table 4.1 Checklist of the mammals of Sriharikota

Sl. no.	Common name	Scientific name	Local name
1	Bonnet macaque	*Macaca radiata* (Geoffroy)	Koti
2	Slender loris	*Loris tardigradus* (Linnaeus)	Dewang
3	Jungle cat	*Felis chaus* (Guldenstaedt)	Jungu Pilli
4	Rusty-spotted cat	*Felis rubiginosa* (Geoffroy)	Mottabala Pilli
5	Small Indian civet	*Viverricula indica* (Desmarest)	Punigi Pilli
6	Common Indian mongoose	*Herpestes edwardsii* (Geoffroy)	Mongoosa, Mentrava
7	Golden jackal	*Canis aureus* (Linnaeus)	Nakka
8	Grey musk shrew	*Suncus murinus* (Linn.)	–
9	Savi's pygmy shrew	*S. etruscus* (Savi)	–
10	Indian flying fox	*Pteropus giganteus* (Brunnich)	Gabadai
11	Short-nosed fruit bat	*Cynopterus sphinx* (Vahl)	"
12	Dusky leaf-nosed bat	*Hipposideros ater* (Templeton)	"
13	Schneider leaf-nosed bat	*H. speoris* (Schneider)	"
14	Greater false vampire	*Megaderma lyra* (Geoffroy)	"
15	Asiatic greater yellow house bat	*Scotophilus heathii* (Horsfield)	"
16	Little Indian bat	*Pipistrellus coromandra* (Gray)	"
17	Three-striped palm squirrel	*Funambulus palmarum* (Linnaeus)	Odata
18	Indian mole rat	*Bandicota bengalensis* (Gray and Hardwicke)	Genu Eluka
19	Indian field mouse	*Mus booduga* (Gray)	Chittu Eluka
20	Long-tailed tree mouse	*Vandeleuria oleracea* (Bennett)	Parssh Eluka
21	Indian gerbil	*Tatera indica* (Hardwicke)	Tela Eluka
22	Common house rat	*Rattus rattus* (Linn)	Int Eluka
23	Bandicoot rat	*Bandicota indica* (Bechstein)	Pandi Kokku
24	Black-naped hare	*Lepus nigricollis nigricollis* (F. Cuvier)	Kuundeli
25	Spotted deer	*Axis axis* (Erxleben)	Jinka
26	Indian wild boar	*Sus scrofa* (Linnaeus)	Aduvi Pandi

4.5 Birds

A total of 215 species of birds were recorded in Sriharikota (Table 4.2). The discovery of three heronries at Beripeta, Karimanal and Madugu is another significant record for the island. Six species of water birds were recorded nesting in these heronries. The largest colony is at Beripeta in the central part of the island and contained around 250 nests of the 'near threatened' painted stork *Mycteria leucocephala*. Prior to the discovery of these heronries, the only known breeding sites for colonial water birds of Pulicat Lake were three heronries on the mainland (Subramanya 2001). Important records were the sighting of 13 white-rumped vulture *Gyps bengalensis*, a critically endangered species (IUCN 2014), but the species probably became locally extinct by 2003, as there were no further sightings from the area.

Table 4.2 Checklist of the birds of Sriharikota

Common name	Scientific name
Grebes	**Podicipedidae**
1. Little grebe	*Tachybaptus ruficollis* (Pallas, 1764)
Pelicans	**Pelecanidae**
2. Spot-billed pelican	*Pelecanus philippensis* (Gmelin, 1789)
Cormorants/shags	**Phalacrocoracidae**
3. Little cormorant	*Phalacrocorax niger* (Vieillot, 1817)
4. Indian shag	*Phalacrocorax fuscicollis* (Stephens, 1826)
Darters	**Anhingidae**
5. Darter	*Anhinga melanogaster* (Pennant, 1769)
Frigatebirds	**Fregatidae**
6. Lesser frigatebird	*Fregata ariel* (G.R. Gray, 1845)
Herons, egrets and bitterns	**Ardeidae**
7. Little egret	*Egretta garzetta* (Linnaeus, 1766)
8. Western reef egret	*Egretta gularis* (Bosc, 1792)
9. Grey heron	*Ardea cinerea* (Linnaeus, 1758)
10. Purple heron	*Ardea purpurea* (Linnaeus, 1766)
11. Large egret	*Casmerodius albus* (Linnaeus, 1758)
12. Median egret	*Mesophoyx intermedia* (Wagler, 1829)
13. Cattle egret	*Bubulcus ibis* (Linnaeus, 1758)
14. Indian pond-heron	*Ardeola grayii* (Sykes, 1832)
15. Little green heron	*Butorides striatus* (Linnaeus, 1758)
16. Black-crowned night heron	*Nycticorax nycticorax* (Linnaeus, 1758)
17. Yellow bittern	*Ixobrychus sinensis* (Gmelin, 1789)
18. Chestnut bittern	*Ixobrychus cinnamomeus* (Gmelin, 1789)
19. Black bittern	*Dupetor flavicollis* (Latham, 1790)
20.Great bittern	*Botaurus stellaris* (Linnaeus, 1758)
Storks	**Ciconiidae**
21. Painted stork	*Mycteria leucocephala* (Pennant, 1769)
22. Asian openbill stork	*Anastomus oscitans* (Boddaert, 1783)
Ibises and spoonbills	**Threskiornithidae**
23. Glossy ibis	*Plegadis falcinellus* (Linnaeus, 1766)
24. Oriental white ibis	*Threskiornis melanocephalus* (Latham, 1790)
25. Eurasian spoonbill	*Platalea leucorodia* (Linnaeus, 1758)
Flamingos	**Phoenicopteridae**
26. Greater flamingo	*Phoenicopterus ruber* (Linnaeus, 1758)
27. Lesser flamingo	*Phoenicopterus minor* (Geoffroy, 1798)
Swans, geese and ducks	**Anatidae**
28. Lesser whistling duck	*Dendrocygna javanica* (Horsfield, 1821)
29. Bar-headed goose	*Anser indicus* (Latham, 1790)
30. Comb duck	*Sarkidiornis melanotos* (Pennant, 1769)
31. Cotton teal	*Nettapus coromandelianus* (Gmelin, 1789)
32. Gadwall	*Anas strepera* (Linnaeus, 1758)

(continued)

Table 4.2 (continued)

Common name	Scientific name
33. Eurasian wigeon	*Anas penelope* (Linnaeus, 1758)
34. Spot-billed duck	*Anas poecilorhyncha* (J.R. Forester, 1781)
35. Northern shoveller	*Anas clypeata* (Linnaeus, 1758)
36. Northern pintail	*Anas acuta* (Linnaeus, 1758)
37. Garganey	*Anas querquedula* (Linnaeus, 1758)
38. Common teal	*Anas crecca* (Linnaeus, 1758)
Hawks, eagles, buzzards, old world vultures, kites, harriers	**Accipitridae**
39. Oriental honey buzzard	*Pernis ptilorhynchus* (Temminck, 1821)
40. Black-shouldered kite	*Elanus caeruleus* (Desfontaines, 1789)
41. Black kite	*Milvus migrans* (Boddaert, 1783)
42. Brahminy kite	*Haliastur indus* (Boddaert, 1783)
43. White-bellied sea eagle	*Haliaeetus leucogaster* (Gmelin, 1788)
44. Indian white-backed vulture	*Gyps bengalensis* (Gmelin, 1788)
45. Short-toed snake eagle	*Circaetus gallicus* (Gmelin, 1788)
46. Crested serpent eagle	*Spilornis cheela* (Latham, 1790
47. Western marsh harrier	*Circus aeruginosus* (Linnaeus, 1758)
48. Pallid harrier	*Circus macrourus* (S.G. Gmelin, 1770)
49. Shikra	*Accipiter badius* (Gmelin, 1788)
50. Besra sparrowhawk	*Accipiter virgatus* (Temminck, 1822)
51. White-eyed buzzard	*Butastur teesa* (Franklin, 1832)
Osprey	**Pandionidae**
52. Osprey	*Pandion haliaetus* (Linnaeus, 1758)
Falcons	**Falconidae**
53. Common kestrel	*Falco tinnunculus* (Linnaeus, 1758)
54. Amur falcon	*Falco amurensis* (Radde, 1863)
55. Peregrine falcon	*Falco peregrinus* (Tunstall, 1771)
Pheasants, partridges, quails	**Phasianidae**
56. Grey francolin	*Francolinus pondicerianus* (Gmelin, 1789)
57. Blue-breasted quail	*Coturnix chinensis* (Linnaeus, 1766)
58. Red spurfowl	*Galloperdix spadicea* (Gmelin, 1789)
59. Grey junglefowl	*Gallus sonneratii* (Temminck, 1813)
Buttonquails/bustard quails	**Turnicidae**
60. Yellow-legged buttonquail	*Turnix tanki* (Blyth, 1843)
61. Common buttonquail	*Turnix suscitator* (Gmelin, 1789)
Rails, crakes, moorhens, coots	**Rallidae**
62. Blue-breasted rail	*Gallirallus striatus* (Linnaeus, 1766)
63. Water rail	*Rallus aquaticus* (Linnaeus, 1758)
64. White-breasted waterhen	*Amaurornis phoenicurus* (Pennant, 1769)
65. Common moorhen	*Gallinula chloropus* (Linnaeus, 1758)
66. Common coot	*Fulica atra* (Linnaeus, 1758)
Bustards	
67. Lesser florican	*Sypheotides indica* (J.F. Miller, 1782)
Jacanas	**Jacanidae**

(continued)

Table 4.2 (continued)

Common name	Scientific name
68. Pheasant-tailed jacana	*Hydrophasianus chirurgus* (Scopoli, 1786)
Painted-snipes	**Rostratulidae**
69. Greater painted-snipe	*Rostratula benghalensis* (Linnaeus, 1758)
Plovers, dotterels, lapwings	**Charadriidae**
70. Pacific golden plover	*Pluvialis fulva* (Gmelin, 1789)
71. Grey plover	*Pluvialis squatarola* (Linnaeus, 1758)
72. Little ringed plover	*Charadrius dubius* (Scopoli, 1786)
73. Kentish plover	*Charadrius alexandrinus* (Linnaeus, 1758)
74. Lesser sand plover	*Charadrius mongolus* (Pallas, 1776)
75. Greater sand plover	*Charadrius leschenaultii* (Lesson, 1826)
76. Yellow-wattled lapwing	*Vanellus malabaricus* (Boddaert, 1783)
77. Red-wattled lapwing	*Vanellus indicus* (Boddaert, 1783)
Sandpipers, stints, snipes, godwits and curlews	**Scolopacidae**
78. Pintail snipe	*Gallinago stenura* (Bonaparte, 1830)
79. Wood snipe	*Gallinago nemoricola* (Hodgson, 1836)
80. Black-tailed godwit	*Limosa limosa* (Linnaeus, 1758)
81. Whimbrel	*Numenius phaeopus* (Linnaeus, 1758)
82. Eurasian curlew	*Numenius arquata* (Linnaeus, 1758)
83. Spotted redshank	*Tringa erythropus* (Pallas, 1764)
84. Common redshank	*Tringa totanus* (Linnaeus, 1758)
85. Marsh sandpiper	*Tringa stagnatilis* (Bechstein, 1803)
86. Common greenshank	*Tringa nebularia* (Gunner, 1767)
87. Green sandpiper	*Tringa ochropus* (Linnaeus, 1758)
88. Wood sandpiper	*Tringa glareola* (Linnaeus, 1758)
89. Terek sandpiper	*Xenus cinereus* (Guldenstadt, 1774)
90. Common sandpiper	*Actitis hypoleucos* (Linnaeus, 1758)
91. Ruddy turnstone	*Arenaria interpres* (Linnaeus, 1758)
92. Great knot	*Calidris tenuirostris* (Horsfield, 1821)
93. Red knot	*Calidris canutus* (Linnaeus, 1758)
94. Little stint	*Calidris minuta* (Leisler, 1812)
95. Temminck's stint	*Calidris temminckii* (Leisler, 1812)
96. Curlew sandpiper	*Calidris ferruginea* (Pontoppidan, 1813)
97. Ruff	*Philomachus pugnax* (Linnaeus, 1758)
Ibisbill, avocets and stilts	**Recurvirostridae**
98. Black-winged stilt	*Himantopus himantopus* (Linnaeus, 1758)
99. Pied avocet	*Recurvirostra avosetta* (Linnaeus, 1758)
Phalaropes	**Phalaropidae**
100. Red-necked phalarope	*Phalaropus lobatus* (Linnaeus, 1758)
Stone-curlew and stone-plovers/thick-knees	**Burhinidae**
101. Stone-curlew	*Burhinus oedicnemus* (Linnaeus, 1758)
Gulls, terns and noddies	**Laridae**
102. Heuglin's gull	*Larus heuglini* (Bree, 1876)
103. Pallas's gull	*Larus ichthyaetus* (Pallas, 1773)
104. Brown-headed gull	*Larus brunnicephalus* (Jerdon, 1840)

(continued)

Table 4.2 (continued)

Common name	Scientific name
105. Black-headed gull	*Larus ridibundus* (Linnaeus, 1766)
106. Gull-billed tern	*Gelochelidon nilotica* (Gmelin, 1789)
107. Caspian tern	*Sterna caspia* (Pallas, 1770)
108. Little tern	*Sterna albifrons* (Pallas, 1764)
109. Whiskered tern	*Chlidonias hybridus* (Pallas, 1811)
110. Black tern	*Chlidonias niger* (Linnaeus, 1758)
Pigeons and doves	**Columbidae**
111. Blue rock pigeon	*Columba livia* (Gmelin, 1789)
112. Oriental turtle dove	*Streptopelia orientalis* (Latham, 1790)
113. Little brown dove	*Streptopelia senegalensis* (Linnaeus, 1766)
114. Spotted dove	*Streptopelia chinensis* (Scopoli, 1786)
115. Red collared dove	*Streptopelia tranquebarica* (Hermann, 1804)
116. Orange-breasted green pigeon	*Treron bicincta* (Jerdon, 1840)
Parakeets and hanging parrots	**Psittacidae**
117. Rose-ringed parakeet	*Psittacula krameri* (Scopoli, 1769)
118. Plum-headed parakeet	*Psittacula cyanocephala* (Linnaeus, 1766)
Cuckoos, malkohas and coucals	**Cuculidae**
119. Pied crested cuckoo	*Clamator jacobinus* (Boddaert, 1783)
120. Red-winged crested cuckoo	*Clamator coromandus* (Linnaeus, 1766)
121. Brainfever bird	*Hierococcyx varius* (Vahl, 1797)
122. Lesser cuckoo	*Cuculus poliocephalus* (Latham, 1790)
123. Indian plaintive cuckoo	*Cacomantis passerinus* (Vahl, 1797)
124. Asian koel	*Eudynamys scolopacea* (Linnaeus, 1758)
125. Small green-billed malkoha	*Phaenicophaeus viridirostris* (Jerdon, 1840)
126. Greater coucal	*Centropus sinensis* (Stephens, 1815)
Barn owls	**Tytonidae**
127. Barn owl	*Tyto alba* (Scopoli, 1769)
Owls	**Strigidae**
128. Collared scops owl	*Otus bakkamoena* (Pennant, 1769)
129. Eurasian eagle-owl	*Bubo bubo* (Linnaeus, 1758)
130. Brown fish-owl	*Ketupa zeylonensis* (Gmelin, 1788)
131. Mottled wood-owl	*Strix ocellata* (Lesson, 1839)
132. Spotted owlet	*Athene brama* (Temminck, 1821)
Nightjars	**Caprimulgidae**
133. Indian jungle nightjar	*Caprimulgus indicus* (Latham, 1790)
134. Common Indian nightjar	*Caprimulgus asiaticus* (Latham, 1790)
135. Franklin's nightjar	*Caprimulgus affinis* (Horsfield, 1821)
Swifts	**Apodidae**
136. Asian palm swift	*Cypsiurus balasiensis* (J.E. Gray, 1829)
137. House swift	*Apus affinis* (J.E. Gray, 1830)
Kingfishers	**Alcedinidae**
138. Small blue kingfisher	*Alcedo atthis* (Linnaeus, 1758)

(continued)

Table 4.2 (continued)

Common name	Scientific name
139. White-breasted kingfisher	*Halcyon smyrnensis* (Linnaeus, 1758)
140. Black-capped kingfisher	*Halcyon pileata* (Boddaert, 1783)
141. Lesser pied kingfisher	*Ceryle rudis* (Linnaeus, 1758)
Bee-eaters	**Meropidae**
142. Small bee-eater	*Merops orientalis* (Latham, 1801)
143. Blue-tailed bee-eater	*Merops philippinus* (Linnaeus, 1766)
Rollers	**Coraciidae**
144. Indian roller	*Coracias benghalensis* (Linnaeus, 1758)
Hoopoes	**Upupidae**
145. Common hoopoe	*Upupa epops* (Linnaeus, 1758)
Barbets	**Capitonidae**
146. Coppersmith barbet	*Megalaima haemacephala* (P.L.S. Müller, 1776)
Woodpeckers	**Picidae**
147. Lesser golden-backed woodpecker	*Dinopium benghalense* (Linnaeus, 1758)
Pittas	**Pittidae**
148. Indian pitta	*Pitta brachyura* (Linnaeus, 1766)
Larks	**Alaudidae**
149. Jerdon's bush lark	*Mirafra affinis* (Blyth, 1845)
150. Ashy-crowned sparrow-lark	*Eremopterix grisea* (Scopoli, 1786)
151. Eastern skylark	*Alauda gulgula* (Franklin, 1831)
Swallows and martins	**Hirundinidae**
152. Common swallow	*Hirundo rustica* (Linnaeus, 1758)
153. Wire-tailed swallow	*Hirundo smithii* (Leach, 1818)
Wagtails and pipits	**Motacillidae**
154. Forest wagtail	*Dendronanthus indicus* (Gmelin 1789)
155. Large pied wagtail	*Motacilla maderaspatensis* (Gmelin, 1789)
156. Citrine wagtail	*Motacilla citreola* (Pallas, 1776)
157. Yellow wagtail	*Motacilla flava* (Linnaeus, 1758)
158. Paddyfield pipit	*Anthus rufulus* (Vieillot, 1818)
Cuckoo-shrikes, flycatcher-shrikes, trillers, minivets, woodshrikes	**Campephagidae**
159. Large cuckoo-shrike	*Coracina macei* (Lesson, 1830)
160. Black-headed cuckoo-shrike	*Coracina melanoptera* (Rüppell, 1839)
161. Ashy minivet	*Pericrocotus divaricatus* (Raffles, 1822)
162. Common woodshrike	*Tephrodornis pondicerianus* (Gmelin, 1789)
Bulbuls and finchbills	**Pycnonotidae**
163. Red-whiskered bulbul	*Pycnonotus jocosus* (Linnaeus, 1758)
164. Red-vented bulbul	*Pycnonotus cafer* (Linnaeus, 1766)
165. White-browed bulbul	*Pycnonotus luteolus* (Lesson, 1841)
Ioras, chloropsis/leafbird, fairy-bluebird	**Irenidae**
166. Common iora	*Aegithina tiphia* (Linnaeus, 1758)

(continued)

Table 4.2 (continued)

Common name	Scientific name
Shrikes	**Laniidae**
Brown shrike	*Lanius cristatus* (Linnaeus, 1758)
167. Bay-backed shrike	*Lanius vittatus* (Valenciennes, 1826)
168. Rufous-backed shrike	*Lanius schach* (Linnaeus, 1758)
Thrushes, shortwings, robins, forktails, wheaters	**Turdinae**
169. Orange-headed thrush	*Zoothera citrina* (Latham, 1790)
170. Indian blue robin	*Luscinia brunnea* (Hodgson, 1837)
171. Oriental magpie-robin	*Copsychus saularis* (Linnaeus, 1758)
172. White-rumped Shama	*Copsychus malabaricus* (Scopoli, 1786)
173. Indian robin	*Saxicoloides fulicata* (Linnaeus, 1776)
174. Black redstart	*Phoenicurus ochruros* (Gmelin, 1774)
Babblers, laughingthrushes, babaxes, barwings, yuhinas	**Timaliidae**
175. White-headed babbler	*Turdoides affinis* (Jerdon, 1847)
Goldcrest, prinias, tesias, warblers	**Sylviinae**
176. Streaked fantail-warbler	*Cisticola juncidis* (Rafinesque, 1810)
177. Franklin's prinia	*Prinia hodgsonii* (Blyth, 1844)
178. Plain prinia	*Prinia inornata* (Sykes, 1832)
179. Blyth's reed warbler	*Acrocephalus dumetorum* (Blyth, 1849)
180. Indian great reed warbler	*Acrocephalus stentoreus* (Hemprich & Ehrenberg, 1833)
181. Thick-billed warbler	*Acrocephalus aedon* (Pallas, 1776)
182. Common tailorbird	*Orthotomus sutorius* (Pennant, 1769)
183. Greenish leaf warbler	*Phylloscopus trochiloides* (Sundevall, 1837)
184. Large-billed leaf warbler	*Phylloscopus magnirostris* (Blyth, 1843)
185. Common lesser whitethroat	*Sylvia curruca* (Linnaeus, 1758)
Flycatchers	**Muscicapinae**
186. Asian brown flycatcher	*Muscicapa dauurica* (Pallas, 1811)
187. Brown-breasted flycatcher	*Muscicapa muttui* (Layard, 1854)
188. Red-throated flycatcher	*Ficedula parva* (Bechstein, 1792)
189. Blue-throated flycatcher	*Cyornis rubeculoides* (Vigors, 1831)
Monarch flycatchers and paradise flycatchers	**Monarchidae**
190. Asian paradise flycatcher	*Terpsiphone paradisi* (Linnaeus, 1758)
191. Black-naped monarch flycatcher	*Hypothymis azurea* (Boddaert, 1783)
Flowerpeckers	**Dicaeidae**
192. Tickell's flowerpecker	*Dicaeum erythrorhynchos* (Latham, 1790)
Sunbirds and spiderhunters	**Nectariniidae**
193. Purple-rumped sunbird	*Nectarinia zeylonica* (Linnaeus, 1766)
194. Purple sunbird	*Nectarinia asiatica* (Latham, 1790)
195. Loten's sunbird	*Nectarinia lotenia* (Linnaeus, 1766)
Munias (estrildid finches)	**Estrildidae**
196. White-throated munia	*Lonchura malabarica* (Linnaeus, 1758)

(continued)

Table 4.2 (continued)

Common name	Scientific name
197. White-rumped munia	*Lonchura striata* (Linnaeus, 1766)
198. Black-headed munia	*Lonchura malacca* (Linnaeus, 1766)
Sparrows and snowfinches	**Passerinae**
199. House sparrow	*Passer domesticus* (Linnaeus, 1758)
200. Yellow-throated sparrow	*Petronia xanthocollis* (Burton, 1838)
Weavers	**Ploceidae**
201. Baya weaver	*Ploceus philippinus* (Linnaeus, 1766)
Starlings and mynas	**Sturnidae**
202. Brahminy starling	*Sturnus pagodarum* (Gmelin, 1789)
203. Rosy starling	*Sturnus roseus* (Linnaeus, 1758)
204. Common starling	*Sturnus vulgaris* (Linnaeus, 1758)
205. Common myna	*Acridotheres tristis* (Linnaeus, 1766)
Orioles	**Oriolidae**
206. Eurasian golden oriole	*Oriolus oriolus* (Linnaeus, 1758)
Drongos	**Dicruridae**
207. Black drongo	*Dicrurus macrocercus* (Vieillot, 1817)
208. Ashy drongo	*Dicrurus leucophaeus* (Vieillot, 1817)
209. White-bellied drongo	*Dicrurus caerulescens* (Linnaeus, 1758)
210. Spangled drongo	*Dicrurus hottentottus* (Linnaeus, 1766)
Woodswallows/swallow-shrikes	**Artamidae**
211. Ashy woodswallow	*Artamus fuscus* (Vieillot, 1817)
Crows, jays, treepies, magpies	**Corvidae**
212. Indian treepie	*Dendrocitta vagabunda* (Latham, 1790)
213. House crow	*Corvus splendens* (Vieillot, 1817)
214. Jungle crow	*Corvus macrorhynchos* (Wagler, 1827)

Another significant sighting was of the water rail *Rallus aquaticus*, this being the southernmost record of the species in India, which till then had not been reported south of Mumbai.

4.6 Herpetofauna

In total, 12 amphibian species were recorded in Sriharikota (Table 4.3). The record of painted kaloula *Kaloula taprobanica* in Sriharikota constitutes the first record of the species in Andhra Pradesh. The reptile diversity included four species of turtles and tortoises, three species of geckos, two agamids, five skinks, one each of chameleon and monitor lizard and 18 species of snakes (Table 4.4). A new species of cat snake, misidentified by us as the Beddome's cat snake *Boiga beddomei*, has recently been described as a new species to science; yellow-green cat snake *Boiga flaviviridis* (Vogel and Ganesh 2013) with specimens also recorded from a few sites in the dry forests of eastern Peninsular India. The checklist of reptiles of Sriharikota constitutes ca. 40% of the reptiles reported from the eastern part of

Table 4.3 Checklist of the amphibians of Sriharikota

Sl. no.	Common name	Scientific name
Family: Bufonidae		
1.	Common Indian toad	*Bufo melanostictus* (Schneider, 1799)
Family: Microhylidae		
2.	Marbled narrow-mouthed frog	Ramanella variegata (Stoliczka, 1872)
3.	Ornate microhylid	*Microhyla ornata* (Dumeril and Bibron, 1841)
4.	Red microhylid	*Microhyla rubra* (Jerdon, 1854)
5.	Marbled balloon frog	*Uperodon systoma* (Schneider, 1799)
6.	Painted kaloula	*Kaloula taprobanica* (Parker, 1934)
Family: Ranidae		
7.	Skittering frog	*Euphlyctis cyanophlyctis* (Schneider, 1799)
8.	Indian pond frog	*Euphlyctis hexadactylus* (Lesson, 1834)
9.	Jerdon's bull frog	*Hoplobatrachus crassus* (Jerdon, 1853)
10.	Indian cricket frog	*Fejervarya limnocharis* (Gravenhorst, 1829)
11.	Indian burrowing frog	*Tomopterna rolandae* (Dubois, 1983)
Family: Rhacophoridae		
12.	Common tree frog	*Polypedates* maculatus (Gray, 1834)

Table 4.4 Checklist of the reptiles of Sriharikota

Sl. No.	Common Name	Scientific Name
Family: Cheloniidae		
1.	Olive ridley turtle	*Lepidochelys olivacea* (Eschscholte, 1829)
Family: Bataguridae		
2.	Indian pond terrapin	*Melanochelys trijuga* (Schweigger, 1812)
Family: Trionychidae		
3.	Indian flapshell turtle	*Lissemys punctata* (Bonnaterre, 1789)
Family: Testudinidae		
4.	Starred tortoise	*Geochelone elegans* (Schoepff, 1795)
Family: Gekkonidae		
5.	Bark gecko	*Hemidactylus leschenaultii* (Dumeril and Bibron, 1836)
6.	Brook's gecko	*Hemidactylus brookii* (Gray, 1845)
7.	Southern house gecko	*Hemidactylus frenatus* (Schlegel in: Dumeril & Bibron, 1836)
Family: Agamidae		
8.	Common garden lizard	*Calotes versicolor* (Daudin, 1802)
9.	Fan-throated lizard	*Sitana ponticeriana* (Cuvier, 1844)
Family: Chamaeleonidae		
10.	Indian chameleon	*Chamaeleo zeylanicus* (Laurenti, 1768)
Family: Scincidae		
11.	Common skink	*Mabuya carinata* (Schneider, 1801)
12.	Little skink	*Mabuya macularia* (Blyth, 1853)
13.	Sand skink	*Mabuya bibronii* (Gray, 1838)
14.	Snake skink	*Lygosoma punctata* (Gmelin, 1799)

(continued)

Table 4.4 (continued)

Sl. No.	Common Name	Scientific Name
15.	White-spotted skink	*Lygosoma albopunctata* (Gray, 1846)
Family: Varanidae		
16.	Common Indian monitor	*Varanus bengalensis* (Daudin, 1802)
Family: Typhlopidae		
17.	Common worm snake	*Ramphotyphlops braminus* (Daudin, 1803)
Family: Boidae		
18.	Russell's earth boa	*Eryx conicus* (Schneider, 1801)
19.	John's earth boa	*Eryx johnii* (Russell, 1801)
20.	Indian python	*Python molurus* (Linnaeus, 1758)
Family: Colubridae		
20.	Common rat snake	*Ptyas mucosa* (Linnaeus, 1758) $
21.	Variegated kukri snake	*Oligodon taeniolatus* (Jerdon, 1853)
22.	Common Indian bronzeback tree snake	*Dendrelaphis tristis* (Daudin, 1803)
23.	Common wolf snake	*Lycodon aulicus* (Linnaeus, 1758)
24.	Checkered keelback	*Xenochrophis piscator* (Schneider, 1799)
25.	Buff striped keelback	*Amphiesma stolatum* (Linn. 1758)
26.	Olivaceous keelback	*Atretium schistosum* (Daudin, 1803)
27.	Common cat snake	*Boiga trigonata* (Schneider in: Bechstein, 1802)
28.	Yellow-green cat snake	*Boiga flaviviridis* (Vogel and Ganesh 2013)
29.	Common green whip snake	*Ahaetulla nasuta* (Anderson, 1898)
30.	Black-headed snake	*Sibynophis subpunctatus* (Dum. & Bibr. 1854)
Family: Elapidae		
31.	Indian cobra	*Naja naja* (Linn. 1758)
32.	Common Indian krait	*Bungarus caeruleus* (Schneider, 1801)
Family: Viperidae		
33.	Russell's viper	*Daboia russelii* (Shaw and Nodder, 1797)
34.	Saw-scaled viper	*Echis carinatus* (Schneider, 1801)

India by (Das 1996; Shanker et al. 2003; Choudhury et al. 2003) and 53% reported in Andhra Pradesh by Sanyal et al. (1993). Interesting records were of the nesting of olive ridley turtles *Lepidochelys olivacea* (around 150 nests) which is placed in the 'Endangered' category by IUCN and Schedule I of the Indian Wildlife (Protection) Act, 1972. Other interesting records were of the Indian chameleon *Chamaeleo zeylanicus* and common Indian monitor *Varanus bengalensis* which are listed under the 'Vulnerable' category of the IUCN and Schedule II of the Indian Wildlife (Protection) Act, 1972.

4.7 Fish

A total of 53 fish species belonging to 36 families and 10 orders were recorded in Sriharikota (Table 4.5). Nineteen species were recorded from freshwater habitats, 38 from fresh-brackish and 39 from brackish-saline wetlands. Two important perennial wetlands in the island, Malliplate Vagu (a fresh-brackish stream) and Urugayya (a brackish-saline lake), serve as nurseries for the young of marine fish and prawn species, including two anguillid eels. The abandoned irrigation ponds are major refuges for two threatened air-breathing species, *Clarias batrachus* and *Anabas testudineus.* An exotic species *Oreochromis mossambicus* and a species native to north India, *Colisa lalia*, were recorded.

Table 4.5 Checklist of the fishes of Sriharikota

Common name	Freshwater	Fresh-brackish wetlands[a]	Brackish-saline wetlands[b]
Family: Elopidae			
Giant herring *Elops machnata* (Forsskal)	–	#	#
Family: Megalopidae			
Oxeye tarpon *Megalops cyprinoides* (Broussonet)	+	+	+
Family: Anguillidae			
Longfin eel Anguilla bengalensis (Gray)	–	+	+
Shortfin eel *Anguilla bicolor* (McClelland)	–	+	+
Family: Ophichthidae			
Paddy snake eel *Pisodonophis boro* (Ham.-Buch.)	–	#	#
Family: Clupeidae			
Bloch's gizzard-Shad *Nematalosa nasus* (Bloch)	–	–	+
Family: Chanidae			
Milkfish Chanos chanos (Forsskal)	–	#	#
Family: Cyprinidae			
Common flying barb *Esomus danricus* (Ham.-Buch.)	+	–	–
Spotfin swamp barb *Puntius sophore* (Ham.-Buch)	+	+	–
Family: Cobitidae			
Malabar loach *Lepidoce phalus thermalis* (Val.)	+	–	–
Family: Bagridae			
Long-whiskered catfish *Mystus gulio* (Ham.-Buch.)	–	+	+
Striped dwarf catfish *Mystus vittatus* (Bloch)	+	–	–

(continued)

Table 4.5 (continued)

Common name	Freshwater	Fresh-brackish wetlands[a]	Brackish-saline wetlands[b]
Family: Clariidae			
Magur *Clarias batrachus* (Linn.)	+	–	–
Family: Heteropneustidae			
Stinging catfish *Heteropneustes fossilis* (Bloch)	+	–	–
Family: Ariidae			
Threadfin sea catfish *Arius arius* (Ham.-Buch.)	–	#	#
Family: Plotosidae			
Canine eel catfish *Plotosus canius* (Ham.-Buch.)	–	#	#
Family: Hemiramphidae			
Congaturi halfbeak *Hyporhamphus limbatus* (Val.)	–	–	+
Family: Belonidae			
Spot-tail garfish *Strongylura strongylura* (V. Hasselt)	–	–	+
Family: Adrianichthyidae			
Carnatic ricefish *Oryzias carnaticus* (Jerdon)	+	+	+
Family: Aplocheilidae			
Dwarf panchax *Aplocheilus parvus* (Raj)	+	+	+
Family: Platycephalidae			
Bartail flathead *Platycephalus indicus* (Linn.)	–	#	+
Family: Latidae			
Barramundi *Lates calcarifer* (Bloch)	–	+	+
Family: Ambassidae			
Commerson's glassy perchlet *Ambassis ambassis* (Cuvier)	–	+	+
Bald glassy perchlet *Ambassis gymnocephalus* (Lacepede)	–	–	+
Indian glassfish *Parambassis ranga* (Ham.-Buch.)	+	–	+
Family: Terapontidae			
Target terapon *Terapon jarbua* (Forsskal)	–	+	+
Family: Sillaginidae			
Silver Silago *Sillago sihama* (Forsskal)	–	#	#
Family: Carangidae			
Six-banded trevally *Caranx sexfasciatus* (Quoy & Gaimard)	–	#	+

(continued)

Table 4.5 (continued)

Common name	Freshwater	Fresh-brackish wetlands[a]	Brackish-saline wetlands[b]
Family: Lutjanidae			
River snapper *Lutjanus argentimaculatus* (Forsskal)	–	+	–
Family: Gerreidae			
Whiptail silver-biddy *Gerres filamentosus* (Cuvier)	–	#	+
Black-tipped silver-biddy *Gerres lucidus* (Cuvier)	–	#	+
Family: Scatophagidae			
Spotted scat *Scatophagus argus* (Linn.)	–	#	#
Family: Cichlidae			
Orange chromide *Etroplus maculatus* (Bloch)	+	+	+
Banded pearlspot *Etroplus suratensis* (Bloch)	–	+	+
Mozambique tilapia *Oreochromis mossambicus* (Peters)	+	+	+
Family: Mugilidae			
Greenback mullet *Liza subviridis* (Val.)	–	+	+
Flathead mullet *Mugil cephalus* (Linn.)	–	+	+
Family: Gobiidae			
Bighead goby *Drombus globiceps* (Hora)	–	–	+
Tropical sand goby *Favonigobius reichei* (Bleeker)	–	–	+
Tank goby *Glossogobius giuris* (Ham.-Buch)	+	+	+
Sharptail goby *Oligolepis acutipennis* (Val.)	–	+	–
Javanese goby *Pseudogobius javanicus* (Bleeker)	–	+	+
Barred goby *Pseudogobius poicilosoma* (Bleeker)	–	–	+
Family: Eleotrididae			
Broadhead sleeper *Eleotris melanosoma* (Bleeker)	–	+	+
Family: Anabantidae			
Indian climbing perch *Anabas testudineus* (Bloch)	+	+	–
Family: Osphronemidae			
Spike-tailed paradise fish *Pseudosphromenus cupanus* (Val.)	+	+	–
Family: Osphronemidae			
Dwarf gourami *Colisa lalia* (Ham.-Buch.)	+	–	–

(continued)

Table 4.5 (continued)

Common name	Freshwater	Fresh-brackish wetlands[a]	Brackish-saline wetlands[b]
Family: Channidae			
Spotted snakehead *Channa punctatus* (Bloch)	+	+	–
Striped snakehead *Channa striatus* (Bloch)	+	+	–
Family: Mastacembelidae			
Striped spiny eel *Macrognathus pancalus* (Ham.-Buch.)	#	–	–
Family: Soleidae			
Oriental sole *Brachirus orientalis* (Bloch & Schn.)	–	#	#
Family: Tricanthidae			
Short-nosed tripod fish *Triacanthus biaculeatus* (Bloch)	–	#	+
Family: Tetraodontidae			
Patoka pufferfish *Chelonodon patoca* (Ham.-Buch.)	–	+	+

In the case of creeks, the salinity increase is primarily due to inflows from the Bay of Bengal (Sateneru-Sidimuthu Kayya) or Pulicat Lake (Perumbakkam Basin)
+ recorded, – not recorded, # reported by fishermen
[a]Freshwater stretches and/or becomes fresh during the peak monsoon; brackish water otherwise (Malliplate Vagu)
[b]Brackish water during the peak monsoon turns saline as summer progresses (Urugayya)

4.8 Conservation Issues

The wilderness and wildlife of Sriharikota though relatively well-protected, due to high-security status of the island and the conservation-oriented officials of ISRO, do face existing threats or those that could result in the future. These are discussed in brief below:

Plantations The trend in India is that formations lacking in timber species are as a rule considered useless and felled or replaced by plantations, little realising that these have rich diversity and are repositories of economic-medicinal plants and natural habitat for wildlife (Meher-Homji 1997). As discussed earlier, Sriharikota too has a history of clearing of native vegetation to raise fast-growing or commercially important species for afforestation, shelter belts, stabilising sand dunes and revenue and employment generation. Studies in Sriharikota have confirmed the deleterious impact of plantations on birds, besides mammals, herpetofauna and butterflies. Fortunately, the earlier practice of clearing the native vegetation to raise plantations has stopped after BNHS representations to SDSC-SHAR, and plantations are now raised in open scrub or sandy area. Another positive outcome has been the ban on raising new eucalyptus plantations.

Invasives Chilean mesquite *Prosopis chilensis* and cane *Calamus rotang* are major invasive plant species in Sriharikota. The Chilean mesquite, an exotic from South America, has proliferated on its own in areas that faced clearing in the past and where the soils are saline (mostly in areas bordering Pulicat Lake). Cane, introduced during the British era, has now spread and engulfed most of the freshwater streams and ponds and their margins eliminating native vegetation. Another invasive that is now seen in the residential and office compounds is *Lantana camara*, which is a major problem in many forest tracts of India. Once established, it forms a dense shrub layer preventing other plants from surviving. Steps must be taken to weed out the species from the island and not introduce it into residential areas, gardens and parks.

Wetlands The problems affecting the aquatic habitats in the island and which impact the fish fauna are:

1. The proliferation of cane in all the freshwater habitats, engulfing the smaller ones and forming impenetrable brakes in streams obstructing the water flow
2. The spread of the exotic aquatic weed water hyacinth *Eichornia crassipes*, which has almost completely covered the surface of many abandoned irrigation ponds and also the streams
3. Siltation, a major problem confronting abandoned irrigation ponds and streams
4. The occurrence of invasive fish species, two recorded being the Mozambique tilapia *Oreochromis mossambicus* and dwarf gourami *Colisa lalia*

Expansion of the Spaceport The developmental activities and expansion plans of the SDSC-SHAR have been making demands on the land. Large tracts of land were taken over by the spaceport for construction of a number of new buildings and facilities and for a new launch pad in recent years. Though acquisition of land for such purposes is unavoidable, measures could be taken up to lessen the impacts on the wildlife and their habitats such as (1) acquiring land dominated by mesquite, eucalyptus and areas largely devoid of vegetation, (2) optimal use of land for expansion plans and adopting landscape designing to retain as much of the native vegetation as possible around new facilities and (3) demarcation of exclusive biodiversity conservation zone(s).

4.9 Conclusion and Conservation Perspectives

Sriharikota has one of the last remaining and largest tract of coastal tropical dry evergreen forest (a forest type exclusive to the Coromandel Coast) in India. Other than this dominant vegetation type, due to the varied ecological conditions and past and recent anthropogenic factors, tracts of scrub jungle, plantations, 'abandoned village forest', grasslands and sand dune vegetation also occur. The aquatic habitats range from brackish-saline water lakes and marshes to freshwater ponds and streams and creeks connected to Pulicat Lake or the Bay of Bengal. Our surveys to document the vertebrate diversity of Sriharikota revealed that in concurrence with the diversity of habitat types, the island also has a rich faunal diversity. The value of this site in terms of conservation potential is further enhanced by the fact that the landscape along the Coromandel Coast has lost most its natural wealth to cultivation and urbanisation.

A visit to the areas bordering Sriharikota easily provides an idea of what would have happened to Sriharikota if ISRO had not taken over the island. It would have ended up as nothing more than a cluster of overgrown fishing villages. However, like all the wilderness areas of India, even those in Sriharikota are not without conservation issues. These issues are primarily related to the expansion and infrastructure development of the spaceport's facilities, resulting in loss and fragmentation of forest cover and their related impacts on the fauna. Another is the raising of plantations carried out as part of afforestation schemes, employment generation for tribals (who now live in a colony in the residential area in the island), meeting fuelwood demands, revenue generation and creation of shelter belts along the sea coast for protection against cyclones. The pauperization of fauna in general as a result of monoculture plantations had been highlighted in our project reports, and fortunately, the authorities have put a stop to further raise eucalyptus plantations and are also planning to remove eucalyptus in phases to allow the native forest to regenerate and get back their pristine glory.

With India's alarming biodiversity loss, especially in recent times, places like Sriharikota, with limited human intrusion, become significant for biodiversity conservation and could become more so in the future. It would not be wrong to assume that very little of the forest or wildlife of Sriharikota would have remained if ISRO had not taken over the island. From the example of Sriharikota, it is obvious that similar organisations holding vast tracts of land and with high-security status (e.g. ISRO, atomic and other power plants and defence establishments) can play a major role in protecting wilderness areas provided there is a pro-conservation attitude with those at the helm of affairs. However, as seen in the case in Sriharikota, even such wilderness havens may face threats in the not too distant future due to expansion programmes, and it will be wise that policies are framed so that these wilderness areas survive into the future or threats are minimised to low levels possible.

Acknowledgements The authors thank the authorities in the Indian Space Research Organisation, Bangalore, for sanctioning the projects and granting permission to undertake the project in Sriharikota. Special thanks to Mr. R. B. Singh, former Head, Conservation and Landscaping Division (C & LD); Dr. Rabbani, present Head (C & LD); Mr. Srinivasulu Reddy (Scientist G); and Mr. Amitav Mohanty (Scientist F) for providing logistics and support at Sriharikota. We thank the resourceful Yanadi tribal assistant Manikala Parandamaiah for helping us in the fieldwork.

References

Blasco F, Legris P (1973) Dry evergreen forest of Point Calimere and Marakanam. J Bombay Nat Hist Soc 70:279–294

Bombay Natural History Society (1977) Birds of Sriharikota Island. A preliminary survey by the BNHS submitted to ISRO. Bombay Natural History Society, Mumbai

Champion HG, Seth SK (1968) A revised survey of forest types of India. Government of India Press, New Delhi
Choudhury BC, Pandav B, Tripathy B, Andrews HV (2003) Sea turtle conservation: eco (turtle) friendly coastal development. A GOI–UNDP Project Manual. Centre for Herpetology/Madras Crocodile Bank Trust, Mamallapuram
Das I (1996) Biogeography of the reptiles of South Asia. Krieger Publishing Company, Malabar
David JP, Murugan BS, Manakadan R (2011) Frugivory by birds and mammals in Sriharikota Island, southern India. J Bombay Nat Hist Soc 108(1):24–40
David JP, Murugan BS, Manakadan R (2012) Seasonality in fruiting of fig and non-fig species in a tropical dry evergreen forest in Sriharikota Island, southern India. Trop Ecol 53(1):1–13
Groves CP (2001) Primate taxonomy. Smithsonian Institution Press, Washington, DC
Jackson P (1998) Conserving the smaller cats, 1–4. ENVIS Newsletter (Wildlife & Protected Areas): vol. 1, no. 1. Wildlife Institute of India, Dehradun
Kannan V, Manakadan S, Sivakumar S, Rao P, Mohapatra KK, Santharam V (2008) The waterbirds of Pulicat Lake, southern India including those in the adjoining wetland and heronries, Nellore district. Andhra Pradesh J Bombay Nat Hist Soc 105(2):162–180
Manakadan R, Sivakumar S (2004a) An ecological account of the faunal diversity of Sriharikota Island and its environs. Final Report: Part I – Birds and Mammals. Bombay Natural History Society, Mumbai
Manakadan R, Sivakumar S (2004b) An ecological account of the faunal diversity of Sriharikota Island and its environs. Final Report: Part III – Fish. Bombay Natural History Society, Mumbai
Manakadan R, Sivakumar S (2004c) Sighting of the Water Rail *Rallus aquaticus* in Sriharikota Island, Nellore district, Andhra Pradesh. Newsl Ornithol 1(1&2):15–16
Manakadan R, Sivakumar S, Rahmani AR (2004) An ecological account of the faunal diversity of Sriharikota Island and its environs. Final Report: Part V: Conservation Issues. Bombay Natural History Society, Mumbai
Manakadan R, Sivakumar S, David JP, Murugan SP (2008) Investigations into some ecological aspects of Sriharikota Island. Final Report (2004–2007). Bombay Natural History Society, Mumbai
Manakadan R, Rao P, Mohapatra KK, Sivakumar S, David JP, Murugan BS, Santharam V (2009a) The land birds of Sriharikota Island, southern India and conservation issues. J Bombay Nat Hist Soc 106(1):15–29
Manakadan R, Rema Devi K, Sivakumar S, Indra TJ (2009b) Fish fauna of the wetlands of Sriharikota Island, southern India and their conservation issues. J Bombay Nat Hist Soc 106(3):313–323
Manakadan R, Sivakumar S, David JP, Murugan BS (2013) The mammals of Sriharikota Island, southern India, with insights into their status, population and distribution. J Bombay Nat Hist Soc 110(2):114–121
Mani S, Parthasarathy N (2009) Tree population and above ground biomass changes in two disturbed tropical dry evergreen forests of peninsular India. Trop Ecol 50:249–258
Meher Homji VM (1997) Biodiversity conservation priorities. Indian J Forest 20(1): 1–7
Mukherjee S (1998) Cats: some large, many small, ENVIS Newsletter (Wildlife & Protected Areas): vol. 1, no. 1. Wildlife Institute of India, Dehradun, pp 5–13
Parthasarathy N, Arthur SM, Udayakumar M (2008) Tropical dry evergreen forests of Peninsular India. Ecology and conservation significance. Trop For Sci 1:89–110
Rao P (1998) The bird communities of the Tropical Dry Evergreen Forests of Sriharikota. Ph.D. thesis, University of Bombay
Reddy PS (1981) The displaced Yanadis of Sriharikota Island. Ph.D. thesis, Sri Venkateswara University, Tirupati
Roonwal ML, Mohnot SM (1977) Primates of South Asia: ecology, sociobiology and behaviour. Harvard University Press, London
Samant JS, Rao P (1996) An ecological investigation of the avian community of Sriharikota Island. Final report. Bombay Natural History Society, Bombay

Sanyal DP, Dattagupta B, Gayen NC (1993) Fauna of Andhra Pradesh. Part-I, Reptilia. Zoological Survey of India, Calcutta

Shanker K, Pandav B, Choudhury BC (2003) Sea turtle conservation: population census and monitoring. A GOI – UNDP Project Manual. Madras Crocodile Bank Trust, Mamallapuram

Sivakumar S, Manakadan R (2004) An ecological account of the faunal diversity of Sriharikota Island and its environs. Final report: part II – herpetofauna. Bombay Natural History Society, Mumbai

Sivakumar S, Kailasanathar R, Manakadan R (2004) An ecological account of faunal diversity of Sriharikota Island and its environs. Final Report: Part 1V- Butterflies. Bombay Natural History Society, Mumbai

Sivakumar S, Manakadan R (2005) An account of the heronries and other breeding waterbirds of Sriharikota, Andhra Pradesh, India. Proceedings of the National Seminar on Bird Ecology and Conservation. NLBW and INCERT, Bangalore, pp 93–96

Sivakumar S, Manakadan R (2007) Record of Boiga beddomei (Wall 1909) from Sriharikota, Andhra Pradesh. India J Bombay Nat Hist Soc 104(1):103–104

Sivakumar S, Manakadan R (2008) Ecology of the slender loris *Loris lydekkerianus (Cabrera)* with special reference to vegetation types. In: Investigations into some ecological aspects of Sriharikota Island. Final Report (2004–2007). Bombay Natural History Society, Mumbai, pp 53–63

Sivakumar S, Manakadan R (2010) Wildlife mortality from vehicular traffic in Sriharikota Island, southern India. J Bombay Nat Hist Soc 107(1):53–55

Sivakumar S, Manakadan R, David P (2011) Nesting surveys of Olive Ridley Turtle (*Lepidochelys olivacea*) along the beaches of Sriharikota Island, Andhra Pradesh, India. Indian Ocean Turtle News Lett 14:12–14. Bombay Natural History Society, Mumbai, India

Subramanya S (2001) Heronries of Andhra Pradesh. Mayura 13:1–27

Suryanarayana B, Rao AS, Rao AM, Veeraraju V (1989) Report on the flora of Sriharikota Island, vol 2. Visvodaya Government College, Venkatagiri and SHAR Centre, Sriharikota

Suryanarayana B, Rao AS, Rao AM, Veeraraju V (1998) Flora of Sriharikota Island. Technical Report: ISRO-SHAR-TR-99-98. Indian Space Research Organisation, Bangalore

Vogel G, Ganesh SR (2013) A new species of cat snake (Reptilia: Serpentes: Colubridae: *Boiga*) from dry forests of eastern Peninsular India. Zootaxa 3637:158–168

Group Size and Composition of Gaur (*Bos gaurus gaurus*) in Relation to Environmental Factors in Parambikulam Wildlife Sanctuary, Western Ghats

5

S. M. Vairavel and P. S. Easa

Abstract

The gaur (*Bos gaurus gaurus*) is a threatened species distributed in the Western Ghats, Central India and Himalayan foot hills. Data on group size, composition and structure of gaur in Parambikulam Wildlife Sanctuary, Kerala, were collected through direct observations. Morphological distinguishing features were used for classification of individuals into different age-sex categories. About 45% of the population was adult females and 18% adult males. The sex ratio of adult male to female was 1: 2.49. The basic unit of the groups was formed by adult females in combination with subadults and juveniles. The mean group size was 6.0398 (loners excluded), and the most frequented groups were of 3, 5 and 7. Solitary bulls formed about 21% of the total sightings. There was seasonal influence on the group size, which is explained mostly by the variations in availability of grass. The births were mostly between the two peak rainfall periods. The observations are discussed in relation to ecological parameters.

Keywords

Gaur · Group composition · Population · Kerala · Distribution

5.1 Introduction

Knowledge on animal population parameters, the structure and the trend are of importance for wildlife management. McCullough (1993, 1994) discussed the significance of studies on group composition, size and structure. Herbivores

S. M. Vairavel
Formerly with Kerala Forest Research Institute, Thrissur, Kerala, India

P. S. Easa (✉)
Anugraham, East Fort Post, Thrissur, 680005 Kerala, India
e-mail: easaelephant@yahoo.com

C. Sivaperuman, K. Venkataraman (eds.), *Indian Hotspots*,
https://doi.org/10.1007/978-981-10-6605-4_5

normally live in groups, and the pattern in group size is influenced by environmental parameters (Eisenberg 1966; Leuthold and Leuthold 1975; McBride 1976; Crook et al. 1976; Rodman 1981; Johnson 1983; Southwell 1984).

Gaur (*Bos gaurus gaurus*) is a threatened species confined to Western Ghats, Central Indian highlands and foothills of Himalayas. In Western Ghats, they occur mostly in the hilly forest areas of Karnataka, Tamil Nadu and Kerala. These populations face the problem of habitat loss, fragmentation or degradation and epidemics. Very few studies have been conducted on gaur, and the available information is from casual and short-term observations and is mostly on the number in different areas (Anon 1993; Belsare et al. 1984; Dwivedi and Shukla 1988; Krishnan 1975; Nair et al. 1985; Schaller 1967; Vijayan et al. 1979). The present paper is based on a study conducted in Parambikulam Wildlife Sanctuary during 1993–1996 (Easa 1998).

5.2 Study Area

Parambikulam Wildlife Sanctuary is in Palakkad district of Kerala, India. The sanctuary is about 285 km^2 in extent and is part of a larger landscape comprising Anamalais, Nelliampathis, High Ranges and Palani Hills. The vegetation comprises southern tropical wet evergreen, southern tropical moist deciduous and southern tropical dry deciduous forests and grasslands. Plantations of teak occupy a good portion and eucalypts are also met with. These forests are interspersed and hence the overall vegetation is mosaic. The area is rich in flora and fauna with a good representation of Indian mammals. The wildlife population estimation indicates that the area has the second largest density of gaur in Kerala.

Though the area receives both southwest and northeast monsoons, the former is the more active one. The first peak of rains occurs between June and July and the second during October and November. Based on the rainfall, three seasons are differentiated, viz. dry (from February to May), first wet (from June to September) and second wet (from October to January) seasons.

5.3 Methods

The sanctuary area was covered every month on foot from March 1993 to February 1996. Care was taken to spend time in all habitats proportionate to the size. The individuals in the groups sighted were counted and classified according to age and sex. The individuals were classified into different age and sex classes following Krishnan (1972) and Schaller (1967) with appropriate modifications. Adult males (AM) were sooty black in colour with enlarged dewlap, well diverged and fully converged tip of horn, prominent dorsal ridge and rotary movement of hump while walking. The females in dark brown and closer to black colour with non-prominent dewlap, less diverged but fully converged tip of horn, less prominent dorsal ridge were considered as adult females (AF). The subadult males (SAM) were black or brownish black in colour with prominent dewlap. But the diverged horn was about to converge. The brownish black (more black in thoracic portion and more brownish colour in the rump portion) females without dewlap and less diverged non-converged horn were subadult females. The juveniles (JUV) were brownish in colour with dagger-like spike horn (approximately 20–25 cm in length) in both sexes. The calves (CA) were approximately 1 m in height and golden brownish in colour and can pass through between the legs of its mother. The white stockings were absent among calves.

Group composition analyses were done using only the completely classified groups. Proportion of different age and sex classes in the population was derived on the basis of all the sightings. The solitaries were not considered for calculating mean group size. The 3-year data were considered for the analyses on changes in the mean group size over time. Calving season was arrived based on the number of calves present in the group and their percentage contribution to the population in different months.

Observations were made during feeding and the food species were identified. Clip-and-weigh method was followed for measuring food availability in terms of biomass (dry weight) (Wiegert 1962). Quadrates of 1 m^2 size for grasses and herbs and 25 m^2 for shrubs were laid using stratified random sampling procedure. All food species within the plots were clipped and weighed in the field for wet weight, and subsamples were later oven dried at 60 °C constant temperature till the samples reached constant weight. Biomass of each individual food species was estimated. Food species were grouped into grass, herb and shrub for the analyses. The total of these three groups was taken as total food available. Data collected for six seasons during 1994–1995 were used for food availability estimates.

5.3.1 Analyses

5.3.1.1 Effect of Age-Sex Categories on Group Size

Group size is considered as a sum of the number of individuals in the different component classes. However, because of the complex association possible among the classes, group size needs not respond in a simple manner to the changes in the number of individuals in the component classes.

Regression analysis was done for studying the effect of age-sex categories on group size. Group size was considered as dependent variable and number of individuals in each age-sex category as independent variable. The regression functions included each category one at a time. In such cases, the regression functions fitted were of the following form.

$$E\left(G^{1/2}\right) = \beta_0 + \beta_1 N \qquad (5.1)$$

where

E = expectation
G = group size

N = number of individuals in a particular age-sex category
β_0, β_1 = parameters to be estimated

Since the variance in the herd size was found related to the mean, the group size was subjected to square root transformation. The parameter estimates were obtained through weighted least squares. The weights were inversely proportional to different powers of the corresponding regressors. The value of the index of power for each equation was obtained through a grid search utilizing the procedure WLS of SPSS/PC+ (Anonymous 1987).

A multiple linear regression equation was fitted including all the categories except the unsexed class in the model to study the combined effect of the different categories on group size. Stepwise regression helped in identifying the prominent components affecting group size.

Pearson's product moment correlation coefficient was worked out for association among the different age-sex categories. Because of the possible intercorrelation among the groups, changes in number of individuals in any category may have indirect effect on the group size through other categories, apart from the direct effects. Such effects were studied by carrying out path coefficient analysis (Wright 1921).

5.3.1.2 Changes in the Mean Group Size Over Time

The data for the analyses consisted of monthly observations on group size for 36 months starting from March 1993. The mean group size for each month was computed, and changes in the mean group size over time were analysed using the following general model.

$$G_t = \mu + \alpha t + \in_t \tag{5.2}$$

where

G_t = mean group size at time t
μ = intercept
t = time in years
α = coefficient of the trend variable
$\in_t$ = error at time t

The effect due to months was superimposed on the above model as

$$G_{t(i)} = \mu + \alpha t + \beta_i + \in_{t(i)} \tag{5.3}$$

where

$G_{t(i)}$ = mean group size at time t belonging to ith month
β_i = effect due to ith month
$\in_{t(i)}$ = residual at time t belonging to the ith month
μ, α, t = as defined earlier

Model (3) assumes a continuous change in the mean group size over time with seasonal fluctuations superimposed. The mean-range plot was used for examining the additivity of the monthly effects. Ordinary least squares (OLS) was used first for estimating the coefficients of the model. The autocorrelation coefficient for the residual was non-significant as shown by Durbin-Watson statistic. However, a plot of residuals against the months showed heteroscedasticity. Since the observations in the wet seasons were less variant compared to those of other months, the coefficients of the model were re-estimated through weighted regression. The weights were roughly proportional to the reciprocal of the variance of residuals for each month. The weights (W) were

$W=^{-\delta}$where

W = standard deviation of the residuals for each month
δ = index of power

The optimal value of δ came to 2.2 through a grid search from 1 to 2.6 using the WLS procedure of SPSS/PC+.

5.3.1.3 Changes in Distribution of Group Size Over Time

The previous analyses dealt with only changes in the mean group size over time. However, considerable variation in group size around the mean was observed, and the pattern of this variation itself could change from time to time. A chi-square test was carried out. The group size distribution over different seasons and years were tested for significance using a chi-square test. The months were grouped into dry, first wet and second wet seasons for analyses. Since most of the cells had expected frequency below 5 making the test ineffective, the analysis was done through loglinear models (Haberman 1978). Groups of size 11 and above were put in a single class. The solitary animals were excluded from the data.

The changes in the herd size distribution over different seasons and years were studied using the following model.

$$\ln f_{ijk}=\mu+\lambda_i^{G}+\lambda_j^{S}+\lambda_k^{Y}+\lambda_{ij}^{GS}+\lambda_{ik}^{GY}+\lambda_{jk}^{SY}+\lambda_{ijk}^{GSY} \quad (5.4)$$

where

f_{ijk} = the frequency in the multiway contingency table corresponding to the *i*th group size, *j*th season, *k*th year
ln = indicates natural logarithm
λ's = are the parameters corresponding to the main effects and interactions due to factors, viz. group size (G), season (S) and years (Y)

The above is a saturated model with full set of parameters in a three-way setup. The non-significant parameters were eliminated through backward elimination using HILOGLINEAR procedure of SPSS/PC+.

5.3.1.4 Mean Group Size in Relation to Food Availability

A multiple regression equation was fitted to understand the relation between mean group size and food availability. Mean group size was taken as dependent variable and components of food availability (grass, herb and shrub) as independent variables. The model was as follows:

$$y = a + b_1 x_1 + b_2 x_2 + b_3 x_3 \quad (5.5)$$

where

y = mean group size
x_1 = grass food
x_2 = herb food
x_3 = shrub food; a,b_1,b_2 and b_3 are fitted constants

A simple linear regression equation was also fitted for finding significant relation between mean group size and total food availability inclusive of grass, herb and shrub.

5.4 Results

5.4.1 Group Composition

There were 241 sightings consisting of 1141 individuals, and these were considered for age and sex classification. Of the 241, 62 sightings were of loners. About 82% of the loners were males. The adult females formed 44.79% of the population followed by adult males (17.97%), subadult females (14.81%) and subadult males (7.71%). The juveniles and calves formed 7.71% and 6.31%, respectively. About 1% of the total number of individuals could not be sexed. The adult male to female sex ratio was 1:2.49.

The highest percentage contribution of calves to the population was in September (10.71%) and the minimum in February (3.7%). The overall calf-adult female ratio observed was 1:7.1. High ratio between calf and adult female was observed during September (1:4.33) followed by October (1:4.6) and April (1:4.6).

Comparison of month-wise percentage of calves in the population with the mean precipitation (obtained from 1991 to 1995) indicates an increase in the number of calves before and after the first peak of precipitation and a decrease after the second peak (Fig. 5.1). The percentage of calves in relation to mean monthly temperature (Fig. 5.2) and to mean monthly temperature difference showed a negative correlation of the number of calves with higher temperature and temperature differences.

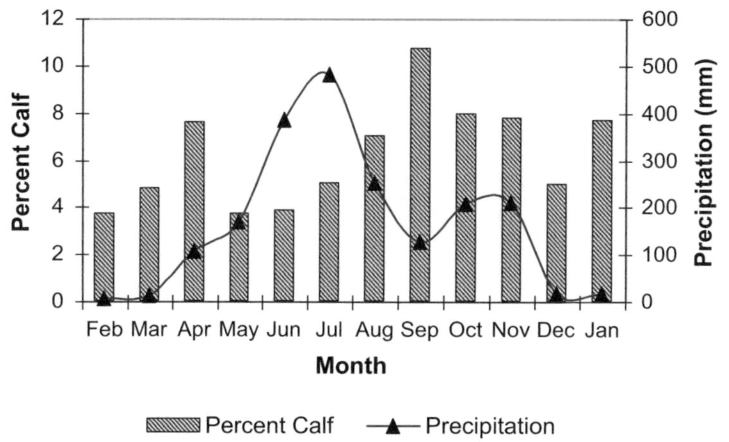

Fig. 5.1 Relation of mean monthly precipitation and percentage of calves

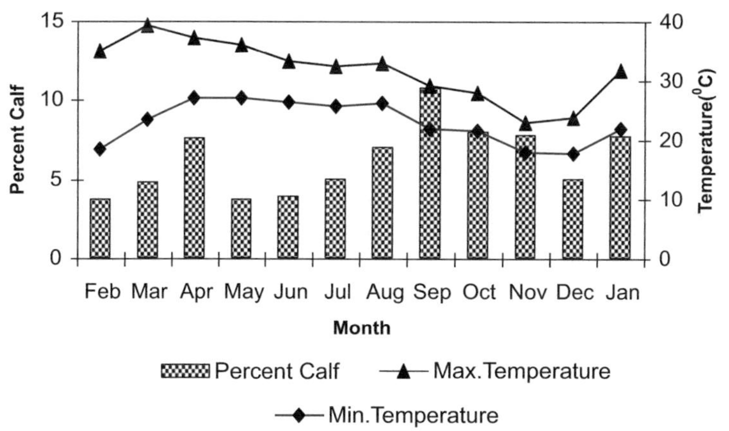

Fig. 5.2 Relation between mean monthly temperature and percentage of calves

5.4.2 Age-Sex Categories and Group Size

The simple linear regression between group size and different age-sex categories (Table 5.1) indicates that the number of adult female had the largest adj. R^2 value individually. This means that a substantial part of variation in group size is due to the adult females in the group. The corresponding regression coefficient (β_1) was 1.7582 indicating that with addition of every adult female, multiple numbers of individuals could be added to the group as the dependent variable is in the square root scale. The variation in group size individually by the subadult males was 47%. The subadult males occur in large groups as indicated by the high value for regression coefficient. The results of stepwise regression analysis confirmed the above

Table 5.1 Regression coefficient associated with simple linear regression of group size on age-sex categories of gaur

Age-sex class	β_0	SE (β_0)	β_1	SE(β_1)	Adj. R^2	Value of δ used for transformation
AM	3.2392	0.9309	2.7523	0.7982	0.0894	1.25
AF	1.0711	0.2931	1.7582	0.1085	0.6043	0.75
SAM	1.0411	1.1176	8.5206	1.0243	0.4701	0.50
SAF	3.8407	0.6183	2.2849	0.3805	0.2520	0.25
JUV	4.5447	1.0953	2.6660	0.8112	0.1406	0.25
CA	4.1051	1.0573	2.8224	0.8464	0.1708	0.75

SE standard error

Table 5.2 Stepwise regression analysis – cumulative effect of various age-sex categories on group size

Age-sex class	Cumulative value of R^2	Partial regression coefficients of the final model(β)	SE(β)
AF	0.6635	0.9713	0.0215
SAM	0.7909	1.0725	0.0573
AM	0.8733	1.0035	0.0377
SAF	0.9079	1.0215	0.0321
JUV	0.9526	1.0006	0.0392
CA	0.9900	1.0058	0.0397
Constant		0.0642	0.0595

Table 5.3 Correlation matrix of different age-sex categories and group size (GS)

	AM	AF	SAM	SAF	JUV	CA	GS
AM	1.000	0.171	0.184	0.274**	0.172	0.095	0.460**
AF	0.171	1.000	0.524**	0.446**	0.298**	0.254**	0.815**
SAM	0.184	0.524**	1.000	0.436**	0.400**	0.301**	0.731**
SAF	0.274**	0.446**	0.437**	1.000	0.120	0.012	0.656**
JUV	0.172	0.298**	0.400**	0.120	1.000	0.226*	0.540**
CA	0.095	0.254**	0.301**	0.012	0.226*	1.000	0.448**

Two-tailed significance. *significant at $P = 0.01$, **significant at $P = 0.001$

findings showing that adult female combined with subadult males explained about 80% of the variation in group size (Table 5.2).

Group size was highly correlated positively with adult female, subadult male and subadult female (Table 5.3). The adult female was associated with all the categories except adult male. The adult female had high association with subadult male (0.524) and subadult female (0.446). The subadult female showed more or less equal extent of association with adult female (0.446) and subadult male (0.437). The juvenile and calf categories were more attached to subadult male (0.400, 0.301). The basic unit of a group is obviously formed by adult females in combination with subadult males, subadult females and juveniles.

Table 5.4 Path coefficient analyses – direct and indirect effects of different age-sex categories on group size

	AM	AF	SAM	SAF	JUV	CA
AM	0.2140	0.0736	0.0344	0.0808	0.0373	0.0197
AF	0.0366	0.4305	0.0981	0.1315	0.0647	0.0531
SAM	0.0393	0.2254	0.1873	0.1287	0.0870	0.0628
SAF	0.0586	0.1921	0.0818	0.2948	0.0260	0.0026
JUV	0.0367	0.1283	0.0750	0.0352	0.2173	0.0472
CA	0.0202	0.1095	0.0563	0.0036	0.0491	0.2089

Note: Direct effects are in main diagonal with underline; off diagonal elements are indirect effects

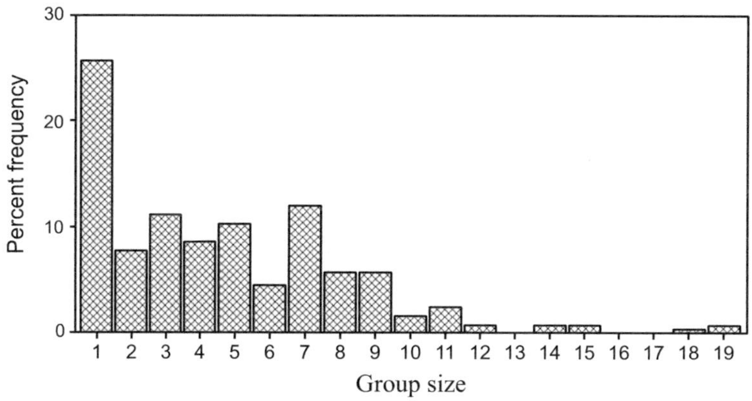

Fig. 5.3 Percentage frequency distribution of group size of gaur

Results of path coefficient analysis are presented in Table 5.4. The estimate of residual variation was 0.01, thereby leaving only a negligible part of the total variation unexplained. The adult females had the maximum positive direct effect on group size (0.4305) followed by subadult females (0.2948). No age-sex class had a negative effect. Adult females had high indirect effect on the size through subadult females. This indicates that variations in these categories are mostly simultaneous in a group. Similarly, the indirect effect of the subadult males, juveniles and calves on the group size was through adult females. The adult males were mostly a standalone group. But these had high indirect effects on the group size through subadult females and adult females implying the possible attractions between the groups.

5.4.3 Changes in Distribution of Group Size Over Time

Group size ranged from 1 to 19 with a positively skewed distribution (Fig. 5.3). Out of 241 sightings, 25.73% were solitaries (mostly bulls). About 12% of the groups were of 7. Group size of 3 and 5 were observed in 11.2% and 10.4%, respectively. Group size of 2 and 4 formed only 7.9% and 8.7%, respectively. The largest group observed was of 19 individuals.

Table 5.5 Results of analysis through loglinear models

Factors	Df	Likelihood ratio χ^2	Probability of χ^2 value
G,S,Y(combined)	13	38.269	0.0003
GS,GY,SY (combined)	40	43.018	0.3433
GSY	36	32.663	0.6281
Y	2	1.611	0.4469
S	2	3.042	0.2185
G (group size)	9	33.616	0.0001

The results of analysis through loglinear models (Table 5.5) showed that two-way and three-way interactions were non-significant. The only factor which came out significant through the likelihood ratio χ^2 was group size. This indicates that the group size distribution, i.e. the pattern of variation around the mean did not undergo changes with the seasons or years. The mean group size between seasons was different. However, the two-way and three-way interactions did not show significant differences in group size distribution around the mean in relation to seasons or years indicating that the group size distribution was not changing over the seasons. Similar was the case with group size-year interaction. The three-way interaction (GSY) was non-significant. Thus there was no change in the interaction between group size and season with change in years. But the percentage of solitaries in dry season (36.45%) was more compared to the first wet (13.70%) and second wet (21.31%) seasons.

5.4.4 Mean Group Size (MGS)

The loners were not considered for calculating MGS. The overall mean group size for the observations in 3 years was 6.0398. The MGS in 1993 was the lowest (5.4706) and the highest (6.3281) was in 1994. There was not much deviation in the annual MGS from the overall MGS. The second wet season had the highest of 6.9167 compared to dry (5.4769) and first west (5.9524) seasons. Though the seasonal difference was significant ($F = 2.7607$, $\alpha = 0.1$), one-way ANOVA for annual variation turned out to be non-significant.

5.4.5 Changes in the Mean Group Size Over Time

The estimates of parameters in model (3) are given in Table 5.6. The coefficient of adjusted multiple determination for the model was 0.8571. There was a slightly increasing trend in the order of 0.2132 per year for the mean group size. While estimating the effects of months, December was kept as a reference with no deviation from the trend line. The mean group size showed significant decrease in March, June and July.

Table 5.6 Parameter estimates for the different effects in the model

Effect	Coefficient(C)	SE(C)	t value	Probability of t value
Constant	7.3524	1.0251	7.172	<0.0001*
Trend	0.2132	0.0637	3.349	0.0028*
Jan	−1.5444	1.0258	−1.506	0.1458 ns
Feb	−1.6112	2.1239	−0.759	0.4555 ns
Mar	−2.9167	1.1880	−2.455	0.0221*
Apr	−0.7478	1.4473	−0.517	0.6103 ns
May	−2.2489	1.9079	−1.179	0.2506 ns
Jun	−3.0467	1.0211	−2.934	0.0066*
Jul	−2.4778	1.0307	−2.404	0.0247*
Aug	−0.1455	1.1856	−0.123	0.9033 ns
Sep	−0.9134	1.4419	−0.633	0.5327 ns
Oct	−1.3978	1.0381	−1.347	0.1913 ns
Nov	−0.7009	1.6053	−0.442	0.6629 ns

ns non-significant
*Significant at $P = 0.05$

5.4.6 Mean Group Size in Relation to Food Availability

Multiple regression analyses for the relation between mean group size and availability of individual food items (grass, herb and shrub) as individual variables turned out to be significant. The model of multiple regression equation fitted using stepwise regression was as follows:

$$\begin{aligned} &\text{MGS} = 3.1063 + 0.1435(\text{Grass}) - 0.0789(\text{Shrub}) \\ &(0.3611)(0.0202)(0.0211)\ \text{Adj.R}^2 = 0.9456 \end{aligned} \tag{5.5}$$

The figures in parentheses are standard errors of the coefficients.

The variation in mean group size explained by herb food item was not significant. Availability of grass explained about 69% of the variation in mean group size. Grass and shrub combined explained 94.56% of variation in the mean group size. The relation between mean group size and total food availability was not significant as evident from the combined regression analysis (Adj.$R^2 = 0.1695$).

5.5 Discussion

Gaur is a gregarious animal and the group is centred around the adult females. Family links in bison family groups exert an important influence on formation and shaping of different groups (Krasinski 1978). Basic unit, as found from the present study, consists of different age-sex individuals except adult males. Adult females had highest influence on the group size followed by subadult females. The adult male gaurs are found to influence the group only through adult and subadult females. A similar pattern is reported in the American (Meagher 1973; Shult 1972;) and

European (Krasinski 1967) bisons and wood bison groups (Larter and Gates 1994). Fuller (1960), Larter (1988) and McHugh (1958) have also reported similar observations.

The solitary bulls in Parambikulam constituted 21.45% of the total sightings, and all of them were healthy adults. Based on the observations on American bison (Soper 1941; Fuller 1960) and European bison (Bojanus 1827), there were conclusions that the solitaries were older bulls. This was later refuted (Krasinski 1978). The tendency of bulls to be solitaries is considered to be a property of males of the genus *Bison* in both America and Europe (Krasinski 1978). Presence of solitaries was observed throughout the year in Parambikulam though in low proportions compared to those in American and European bisons. However, further observations are required to explain the occurrences of females forming about 18% of the observed solitaries. Unlike the *Bison* sp. (Krasinska and Krasinski 1995), there was no bull groups observed in Parambikulam. Though there were seasonal differences in the proportion of solitaries, proportion of adult males within the group was almost constant. This explains the non-significant influence of adult males on group size. The solitary gaur bulls in the study area were seen moving from group to group as in the case of American and European bisons (McHugh 1958) ensuring exchange of genes.

The sex ratio of gaur in Parambikulam is distorted favouring females. Similar observation was also reported by Schaller (1967). Alteration of the age ratios in North American deer is due to the decreased fawn survival associated with limitations in quality and quantity of available food among several other factors (Klein,1970). Information on the sex ratio of gaur at birth is not available. However, assuming an equal sex ratio at birth, the distorted sex ratio could be due to differential sex mortality. Robinette et al. (1957) has reviewed such differential sex mortality among mule deer. Krasinski (1978) reported a male mortality of about 69% in European bison of which 23% was of calves of less than 1 year. Mortality of males due to poor nutrition, particularly during the first year of life, was greatly accentuated. Longhurst and Douglas (1953) and Taber and Dasmann (1954) have reported similar observations on black-tailed deer. Male appears to be more susceptible in the wake of food scarcity (Klein 1970). This could be attributed to their higher metabolic rate resulting from a greater rate of growth, activity, curiosity and independence than the females. Such factors could also be responsible for the distorted sex ratio among gaur in Parambikulam.

The mean group size of 6.0398 and the most frequent group size of 3, 5 and 7 observed in Parambikulam agree with reports of Dwivedi and Shukla (1988), Schaller (1967) and Belsare et al. (1984). However, large groups were reported by Belsare et al. (1984) in Kanha and Nair et al. (1985) in Periyar. These could be aggregations of smaller units probably due to the environmental factors in these areas. Casual assemblages of individuals are also reported by Dwivedi and Shukla (1988). Krasinski (1978) attributed similar aggregations in European bison to constant supply of supplementary food. Habitat conditions and group size are reported to be related (Shackleton 1968). The mosaic nature of habitat with very few large open areas could be the reason for large number of smaller groups of gaur in Parambikulam.

Group size is an optimal response to the environment. Mean group size in primates is also affected by ecological factors (Crook 1972; Denham 1971) and in the African antelopes (Jarman 1974). Seasonal influence on mean group size of gaur in Parambikulam indicates direct relation with food availability and also the proportion of calves. The second wet season had higher food availability and also the highest proportion of calves in the population. However, variation in mean group size is explained largely by the combination of grass and shrubs and is highly influenced by the availability of grass alone. The large-sized groups in gaur, though with less cohesion, are explained by the higher proportion of calves during the second wet season.

The general objective strategy of ungulates is to drop their young ones in the beginning of the time most favourable for ultimate survival, or after any particular unfavourable condition. Warm temperature prevails year-round at lower elevation in the tropics. The animals are subjected to different selective regimes. The survival of the newborn is affected by the local monsoons or droughts. The higher number of births between the two peak rainfall months in Parambikulam could be considered as a strategy of gaur to ensure successful calf birth and survival.

Tropical mammals have a very small range of temperature tolerance. Metabolic rate also increases with the fall below a critical minimum temperature (Scholander et al. 1950). Most of the births in tropics are during a period of optimum fawn survival possibility (Delany and Happold 1979; Eltringham 1979; Estes 1976; Skinner et al. 1973). However, a few births have also been reported throughout the year. There had been a few calves in April in Parambikulam with a sudden fall in in the percentage of calves in May, June and July. This explains the low survival rate of calves born during the month of April. These could be due to the adverse effect of higher temperature and daily fluctuation. The newborn mammals have a poorly developed homoeothermic mechanism (Brody 1945). Breeding season of gaur is reported to vary considerably in different areas (Sanderson 1912; Stebbing 1911). The present observations on calving in Parambikulam confirm the variation in calving seasons.

Higher availability of food during the second wet season ensures the additional nutritional requirement of pregnant and lactating females (Sadlier 1969), thereby increasing the chances of survival of newborn calves from August to January. Similar observations of increased fawns during abundant food season are reported in the case of deer in temperate ranges (Linsdale and Tomich 1953; Mitchell et al. 1977; Sadlier 1969; Southern 1964). Klein (1970) observed decreased production and survival of deer fawns in situations where food quality deteriorates due to various reasons. Eltringham (1979) and Phillipson (1975) suggested close association between forage production and breeding cycle with the annual rainfall cycle in tropics.

Parturition was considerably less synchronized and the length of birth season did not fit the pattern typical of ungulates exhibiting a follower strategy of mother-young spatial relations. Gaur, like *Bison* sp. (Green and Rothstein 1993), seems to differ from typical follower species. The calf concealed itself in hiding places for a short period before moving with the group continuously. The golden brown colour of the calf makes it highly conspicuous in the company of dark-coloured adults. This concealment helps reduce calf mortality to some extent.

Gregariousness is an essential adaptation for life in the open (Estes 1974). Adaptations to open habitats favour group formation (Eisenberg 1966). Grouping tendencies are shown by many solitary bovids when attracted to the open onto neutral ground. Small size, solitary habits and concealment behaviour are interrelated elements in an antipredator strategy (Eisenberg 1966). Several large bovids live in cover. In gaur, the disadvantages of being conspicuous due to large size are compensated by the groups.

The females with/without young and solitary adults, two of the three social classes universal in gregarious bovids (Estes 1974), are apparent in gaur. Considering the preference for open grasslands surrounded by forest, grazing feeding habits, less conspicuous sexual dimorphism, well-developed horns in females as well, precocial young, group defence and the small- to medium-sized groups, gaur exhibits the major features of bovine social organization.

References

Anon (1993) Wildlife census – Kerala 1993. Kerala Forest Research Institute, Peechi

Anonymous (1987) SPSS/PC+ trends for the IBM PC/XT/AT and PS/2. SPSS Inc, Chicago

Belsare DK, Chandiramani SC, Dutta JJ (1984) Composition and behaviour of gaur group in Kanha National Park. Cheetal 26:28–31

Bojanus HL (1827) De Uro nostrate eiusque sceleto commentatio. Nova Acta Phys Acad Caes Leopold Carol Nat Curios 23:413–478

Brody S (1945) Bioenergetics and growth. Hafner Press, New York

Crook JH (1972) Sexual selection, dimorphism and social organization in the primates. In: Campbell B (ed) Sexual selection and the descent of man. Aldine, Chicago, pp 1871–1971

Crook JH, Ellis JE, Goss-Custard JD (1976) Mammalian social systems; structure and function. Anim Behav 24:261–274

Delany MJ, Happold HCD (1979) Ecology of African mammals. Longmans, London

Denham WW (1971) Energy relations and some basic properties of primate social organization. Am Anthropol 73:77–95

Dwivedi GD, Shukla R (1988) The Pench bison – an ecological and behavioural study. Cheetal 29:28–31

Easa PS (1998) Habitat utilization of animals in Parambikulam Wildlife Sanctuary with special reference to gaur. Research Report No. 155. Kerala Forest Research Institute, Peechi, Thrissur

Eisenberg JF (1966) The social organisation of mammals. Hdb d Zool 8:1–92

Eltringham SK (1979) The ecology and conservation of large African mammals. McMillan Press, London

Estes RD (1974) Social organization of the African bovidae. In: Geist V, Walther F (eds) The behaviour of ungulates and its relation to management, vol I. IUCN, Morges

Estes RD (1976) The significance of breeding synchrony in the wildebeest. E Afr Wildl J 14:135–152

Fuller WA (1960) Behaviour and social organization of the wild bison of wood buffalo National Park. Canad Arctic 13:1–19

Green WCH, Rothstein A (1993) A synchronous parturition in bison: implications for the hider – follower dichotomy. J Mammal 74:920–925

Haberman SJ (1978) Analysis of qualitative data, vol 1. Academic, London

Jarman PJ (1974) The social organization of antelope in relation to their ecology. Behaviour 48:216–267

Johnson CN (1983) Variations in group size and composition in red and western grey kangaroos *Macropus ruful* (Desmarest) and *M. fuliginosus* (Desmarest). Aust Wildl Res 10:25–31

Klein DR (1970) Food selection by north American deer and response to over utilization of food plants species. In: Watson A (ed) Animal population in relation to their food resources. Blackwell Scientific Publications, Oxford

Krasinska M, Krasinski ZA (1995) Composition, group size and spatial distribution of European bison bulls in Bialowieza Forest. Acta Theriol 40:1–21

Krasinski ZA (1967) Free living European bison. Acta Theriol 12:391–405

Krasinski ZA (1978) Dynamics and structure of the European bison population in the Bialowieza primeval Forest. Acta Theriol 23:3–48

Krishnan M (1972) An ecological survey of the larger mammals of Peninsular India. The Indian elephant. J Bombay Nat Hist Soc 69:297–315

Krishnan M (1975) India's wildlife in 1959–70 – an ecological survey of the larger mammals of Peninsular India. Bombay Natural History Society, Bombay

Larter NC (1988) Diet and habitat selection of an erupting wood bison population. Unpublished M.S. thesis. University of British Columbia, Vancouver

Larter NC, Gates CC (1994) Home range size of wood bison: effects of age, sex and forage availability. J Mammal 75:142–149

Leuthold W, Leuthold BM (1975) Pattern of social grouping in ungulates of Tsavo National Park, Kenya. J Zoo Lond 175:405–420

Linsdale JN, Tomich PQ (1953) A herd of Mule deer. University of California Press, Berkely

Longhurst WM, Douglas JR (1953) Parasite interrelationships of domestic sheep and Columbian black-tailed deer. Trans N Am Wildl Nat Resour Conf 33:181–192

McBride G (1976) The study of social organization. Behaviour 59:96–115

McCullough DR (1993) Variation in black-tailed deer herd composition counts. J Wildl Manag 57:890–897

McCullough DR (1994) What do herd composition counts tell us? Wildl Soc Bull 22:295–300

McHugh TS (1958) Social behaviour of the American buffalo (*Bison bison bison*). Zoologica 43:1–10

Meagher MM (1973) The bison of Yellowstone National Park. Natl Park Serv Sci Monogr Ser 1:1–161

Mitchell B, Staines BW, Welch D (1977) Ecology of red deer. A research review relevant to their management in Scotland. Banchory Institute of Terrestrial Ecology

Nair PV, Ramachandran KK, Vijayan VS, Easa PS, Balakrishnan PV (1985) An ecological study in Periyar Tiger Reserve with special reference to wildlife. KFRI Research Report No. 28. Kerala Forest Research Institute, Peechi

Phillipson J (1975) Rainfall, primary production and carrying capacity of Tsavo National park (East) Kenya. E Afr Wildl J 13:171–201

Robinette WL, Gashwiler JS, Low JB, Jones DA (1957) Differential mortality by sex and age among mule deer. J Wildl Manag 21:1–16

Rodman PS (1981) Inclusive fitness and group size with a reconsideration of group sizes in lions and wolves. Anim Nat 118:275–283

Sadlier RMFS (1969) The ecology of reproduction in wild and domestic animals. Methuen, London

Sanderson G(1912) Thirteen years among the wild beasts of India. Edinburgh

Schaller GB (1967) The deer and the tiger: the study of wildlife in India. The University of Chicago Press, Chicago

Scholander PF, Hock R, Walters V, Johnson F, Irving L (1950) Heat regulation in some arctic and tropical mammals and birds. Biol Bull 99:237–258

Shackleton DM (1968) Comparative aspects of social organization of American bison. The University of Western Ontario, London

Shult MJ (1972) American bison behavior patterns at Wind Cave National Park, Ph.D. thesis, Iowa State University, Iowa

Skinner JD, Vanzyl JHM, Heerden JAH (1973) The effect of season on reproduction in the black wildebeest and hartebeest in South Africa. J Reprod Fertil Suppl 19:101–110

Soper JD (1941) History, range, and home life of the northern bison. Ecol Monogr 11:348–412

Southern HN (1964) The handbook of British mammals. Blackwell Scientific Publications, Oxford

Southwell CJ (1984) Variability in grouping in the eastern grey kangaroo, *Macropus giganteus*. Group density and group size. Aust Wildl Res 11:423–435

Stebbing E (1911) Jungle by-ways in India. London

Taber RD, Dasmann RF (1954) A sex difference in mortality in young Columbia black-tailed deer. J Wildl Manag 18:309–315

Vijayan VS, Balakrishnan M, Easa PS (1979) Periyar tiger reserve – a reconnaissance report. Kerala Forest Research Institute, Peechi

Wiegert RG (1962) The selection of an optimum quadrat size for sampling the standard biomass of grass and forbs. Ecology 43:125–129

Wright S (1921) Correlation and causation. J Agric Res 20:557–585

Diversity, Distribution and Conservation of Freshwater Fishes in Andaman and Nicobar Islands

6

P. T. Rajan, P. Vijay, and Jayasimhan Praveenraj

Abstract

From the present work, 24 species of freshwater fishes belonging to six orders, 12 families and 19 genera are reported. The family Gobiidae constituted six genus and eight species, followed by Eleotridae with four genus and seven species. Seven species are known to be diadromous and share distribution elsewhere; three species, namely, *Channa* sp., *Rasbora* sp. and *Aplocheilus* sp., are distinct from their known congener and of primary freshwater origin. Three species are endemic to Andaman Islands. Six species of alien fishes are known to be introduced for freshwater aquaculture. Threat to the indigenous fish fauna in the form of alien fish introduction and anthropogenic activity is discussed. The present work calls for conservation of freshwater ecosystem in the island groups with a risk assessment of fragile biotopes. It is also concluded that further exploratory streams may yield more species new to science.

Keywords

Andaman · Fish · Freshwater · India · Tropical Islands

P. T. Rajan (✉) · P. Vijay
Zoological Survey of India, Andaman and Nicobar Regional Centre, Port Blair, India
e-mail: rajanpt537@gmail.com

J. Praveenraj
ICAR-Central Island Agricultural Research Institute, Port Blair 744101, India

C. Sivaperuman, K. Venkataraman (eds.), *Indian Hotspots*,
https://doi.org/10.1007/978-981-10-6605-4_6

6.1 Introduction

Fish are the most diverse of the vertebrate taxa, distributed in a range of aquatic environments. Freshwater fishes constitute over 40% of the total recorded species of fish. Fish diversity known from the freshwater and marine water of India constitutes 9.7 % of the total number of about 34,190 species of fish known from the world (Eschmeyer and Fong 2016). About 11,952 species live in freshwater lakes and rivers that cover only 1 % of the earth's surface and account for a little less than 0.01% of its water. A total of 2936 native fishes are reported from India which includes 936 freshwater fishes, 113 brackish water and 1887 marine species belonging to 44 orders, 252 families and 1069 genera (NBFGR 2015–2016). The Andaman and Nicobar islands (ANI) lying off Southeast Asia form part of the Indo-Burma biodiversity hotspot (Myers 1990), which harbours a rich biodiversity of Indo-Malayan affinity. Freshwater fishes are little known, and some of the earliest studies were by Day (1870, 1875–1878), Hora (1925), Mukerji (1935), Herre (1939, 1941), Koumans (1940) and Sen (1975). Herre (1939) conducted the most comprehensive survey recording 112 species of freshwater and littoral fish. The most recent study was of Vijay and Davidar (2009), Rema Devi (2010) and Rajan and Sreeraj (2013, 2014) listing 25 species, of these only 10 are primary freshwater fishes and 39 are freshwater visiting species. It is known that new records of freshwater fishes are reported now and then, which reveals the unexplored status of Andaman Islands. The current study gives an updated checklist of the fishes found in the Andaman Islands comprising of native primary, secondary fishes and introduced alien fauna.

6.2 Freshwater Fishes of ANI

The Indian subcontinent, occupying a position at the confluence of three biogeographic realms, viz. the Palearctic, Afro-Tropical and Indo-Malayan, exhibits a great variety of ecological habitats and freshwater inhabitants, and 1570 are marine. The Indian freshwater fish species represent about 8.9% of the known families and 16 orders, including both primary and secondary freshwater fishes from India, Bangladesh, Myanmar, Nepal, Pakistan and Sri Lanka. The checklist of Menon (1999) lists 446 primary freshwater species under 33 families and 11 orders from the Indian region alone. Of the primary freshwater species, 68 % are constituted by the cyprinids, 18 % by siluroids and 14 % by other groups. Rema Devi and Indra (2000) list 667 species which are grouped under 12 orders, 35 families and 149 genera. The present proportion of the main groups of primary freshwater fishes is 62 % cyprinids, 26% siluroids and 12% other groups. From Andaman Islands, Herre (1939, 1941) listed 112 and 490 species including shoreline, coral reef fishes and very few freshwater fishes from the streamlets of Andaman Islands. Rema Devi (2010) listed 23 species, of these only 8 are primary freshwater fishes and others are secondary freshwater fishes encountered in freshwater habitats. Though the numbers of primary freshwater species are few, there are three species endemic to Andaman waters, viz. the syngnathid *Microphis insularis* (Hora), the eleotrid *Eleotris andamensis* (Hora), and *Schismatogobius* sp. (de Beaufort). The hill stream gobiid species *Sicyopterus microcephalus* (Bleeker) is a synonym of *Sicyopterus garra* (Hora), recorded from Andaman Islands.

6.3 Introduced Fishes of ANI

Currently there are 2500 freshwaters ponds operating in Andaman Islands, in addition to tanks, reservoirs and rivers. Many ponds which were dug for irrigational purpose are being used for fish culture. Many species has been introduced into ANI for consumption purposes including Indian major carps (*Catla catla*, *Labeo rohita* and *Cirrhinus mrigala*), Chinese carps (*Cyprinus carpio*, *Hypophthalmichthys molitrix*, *Ctenopharyngodon idella*) and catfishes (*Heteropneustes fossilis*, *Clarias batrachus*), some of which are well established in the freshwater bodies of ANI. Alien species has played an important role in development of freshwater aquaculture in Andaman and Nicobar Islands. In fact globally the alien fish introductions are mainly for aquaculture purpose.

Alien fishes have been introduced for only two motives in Andaman and Nicobar Islands, that is, for aquaculture and biological control. Their impact in the aquatic ecosystem is not yet studied, and at the instant, it is believed that they possess a great threat to the native freshwater fishes.

6.4 Methods

The study was undertaken to assess the species diversity and composition of freshwater fishes in rivers and streams of Andaman and Nicobar Islands. North Andaman is the northernmost island of the Andaman region. The region lies 285 km south of Myanmar and is located between 12°95′ N latitude and 92°86′ E longitude, covering an area of 1458 km^2. Topographically these islands are hilly and have rugged terrain with numerous mountains, peaks, ridges, hill slopes and valleys. Saddle Peak is the highest point, 732 msl (Anon. 2003). North Andaman has several large perennial streams which are the source of drinking water. Majority of the streams were characterized by low gradient with pebbles, forming the predominant substrate. The cultivation along the stream courses of cleared riparian forest has resulted in the drying up of lower reaches of many of streams during summer months. Most of the study streams were located in the centre of the island with a few towards the north. Middle Andaman (12°15′ N to 13° N latitude and 92°30′ E′ to 93° E longitude) is the largest island in the archipelago, with a total area of 1536 km^2. Austin Strait, a creek, separates Middle Andaman from North Andaman in the North and Middle Strait from the Baratang Island in the south (Anon. 2003). In general, the streams of Middle Andaman are characterized by low gradient, and pebbles form the main substrate. The study areas were fairly well distributed within the Middle Andaman with many streams in the north and the south being sampled but not much in the centre due to the Jarawa Tribal Reserve. South Andaman (11°55′ N latitude and 92°37′ E longitude) is a densely populated island with a total geographical area of 1456 km^2. Port Blair is the capital in the South Andaman. Topography of this island is hilly with numerous mountains and valleys. The highest point is Mt. Harriet at 366 msl (Anon. 2003). Majority of streams were small and medium sized. These streams were characterized by high gradient, having rock and bedrock as stream substrate. Lower reaches of most of these streams were subjected pollution from solid waste and agricultural run-off. The sampled streams were located in the southern part of the island due to the Jarawa Tribal Reserve where it was not possible to enter. Rutland is a large-sized island located south of South Andaman. This island falls under reserve forest area. The vegetation is a unique stunted formation of southern hill top evergreen forest dominated by *Dipterocarpus costatus* (with an average height below 10 m) (Anon. 2003). Apart from this, patches of moist deciduous species, dry deciduous and bamboo stands comprise the forest vegetation. There are several seasonal and perennial streams. These streams are relatively small compared to the streams in the other four study sites. Streams are characterized by high gradient, rocky substratum, riffles and waterfalls. Most of the streams are less disturbed and free from anthropogenic activities. The streams sampled covered a large area of Rutland. Little Andaman is the southernmost island of Andaman group lying between 10°40′ N latitude and 92°45′E longitude with the total area of 710 km^2. Topographically the island is flat with maximum altitude at 187 msl (Anon. 2003). There are several perennial streams flowing on either sides of the island. Streams on the western side are inaccessible due to terrain, and it falls under the protected tribal reserve. On the other hand, those streams on the eastern side are subjected to high levels of disturbance, such as pollution from agricultural run-off, solid waste, etc., and introduced fish species. The streams are low gradient and characterized by wide range of substrate types, viz. rocky, pebble and limestone. The streams sampled were located mostly in the centre and the south of the island. Mount Harriet national park lies between 10°43′57″N latitudes and 92°43′41″ to 92°47′11″ E longitudes. At each site GPS co-ordinates were recorded on a Gramin series GPS unit. Cast nets and handline were used for fish sampling, which was done from 8.00 a.m. to 1.00 p.m.; in addition to this, bamboo traps were used in the night to collect freshwater eels. The collected fishes were stored in water-filled buckets and photographed in fresh conditions. All the material studied has been deposited in the reference collections of Zoological Survey of India at Port Blair.

Table 6.1 List of species grouped into orders and families from Andaman and Nicobar Islands

Order	Family	Species
Elopiformes	Megalopidae	*Megalops cyprinoides* (Broussonet, 1782)
Anguilliformes	Anguillidae	*Anguilla marmorata* (Quoy and Gaimard, 1824)
		Anguilla bicolor (McClelland, 1844)
Cypriniformes	Cyprinidae	*Rasbora* sp.
Siluriformes	Claridae	*Clarias magur* (Hamilton, 1822)
	Heteropneustidae	*Heteropneustes fossilis* (Bloch, 1794)
Cyprinodontiformes	Aplocheilidae	*Aplocheilus* sp.(Hamilton, 1822)
Perciformes	Kuhliidae	*Kuhlia mugil* (Forster, 1801)
		Kuhlia rupestris (Lacepede, 1802)
	Cichlidae	*Oreochromis mossambicus* (Peters, 1852)
	Gobiidae	*Awaous grammepomus* (Bleeker, 1849)
		Exyrias puntang (Bleeker, 1851)
		Glossogobius giuris (Hamilton, 1822)
		Sicyopterus microcephalus (Bleeker, 1854)
		Stenogobius gymnopomus (Bleeker, 1853)
		Redigobius tambujon (Bleeker, 1854)
		Redigobius bikolanus (Herre, 1927)
		Glossogobius aureus (Akihito and Meguro, 1975)
	Eleotridae	*Belobranchus segura* (Keith et al., 2012)
		Belobranchus belobranchus (Valenciennes, 1837)
		Butis butis (Hamilton, 1822)
		Butis amboinensis (Bleeker, 1853)
		Eleotris andamensis (Herre, 1939)
		Eleotris fusca (Bloch and Schneider, 1801)
		Giuris margaritacea (Valenciennes, 1837)
	Anabantidae	*Anabas testudineus* (Bloch, 1792)
	Channidae	*Channa* sp. (Scopoli)
		Channa punctata (Bloch, 1793)
		Channa striata (Bloch, 1793)

6.5 Results

A total of 24 species belonging to six orders, 12 families and 19 genera were collected from the rivers and streams of Andaman and Nicobar Islands (Table 6.1, Images 6.1 and 6.2). Gobiidae constituted the largest family, possessing eight species followed by Eleotridae with six species. Three species, *Rasbora*, *Aplocheilus* and *Channa* were new, and their characters did not match with their known congener. In our surveys we encountered breeding population of *Channa striata*, *Anabas testudineus* and *Clarias batrachus* in the ponds, creeks and dams of South Andaman. Three species, namely, *Schismatogobius* sp., *Microphis insularis* and *Eleotris andamensis*, reported in the previous surveys by other authors are considered to be

Image 6.1 Native inland fishes of Andaman and Nicobar Islands

Oryzias sp.

Redigobius bikolanus

Sicyopterus sp.

Sicyopterus microcephalus

Belobranchus belobranchus

Megalops cypriniodes

Aplocheilus sp.

Microphis insularis

Belobranchus segura

Giuris margaritacea

Image 6.1 (continued)

Awous grammepomus

Exyrias puntag

Glossogobius giurus

Channa sp.

Stenogobius gymnopomus

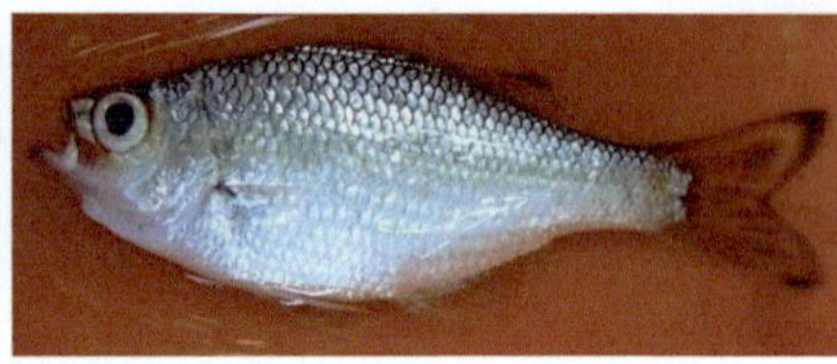

Khulia mugil

Khulia rupestris

Image 6.1 (continued)

endemic, but from the present survey, we believe that three more species, namely, *Rasbora* sp., *Aplocheilus* sp. and *Channa* sp., may be added to the endemic status. As most of the island hill streams, rivulets and ponds are perennial, there is a risk of this species invading all water bodies and completely eradicating the native aquatic biodiversity. A status on the origin of freshwater fishes is presented in Table 6.2. From the present study, it is concluded that more new species can be known from the jungle streams passing through the tribal reserve area.

6.6 Discussion

Aplocheilus sp. and *Rasbora* sp. were very common across the study area. At least four other species, *Sicyopterus microcephalus*, *Glossogobius giuris*, *Glossogobius aureus* and *Channa* sp. were common. Three gobies, *Redigobius tambujon*,

Image 6.2 Introduced freshwater fishes of Andaman Islands

Table 6.2 Native, exotic and endemic freshwater species and their status

Species	Native	Exotic	Endemic	Status
Rasbora sp.	√			Very common
Clarias magur		√		Common
Heteropneustes fossilis		√		Common
Aplocheilus sp.	√			Very common
Oreochromis mossambicus		√		Common
Awaous grammepomus	√			Rare
Glossogobius giuris	√			Common
Schismatogobius sp.	√		√	Rare
Sicyopterus microcephalus	√			Very common
Sicyopterus sp.	√		√	Rare
Stenogobius gymnopomus	√			Very rare
Belobranchus segura	√			Very rare
Eleotris andamensis	√		√	Rare
Eleotris fusca	√			Common
Giuris margaritacea	√			Common
Anabas testudineus	√			Very rare
Channa sp.	√			Common
Channa punctata	√			Common
Channa striata		√		Rare

Redigobius bikolanus and *Awaous grammepomus* were rare, while *Acentrogobius caninus*, *Stenogobius gymnopomus*, *Microphis insularis*, *Schismatogobius* sp. and *Sicyopterus* sp. were very rare. Among non-native species, *Oreochromis mossambica* were very common in streams in which they occurred. Other species such as *Heteropneustes fossilis*, *Clarias batrachus*, *Anabas testudineus* and *Channa striata* were rare. Of the 24 species recorded, 12 were native freshwater fishes, including 2 endemic fishes. Though the numbers of primary freshwater species are few, there are three species endemic to Andaman waters, viz. the syngnathid *Microphis insularis* (Hora), eleotrid *Eleotris andamensis* (Herre) and a*Schismatogobius* sp. (De Beaufort), which is believed to be new and endemic and was not encountered in our survey.

Five species of freshwater fish have been introduced deliberately or accidentally in the freshwater streams of Andaman Islands. Streams are altered by water diversion, channel modification, introduced species and water quality degradation. Compared to streams of Wimberly Gunj and Naya Puram, the streams of Mannarghat, Kalatang and Shoal Bay in South Andaman are pristine. The Andaman Islands, with freshwater streams ranging from the relatively pristine to the highly degraded, offer an opportunity to examine the impacts of human disturbance on native stream communities. For example, urbanization is often accompanied by stream-channel modification and reduced canopy cover, resulting in higher water temperatures and greater fluctuations in daily temperature. Even in relatively pristine streams, diversions can result in decreased flow velocity and water depth, reducing habitat availability. Many non-native aquatic species are better adapted than native species to degraded habitats; once established in these habitats, they can cause further reduction in native populations through competition, predation and the introduction of parasites or diseases. Poisoning, dumping of solid waste and diverting water flows to encroach land are the major threats. Creating awareness, controlling poisoning fishing and protecting the breeding grounds of fishes are some of the measures recommended to counter these threats. The most urgent initiative appears to be a national reserve system 'gap analysis' which would identify those ecosystems most at risk. A comprehensive national assessment of the conservation status of freshwater ecosystems should be undertaken immediately. Such a study would provide a platform for the systematic expansion of the nation's freshwater protected areas, as well as a catalyst for innovative 'bottom-up' conservation approaches driven by local stakeholders. The new findings have indicated that more new species can be found in undisturbed regions. However, many of the native species are threatened due to habitat loss and invasive species.

Acknowledgements We thank Dr. Kailsh Chandra, Director, Zoological Survey of India, Kolkata, and Officer-in-Charge, ZSI, Port Blair for the facilities and encouragement to undertake this study. Thanks to Dr. Rema Devi, Zoological Survey of India, Chennai; Dr. Philippe Keith, Assistant Director and Curator of Fishes, Muséum National d'Histoire Naturelle, France; and Dr. Helen K. Larson, Curator Emeritus, Fishes Museum and Art Gallery of the Northern Territory Australia for fruitful discussions on related topics and to Dr. S.S. Misra, Zoological Survey of India, Kolkata, for comments. We also thank CA Rahaman former DFO, PT Alexander ACF (Retired) and Hussain, Forester of Forest Division, South Andaman, for providing logistics support.

References

Anonymous (2003) Indian Institute of Remote Sensing (IIRS)-Department of Environment and Forests (Andaman and Nicobar Islands) Report. Indian Institute of Remote Sensing, Uttaranchal, 325p

Day F (1870) On the fishes of the Andaman Islands. Proc. Zoo. Soc. London, pp 677–705

Day F (1875–1878) The fishes of India being a natural history of the fishes known to inhabit the seas and freshwaters of India, Burma and Ceylon. Text and Atlas in 4 parts. London, xx +778 pp, 195 pls

Eschmeyer WN, Fong JD (2016, May 2) Species by family/subfamily

Herre AWCT (1939) On a collection of littoral & freshwater fishes from the Andaman Islands. Rec Indian Mus 41(4):327–372

Herre AWCT (1941) List of the fishes known from the Andaman Islands. Mem Indian Mus 13(3):331–403

Hora SL (1925) Doryichthys insularis. In: Annandale N, Hora SL (eds) The freshwater fishes from Andaman Islands. Rec Indian Mus 27(2):33–41

Koumans FP (1940) The fishes of the Indo-Australian Archipelago. X, E.J. Brill Ltd, Leiden, 423 pp.

Menon AGK (1999) Check list – fresh water fishes of India. Records of Zoological Survey of India, Miscellaneous Publication, Occasional Paper (175):1–366

Mukerji DD (1935) Notes on some rare and interesting fishes from the Andaman Islands, with description of two new freshwater gobies. Rec Indian Mus 37(3):250–279

Myers N (1990) The biodiversity challenge: expanded hot-spots analysis. Environmentalist 10(4):243–256

NBFGR (2015-2016) National bureau of fish genetic resources. Annual report 2015–2016

Rajan PT, Sreeraj CR (2013) Diversity, distribution and conservation of freshwater fishes in Mount Harriet National Park, Andaman and Nicobar Islands. Rec Zool Surv India 113(Part-4):35–55

Rajan PT, Sreeraj CR (2014) New record of two species of *Belobranchus* (Teleostei: Gobioidei: Eleotridae) from Andaman Islands. Rec Zool Surv India 114(Part-1):155–158

Rema Devi K (2010) Freshwater fishes of Andaman Islands. In: Ramakrishna, Raghunathan C, Sivaperuman (eds) Recent trends in Biodiversity of Andaman and Nicobar islands, pp 329–339 (Published by the Director, Zool. Sur. India, Kolkata

Rema Devi K, Indra TJ (2000) Freshwater ichthyofaunal resources of Tamil Nadu, pp 77–97. In: Ponniah AG, Gopalakrishnan A (eds) Endemic fish diversity of Western Ghats. NBFGR-NATP Publication. National Bureau of Fish Genetic Resources, Lucknow, vol 1, 347 p

Sen TK (1975) Further light on freshwater fish fauna of Andaman Islands. Sea Food Exp J 7(2):31–33

Vijay P, Davidar P (2009) A survey of freshwater fishes of Andaman Islands. J Bombay Nat Hist Soc 106(1):11–14

Asian Openbill Stork (*Anastomus oscitans*), Not a "Nutcracker": A Study from Kole Wetlands of Thrissur, Kerala

7

P. Greeshma and E. A. Jayson

Abstract

Birds are considered as an indicator of environmental health. The Asian openbill stork (*Anastomus oscitans*) is a large wading bird and belongs to the family Ciconiidae under the order Pelecaniformes. This paper deals with the food and feeding behavior of Asian openbill storks in Kole Wetlands of Thrissur, Kerala, India. Observations on feeding behavior were made with the help of spotting scope (10–45×), stopwatch, Sony HDR videocam, and binocular (7×50). The observation was made between 06.30 and 18.30 h, and the entire day was divided into four time intervals: morning (06.30–09.30), midday (09.30–12.30), afternoon (12.30–15.30), and evening (15.30–18.30). Focal-animal sampling method and direct observation method were used for studying the food and feeding of Asian openbill stork. September–January was the most active season for openbill storks. Feeding activity was highest during morning (06.30–09.30) and evening (15.30–18.30) hours. Asian openbill stork feeds on molluscs mainly snail *Pila globosa* (genus *Pila*) and freshwater mussel (genus *Unio*). Prey-capturing success was more through multiple probing. Detachment of *Pila* takes place underwater, but in the case of *Unio*, it occurs in land. Crushing of shells and feeding on fishes by openbill stork were not supported from our observations from Kole Wetlands.

Keywords

Asian openbill stork · Feeding · Kole Wetlands · Kerala

P. Greeshma (✉) · E. A. Jayson
Wildlife Department, Kerala Forest Research Institute, Thrissur, Kerala, India
e-mail: greeshmap@kfri.res

C. Sivaperuman, K. Venkataraman (eds.), *Indian Hotspots*,
https://doi.org/10.1007/978-981-10-6605-4_7

7.1 Introduction

Birds are considered as an indicator of environmental health. India is rich in avifaunal diversity with passerines and nonpasserines. Storks belong to the family Ciconiidae under the order Pelecaniformes. Nineteen species of storks (Ali and Ripley 1983) are found globally, and out of these, nine species are found in India and six stork species in Kerala (Praveen 2015). South and Southeast Asia have the richest stork diversity with 11 species: 8 residents and 3 migratory (Anam et al. 2016). The Asian openbill stork is a large wading bird with grayish or white with glossy black wings and tail characterized by long legs, neck, and bill, which meets together only at the tip (Hancock et al. 1993). Collection of basic data regarding food availability, habitat use, and foraging techniques is needed for the effective implementation of conservation strategy. The peculiar name "openbill" is derived because of the presence of the distinctive gap formed between the recurved lower and arched upper mandible of the beak in adult storks. Young storks do not have this gap. The fine brushlike structure at the cutting edges of the mandible gives them better grip while holding the snails (Gosner 1993). The food and feeding behavior of openbill storks, *Anastomus lamelligerus* and *Anastomus oscitans*, were studied in Africa, India, Ceylon, and Thailand by Kahl (1971). The general and popular notion about the openbill is that the openbill functions in the fashion of a "nutcracker." This paper deals with the food and feeding behavior of Asian openbill storks in Kole Wetlands of Thrissur, Kerala, India.

7.2 Study Area

The Kole Wetlands is one of largest, highly productive, and threatened wetlands in Kerala and has been declared as Ramsar site for protection (2002), and it comes in the Central Asian Flyway of migratory birds. The Malayalam word "Kole" indicates bumper yield, which refers to a particular type of cultivation method adopted in backwaters from December to April. The Kole Wetlands lies between 10° 20′ and 10° 40′ N latitudes and 75° 58′ and 76° 11′ E longitudes. The Kole Wetlands are low-lying tracts located 0.5–1 m below mean sea level (MSL) and remain submerged for about 6 months in a year. Jayson and Sivaperuman (2005) and Sivaperuman and Jayson (2000) during their avifaunal studies in various regions of Thrissur District concluded that the highest number of birds was reported from Kole Wetlands. This study was carried out in the Kole Wetlands of Thrissur District of Kerala from 2015 January to 2016 November.

7.3 Methods

Observations on feeding behavior were made with the help of spotting scope (10–45×), stopwatch, Sony HDR videocam, and binocular (7×50). The observation was made between 06.30 and 18.30 h, and the entire day was divided into four time

Fig. 7.1 Openbill stork in dewatered land

intervals: morning (06.30–09.30), midday (09.30–12.30), afternoon (12.30–15.30) and evening (15.30–18.30). Activity data were collected according to this time period. The parameters taken for the study of feeding ecology of openbill stork were food items consumed, size of the prey, time for feeding, flock size, and inter-bird distance and techniques adopted for feeding. As openbill storks forage in open wetlands, it was so easy to observe the feeding behavior. Most of the observations were taken from a distance of 50–200 m with the help of spotting scope and recorded with videocam. Focal-animal sampling method and direct observation method (Altmann 1974) were used for studying the food and feeding. The water depth, where the foraging occurred, was also measured. Identification of prey species was made by direct observation.

7.4 Results

In Kerala, Asian openbill stork (*Anastomus oscitans*) is known as "*Njhaunipottan*" (one who cracks snail) and "*Cherrakokkan*" (without a closed bill). Openbill storks forage in Kole Wetlands in different microhabitats like mudflats (Fig. 7.1), shallow water streams, paddy fields (Fig. 7.2), small ditches, and along the bank of small canals characterized by diurnal feeding.

In response to habitat conditions, openbill stork displayed local movements. Large population of openbill stork was seen during the post-monsoon period and least during monsoon. With the advent of post-monsoon, dewatering of paddy fields started, and openbill storks were seen in a flock of 63–378 individuals. September–January is the most active season for openbill storks. Kole lands after dewatering and harvested paddy fields are the favorite foraging ground for the storks. During the dewatering time, it was seen that openbill storks reached the feeding ground after 10–25 min after sunrise. They arrived to the Kole lands as solitary and in groups of three to seven individuals and immediately started feeding. During the

Fig. 7.2 Openbill stork in paddy field

Fig. 7.3 Openbill stork on a raised mudflat

month of December and January, it was seen that openbill storks arrive as one by one to the harvested paddy field and they stood idle on a heap of hay or in a raised mudflat by drooping their head and beak (Fig. 7.3).Usually the individuals stood in a group (inter-bird distance of 2.0–10.0 m) (Fig. 7.4), and they start feeding only after 30–55 min, one by one individually. Moreover they always keep a distance and get scattered away to choose different feeding areas within the same habitat.

Feeding activity was highest during morning (06.30–09.30) and evening (15.30–18.30) hours. In between, the activity was less, and most of the individuals went for roosting in the nearby trees in the bund (Fig. 7.5). It had been observed that after morning feeding, they exhibit preening, basking (Fig. 7.6), and soaring behavior (Fig. 7.7) and they change the feeding ground. The selection of feeding ground mainly depends on the foraging success (prey-capturing success and with least disturbance). Asian openbill storks feed on molluscs mainly snail *Pila globosa* (genus *Pila*) and freshwater mussel (genus *Unio*). The favorite food item was *Pila globosa* which are abundantly distributed throughout the wetlands. The shells of *Pila* and

Fig. 7.4 Openbill storks in paddy field

Fig. 7.5 Roost of openbill stork

Unio discarded by the storks were collected from the Kole lands and measured using vernier calipers (Fig. 7.8), and it was seen that *Pila* shell size ranges from 2.53 to 83 cm in length and 4.83 to 3.92 cm in breadth (n = 206) and *Unio* ranges from 4.7 to 6.15 cm in length and 2.94 to 3.36 cm in breadth (n = 113).

Feeding on crab by openbill stork was recorded only in a single occasion. Openbill stork feeding on fishes was not supported from my observations from Kole lands. *Anastomus oscitans* forage in groups as well as in single and also with mixed flock consisting of painted storks (*Mycteria leucocephala*), black-headed ibis (*Threskiornis melanocephalus*), gray heron (*Ardea cinerea*), purple heron (*Ardea purpurea*), Indian pond heron (*Ardeola grayii*), little egret (*Egretta garzetta*), intermediate egret (*Mesophoyx intermedia*), and great egret (*Ardea alba*). Openbill stork was a long-distance walker and generally walks forward in a linear fashion and thoroughly wades in the substrate. It walks continuously probing in the substrate, up to a distance of more than 60–110 m, touches the end of the paddy field, and returns back after moving straight through the vegetation side along the bunds. This zigzag movement pattern is found to be a simple strategy to find almost every patch of food (Fig. 7.9).

Fig. 7.6 Openbill stork basking in paddy field

Fig. 7.7 Openbill stork soaring in Kole lands

While foraging in dewatered paddy fields, openbill storks always keep a minimum distance (>2 m). They also follow linear movements while foraging on the banks of small streams. Openbill storks were also seen feeding in a tightly packed group when they are foraging in areas with a water depth of 15–26 cm. The individuals were well structured in the group as one individual will move in front and lead the group. Here the probing technique was different as they dip their entire bill up to the eye into the water. Multiple probing (34–129/ min) (n = 112) increased with increased water depth. Usually, the stork slightly opens its bill and stabs through the water. While openbill storks walk, they probe only once in a spot, and when they felt the presence of food item, they stopped walking and undergo multiple probing. During multiple probing, they insert more than half of their bill into the substrate in various directions by twisting their neck, but during single probe, only one third of the bill is gone inside the substrate. In the paddy fields, the probing intensity was inversely proportional to the footstep rate (n = 293), whereas adjacent to floating vegetation and rice stalks, footstep rate is inversely proportional to the probing intensity (n = 149). Feeding techniques employed by openbill stork were

Fig. 7.8 Measuring shell size with vernier calipers

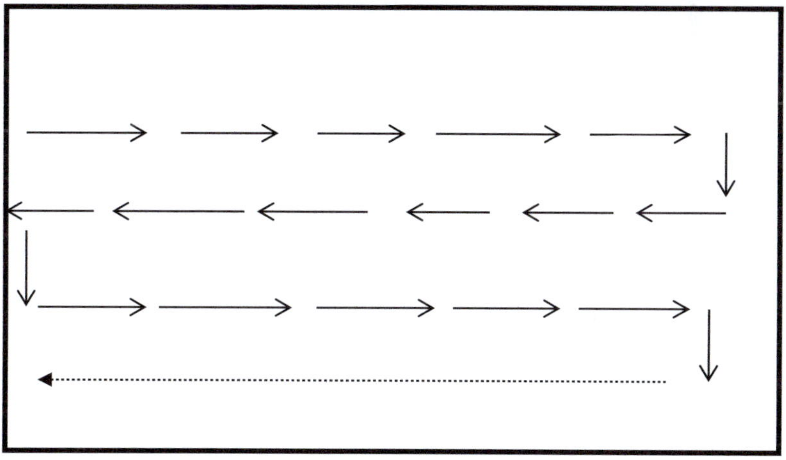

Fig. 7.9 Diagrammatic sketch of foraging path of openbill stork in Kole Wetlands

found to be a successful one. Prey-capturing success was more through multiple probing. When the stork captured a *Pila* of small size (2.53 cm in length and 4.83 cm in breadth) (Fig. 7.10), it picked up with the tip of the bill, inserted the lower mandible into the opening, held tightly using the upper mandible, and shook the bill toward the right and left side (six to nine times) very fast to detach the flesh from the shell. But when they captured large *Pila* (4.83 cm in length and 3.92 cm in breadth) (Fig. 7.11), they picked up with the tip of the bill and rolled up to the middle of the bill and held it for some time (36–129 s).Usually they feed from the same spot where they captured their prey, but sometimes in order to avoid snatching, they kept the prey inside the bill and flew to a distant area from other storks.

After reaching a suitable location, the stork rolled down the *Pila* to the tip or to the ground and pushed the tip of the lower mandible to open the operculum. It dipped the *Pila* (two to four times) in water in order to open the operculum easily. Once the operculum was opened, the stork then inserted its mandibles, lifted up

Fig. 7.10 Openbill stork with small *Pila globosa*

Fig. 7.11 Openbill stork with large *Pila globosa*

Pila, shook its bill (5–14 times) to free the snail's body from its shell, swallowed the flesh by tossing back the head (4–6 s), and discarded the shell. Cracking of shell was not supported from my observations. In summer season, openbill stork was seen foraging in shallow water in streams in search of freshwater mussel. Through multiple probing, they picked up the mussel (Fig. 7.12) with the bill tip, held it, and

Fig. 7.12 Openbill storks foraging in shallow waters, holding freshwater mussel

Fig. 7.13 Openbill stork inserting lower mandible inside the freshwater mussel shell

moved from the water to the mudflat. While feeding on freshwater mussels, the lower mandible was inserted between the two halves of the mussel shell (Fig. 7.13) and forced to break the hinge, where the two halves were connected. The whole mass was consumed immediately after the shell was opened (Fig. 7.14) by tossing back the head, and the shell was left open in the mudflat. Snails were found to be the most favorite food ($n = 397$), then freshwater mussel ($n = 79$), and then crab ($n = 1$). Percentage of prey consumption is plotted in Figs. 7.15 and 7.16.

7.5 Discussion and Conclusion

Food play an inevitable part in the life cycle of birds, especially related to reproductive activities. Kahl (1966) also mentioned about the role of food availability. Major food items consumed by openbill storks were apple snail and freshwater mussels, which are available throughout the season in the Kole Wetlands. The movement pattern and quick movements in mudflats and shallow water were to cover the whole

Fig. 7.14 Openbill storks tossing back the head and feeding freshwater mussel

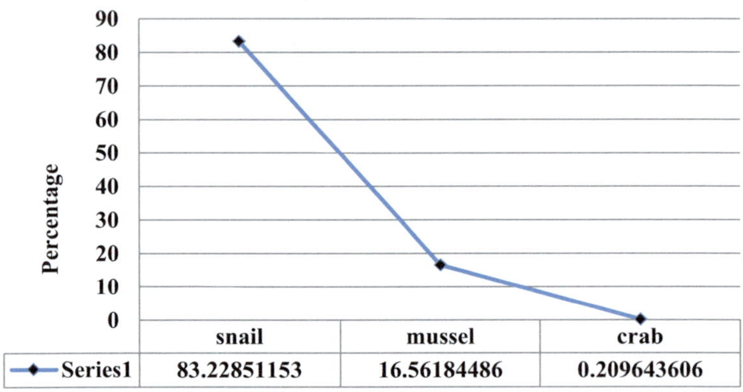

Fig. 7.15 Percentage of prey consumed by openbill stork

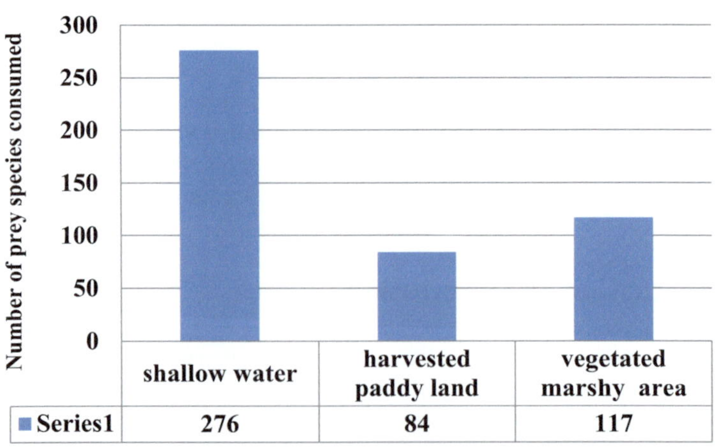

Fig. 7.16 Feeding habitat of openbill stork

food patch. Like egrets, openbill storks also wade very fast that makes their prey capturing a success. This point was substantiated with the findings of Meyerriecks (1962) and Kushlan (1976) that wading birds of family Ciconiiformes walk fast or running behavior was displayed mainly for capturing of moving prey and spotting the prey better. It was observed that openbill storks and black-headed ibises spent more time in vegetated areas and floating vegetation. The reason may be that, when predator pressure is more, the prey species move to safer places. The "openbill" makes the stork to invade a variety of habitat and to handle different prey species very easily. The mandibles firmly held the prey underwater without being washed off, which helps to roll up and roll down the snails and to extract body mass from the shells. Kahl (1971) also supported that the snails were handled in the distal part of the bill and are not crushed. Feeding on fishes by openbill stork as described by Anam (2016) was not supported from our observations from Kole Wetlands.

Acknowledgements The authors are grateful to the Director of KFRI for the facilities and infrastructure. We wish to thank Kerala State Council for Science, Technology and Environment for providing necessary financial support and Mr. Manoj Karingamadathil, Mr. Joy Vincent, Dr. Nishad PM, Mr. Jinoy for their field assistance.

References

Ali S, Ripley SD (1983) Handbook of the birds of India and Pakistan. Oxford University Press, Delhi, p 737

Altmann J (1974) Observational study of behavior: sampling methods. Behaviour 49:227–267

Gosner KL (1993) Scopate tomia: an adaptation for handling hard-shelled prey? Wilson Bull 105(2):316–324

Hancock JA, Kushlan JA, Kahl MP (1993) Storks, ibises and spoonbills of the world. Academic, London

Jayson EA, Sivaperuman C (2005) Avifauna of Thrissur district, Kerala, India. Zoo's Print J 20(2):1774–1783

Jinnath A, Mansur A, Saikia MK, Saikia PK (2016) Food and feeding behaviour of Openbill stork (*Anastomus oscitans*) in Assam, India. J Glob Biosci 5(6):4188–4196

Kahl MP (1966) A contribution to the ecology and reproductive biology of the Marabou stork (*Leptoptilos crumeniferus*) in East Africa. J Zool 148:289–311

Kahl MP (1971) Food and feeding behavior of Openbill storks. J Ornithol 112:21–35

Kushlan JA (1976) Feeding rhythm in nesting White This. Wilson Bull 1:656–657

Meyerriecks AJ (1962) Diversity typifies heron feeding. Nat Hist 71(6):48–59

Praveen J (2015) A checklist of birds of Kerala, India. J Threatened Taxa 7(13):7983–8009

Sivaperuman C, Jayson EA (2000) Birds of Kole wetlands, Thrissur, Kerala. Zoo's Print J 15(10):344–349

Avifauna of North West Himalaya

8

Anil Kumar

Abstract

Northwestern Himalaya is known for a highly heterogeneous physiography and climate and an amazing biodiversity. This chapter is based on the published literature, surveys conducted by Zoological Survey of India during the last two decades, and studies undertaken by author during the last 7 years. A total of 768 species of birds belonging to 21 orders and 95 families are enlisted in the present study including some rare/interesting records. About 48% species of birds of this region are migratory and exhibit different types of migration. Among them, 16.02% are long-distance winter migrants followed by 6.51% passage migrants and 14.58% summer migrants. The rest of the birds (483 spp.) are resident including 81 local migrants. Out of 768 species, 61 (7.94%) belong to threatened categories. Four species are 'critically endangered', while five are 'endangered'. Twenty-three species belong to 'vulnerable' category followed by 29 species in 'near-threatened' category. The major threat to avian diversity is loss, degradation and fragmentation of habitats owing to unorganized development, agriculture, forest fire and excessive extraction of natural resources.

Keywords

Avifauna · Distribution · Himalaya · Status

A. Kumar (✉)
Northern Regional Centre, Zoological Survey of India, Dehra Dun, Uttarakhand, India
e-mail: anilsonta@gmail.com

C. Sivaperuman, K. Venkataraman (eds.), *Indian Hotspots*,
https://doi.org/10.1007/978-981-10-6605-4_8

8.1 Introduction

The northwestern Himalayan region with a highly heterogeneous geography, wide range of altitudes, varied topography and great climatic variability is known for its amazing biodiversity richness. The western Himalaya extends along the mountain chain from western Nepal (west of the Kali Gandaki valley) through Himachal Pradesh, Jammu and Kashmir in northwest India and northern Pakistan and then southwest along the mountains in the border region between Pakistan and Afghanistan (Islam and Rahmani 2004). In Indian Territory, three states, namely, Jammu and Kashmir, Himachal Pradesh and Uttarakhand, cover major area. The total geographical area of this region (all three states) is 3,31,392 km^2 (ISRF 2013). About 27% (91,914 km^2) is recorded forest area, which can be further classified into reserved forests (44,184 km^2), protected forests (45,566 km^2) and unclassified forests (2164 km^2). The faunal wealth of northwest Himalaya is quite rich. On the basis of altitude, the forests are mainly classified as subtropical forests, temperate forests, subalpine forests and alpine vegetation. The region's climatic zones contain a fairly rich diversity of species and ecosystems that exist along a pronounced humidity gradient. Vegetation differs from subtropical semidesert and thorny scrub/meadows in the northwest to broadleaf/deciduous forests in the south-eastern part (Islam and Rahmani 2004; Schickhoff 2005).

Jammu and Kashmir (32° 17′ – 37° 05′ N and 72° 31′ – 80° 20′ E) are located in western Himalaya. On the north, it is bounded by international border with China and on the east by Tibet. Southern boundary lies with Himachal Pradesh and Punjab, while on the west by Pakistan and Afghanistan. It is divided into three geographical regions, namely, Kashmir Valley, Ladakh region and Jammu region (Narwade et al. 2006). The higher regions are covered by Pir Panjal, Karakoram and the inner Himalayan ranges.

Kashmir Valley is covering an area of 15,948 km^2 and bounded by lofty mountains of the Pir Panjal in the south and southwest and by the Great Himalayan range in the north and east (Dar 2008). It is overwhelmed with varied habitats and vegetation, such as freshwater swamps and marshes (dominated by reeds, rushes, and sedges species, such as *Phragmites australis*, *Juncus* spp., *Scirpus* spp. and some other plants, namely, *Butomus umbellatus*, *Typha*, *Equisetum* and *Alisma*), rocky gorges (mainly comprised of plant species *Isodon rugosus, Viburnum foetens* and *Clematis* spp.) and moist alpine meadows which harbour open grassland plants such as *Aconitum*, *Aquilegia*, *Delphinium*, *Callianthemum*, *Aster*, *Inula* and *Salvia* (Dar 2008). Jammu region is characterized by subtropical to alpine vegetation and harbours a rich array of flora and fauna manifested by diverse habitat conditions (Sharma 2008). Contrary to Kashmir and Jammu region, Ladakh region has sparse vegetation cover (Rawat 2008). It is spread over 96,700 km^2 area. The region exhibits typical biophysical features of cold deserts and can be divided into two provinces, namely, Ladakh Mountains (comprised of rugged mountain ranges and valleys) and Eastern Plateau (undulating elevated landscape) (Rodgers and Panwar 1988). The area mainly comprised of moist meadows (dominated by *Festuca kashmiriana*, *Oryzopsis munroi* and *Melica persica*), marsh meadows (mainly located in

Changthang and dominated by sedges, namely, *Carex*, *Blysmus*, *Kobresia* and *Eleocharis*; grasses such as *Calamagrostis holciformis*, *Poa* spp. and *Puccinellia* spp.; and aquatic species such as *Potamogeton pectinatus*, *Myriophyllum verticillatum*, *Hippuris vulgaris*, *Ranunculus natans* and *R. trichophyllus*) and some rocky crags, riverine scrubs (represented by *Hippophae* spp.) and steppe slops dominated by *Caragana versicolor* (Rawat 2008).

Himachal Pradesh (30° 22′ – 33° 13′ N and 75° 36′ – 79° 02′ E) is located in the northwest of India in the Himalayan ranges (Narwade et al. 2006). In the east, it forms India's international boundary with Tibet (China). It is bounded by Jammu and Kashmir in the north, Uttarakhand in the south-east, Haryana in the south and Punjab in the west. The state is mountainous (ranging between 460 and 6600 m ASL), drained by a number of snow-fed perennial rivers. It has a complex geography and habitats (Narwade et al. 2006) and encompasses a rich temperate flora and fauna. Uttarakhand (28° 33′ – 31° 27′ N and 77° 34′ – 81° 02′ E) is located in the northern part of India and known for its natural wealth including rich flora and fauna. It borders the Tibet (China) on the north, Nepal on the east and the Indian states of Uttar Pradesh to the south and Himachal Pradesh to the northwest. Uttarakhand has a total area of 53,484 km^2, of which 93% is mountainous and 65% is covered by forests. Most of the northern part of the state is covered by high Himalayan peaks and glaciers (Narwade et al. 2006).

The lower Himalaya and Shiwalik hills mainly comprised of mixed and deciduous forests. Simbalbara Wildlife Sanctuary, Pong Lake Wildlife Sanctuary, Bandli Wildlife Sanctuary and Govind Sagar and Naina Devi Wildlife Sanctuary are located in this area. In Pong Dam area, prominent tree species are *Acacia* sp., *Syzygium cumini*, *Dalbergia sissoo*, *Mangifera indica, Morus* sp., *Ficus* sp., *Bauhinia variegata*, *Prunus* sp. and *Phyllanthus emblica* (Kumar and Paliwal 2015). In Chail Wildlife Sanctuary and adjacent areas, the dominant forest tree is ban oak (*Quercus incana),* mixed with chir pine (*Pinus roxburghii)* at lower altitudes. Rhododendron (*Rhododendron arboreum)* forms pure stands in places, and cedar (*Cedrus deodara)* and blue pine (*Pinus wallichiana)* have been planted in some areas. There is little mature forest and much secondary growth due to disturbance (Gaston and Singh 1980). In middle Himalaya, the dominant forest types are Montane Broadleaf Deciduous Forests and Mixed Broadleaf Coniferous Forests. The Kalatop and surrounding area contains thick patches of cedar (*Cedrus deodara)* followed by ban oak (*Quercus incana)* and blue pine (*Pinus wallichiana).* Khajjiar is a meadow surrounded by cedar (Islam and Rahmani 2004). The vegetation of Tirthan Valley is dominated by blue pine (*Pinus wallichiana)* and higher up by a diverse deciduous broadleaf forest on moderately sloping areas and fir (*Abies pindrow*) on steep areas. Tirthan Valley also supports small areas of oak forest (*Quercus* sp. and *Q. incana*). The southerly aspects are generally more open; stands of cedar (*Cedrus deodara)* are interspersed with grassy and shrub-clad hillsides, with a zone of Kharsu oak (*Q. semecarpifolia)* forest above 2800 m (Gaston et al. 1994). In high altitude areas such as Kibber Wildlife Sanctuary and Pin Valley National Park, the vegetation mostly comprises small shrubs of junipers and grasses such as *Poa* and *Agropyron* spp. The valley areas are characterized with scattered patches of some thorny trees and bushes such as seabuck thorn (*Hippophae rhamnoides*), willow (*Salix* spp.) and poplar (*Populus balsamifera*) (Islam and Rahmani 2004).

In the foothills of Uttarakhand, forests are tropical moist deciduous and dominated by tree species like Sal *Shorea robusta*, *Machilus odoratissima*, *Syzygium cumini* and *Litsea* sp. along with the plantation of *Ailanthus excels*, *Tectona grandis* and *Haplophragma* in some areas (Pandey et al. 1994). Forest types of the Kumaon Himalaya comprised of sal forests (found up to 1200 m), pine forests (1200–2400 m), oak forests (1300–3200 m), mixed broadleaf forest (foothills to 3300 m) and *Betula utilis* forest (3200–3500 m). The major tree species included *Quercus leucotrichophora*, *Q. lanata*, *Q. floribunda*, *Q. semecarpifolia*, *Tsuga dumosa* and *Rhododendron arboreum*, in association with *Viburnum* species, *Myrica esculenta*, *Alnus nepalensis*, *Swida oblonga*, *Lyonia ovalifolia*, *Persea duthiei* and *Lindera pulcherrima* (Sultana and Khan 2000). The high altitude areas of Uttarakhand, such as Kedarnath Wildlife Sanctuary and Valley of Flowers National Park characterized by moist meadows above tree line, have a large number of plant species such as *Ranunculus hirtellus*, *Bistorta affinis*, *B. vivipara*, *Anemone rivularis*, *Geranium wallichianum*, *Potentilla atrosanguinea*, *P. argerophylla* and *Geum elatum,* while temperate oak forests found at relatively lower altitude are dominated by *Quercus leucotrichophora*, *Rhododendron arboreum* and *Lyonia ovalifolia* (Singh and Dutt Rai 2008).

8.1.1 Ornithological History of the Region

Ornithological history of the region dates back to 19th sanctuary. Andrew Adams joined army in 1848 as a physician and worked in Himachal and Ladakh. He wrote 'The birds of Cashmere and Ladakh', and his notable contribution was the discovery of orange bullfinch (*Pyrrhula aurantiaca*). He recorded the first breeding site of brown-headed gull (*Larus brunnicephalus*) in the Tibet Plateau. Black-winged snowfinch (*Montifringilla adamsi*) was named after him. He also wrote 'Wanderings of a Naturalist in India, the Western Himalayas and Cashmere in 1867'. Bates wrote 'Breeding birds of Kashmir' and described the behaviour and habitats of over 150 avian species. He was a notable photographer also. Theobald (1862) travelled from Shimla to the Spiti Valley and Chomoriri (Tsomoriri) Lake in 1861 and made observations on the birds encountered en route. Tytler (1868) recorded bird species during a march from Shimla to Mussoorie. Dodsworth (1910, 1912a, b, 1913a, b, c, 1914) made detailed observation on the avifauna of Shimla and adjacent areas. Sir Norman Frederick Frome also studied the birds of this region and contributed notes on Shimla and adjacent localities, published in Journal of Bombay Natural History Society. Allan Octavian Hume visited various parts of the country, including western Himalaya, and contributed tremendously. He started the quarterly journal, 'Stray Feathers' in 1872, and published 12 volumes. He wrote a number of valuable and novel papers. Koelz (1895–1989), Ludlow (1885–1972) and Marshall (1841–1927) were other notable contributors to the ornithological history of the region.

During 1933, Walter Koelz extensively surveyed the Spiti Valley (including Kibber and Pin Valley areas) and reported the occurrence of over 95 species (Koelz 1937). Ludlow contributed to the natural history of a number of species and collected about 7000 specimens of birds. The brown-throated fulvetta (*Alcippe ludlowi*) is named after him. Marshall wrote a number of articles, including 'The Game Birds of India, Burmah and Ceylon' in three volumes. 'The Wildlife of Dehradun', a pioneering piece of work on the birds of the region, was contributed by Bertram Beresford Osmaston (1868–1961). Ripley (1913–2001) took walking tour to Ladakh and Western Tibet and contributed significantly. Later on, he worked with Dr. Salim Ali, and decades-long milestone work of both pioneers established the very strong foundation of Indian ornithology. They produced 'Handbook of the birds of India and Pakistan' in ten volumes, including a number of other books and articles. Ali and Ripley (1983) in their monumental work presented a summarized account of the avifauna of Himachal Pradesh particularly from areas like Shimla, Dalhousie, Dharamshala, etc. Stoliczka (1838–1874) worked with Hume and contributed accordingly. He made a large collection of birds from Sutlej Valley (Stoliczka 1868). He covered Spiti Valley also. Later on the valley was surveyed by Huge Whistler (1889–1943) during July 1922. He covered Spiti Valley up to Kibber and reported over 40 species including records by Stoliczka (Whistler 1923). He wrote 'Popular Handbook of the Birds of India'. Several species/subspecies of birds are named after him. He published an account of birds of Kangra and Kullu districts in erstwhile Punjab state (Whistler 1926a, b). Frome (1946) studied avifauna of Mahasu-Narkanda-Baghi area, and perhaps, it was Jones (1947a, b, 1948) who made available a comprehensive account on pre-independence records of birds from Shimla hills and presented a list of as many as 199 species.

During the last few decades, a number of studies have been carried out by various workers on various aspects of avifauna of the region like geographical and altitudinal distribution pattern (Narang 1989; Mahabal 1992a, b, 1996; Mahabal and Mukherjee 1991; Mahabal and Sharma 1992; Mahabal 2005; Sharma and Mahabal 1997; Suyal 1992; Thakur et al. 2008) and seasonal changes in diversity (Sharma and Mahabal 1997). Avifauna of Uttarakhand has been documented by some workers (Pandey et al. 1995; Sankaran 1995; Sultana and Khan 2000; Kumar and Bhatt 2000; Tak and Kumar 1987; Bhattacharya and Sathyakumar 2007; Mohan and Sondhi 2014). Avifauna of conservation areas like Wildlife Sanctuaries and National Parks of Himachal Pradesh have also been studied to some extent by Pandey (1989a, b), Mahabal (2000a, b), Mahabal and Sharma (1993), Thakur et al. (2002), Bhargav et al. (2007), Tak and Paliwal (2008) and Sharma et al. (2009). Thakur and Mattu (2011) also listed their observations from Kaza area of Spiti Valley. During recent years, I surveyed Kalatop Khajjiar area, Pong Dam, Kibber Wildlife Sanctuary, Pin Valley National Park, Spiti Valley, Chandra Valley, Chail Wildlife Sanctuary and Chandratal area (Kumar et al. 2014; Kumar and Paliwal 2015; Kumar 2015).

8.2 Methods

This chapter is based on the published literature and available information generated through field surveys of various ecosystems, protected areas and wetlands undertaken by scientists of Zoological Survey of India during the last two decades and surveys undertaken by the author during the last 7 years. The author covered Kalatop Khajjiar area, Pong Dam wetlands, Kibber Wildlife Sanctuary, Pin Valley National Park, Spiti Valley, Chandra Valley, Chail Wildlife Sanctuary and Chandratal area in Himachal Pradesh. During 2015–16, I covered some high altitude areas (namely Nubra Valley, Pangong Tso, Puga, Tso Kar, Tsi Gul Tso, Lal Pahari, Hanle, Tso Moriri and Walna Valley) of Ladakh and some areas (i.e. Valley of Flowers, Sonanadi Wildlife Sanctuary and Chilla Range of Rajaji National Park) in Uttarakhand. During past (1997–2004), I covered Dehradun area, Mussoorie hills, Srinagar Garhwal, Bhimgoda Barrage, Asan Barrage and Rajaji National Park in Uttarakhand state. During the field work, observations on birds were made every day during 6.00 am to 4.00 pm (with few exceptions), with the help of prismatic field binocular (10 × 50), and identification of species was carried out with the help of field guide of Birds of India written by Kazmierczak and Perlo (2000) and a pocket guide to the birds of the Indian Subcontinent by Grimmett et al. (2003).

The taxonomic order and nomenclature follow Clements 6th edition updated in 2014. The checklist of the birds of northwest Himalayas (Clements et al. 2014) was downloaded from the Avibase website (http://avibase.bsc-eoc.org/checklist.jsp?region=inwh&list=clements). It was used as a base for addition/deletion of species and information on distribution, conservation status and occurrence. 'Birds of South Asia: The Ripley Guide' (Rasmussen and Anderton 2005) and IOC World Bird List (Gill and Donsker 2014) were also considered. The checklist was prepared accordingly. However, it is difficult to prepare an accurate checklist of birds of western Himalaya, as some species occur in Gangetic plain of Uttarakhand but occasionally seen in the foothills of Himalaya (Shiwalik hills, Bhabar and Tarai area) owing to edge effect. Occurrence of some species was doubtful as they do not have distribution in Himalaya, while some species were reported in the past but not seen since decades. I tried to exclude such species from the list. Conservation status of the species has been incorporated in the study (IUCN 2014). On the basis of migratory status, birds were categorized into four groups, namely, winter migrants (long-distance winter migrants in the region, often stay as passage migrants during onward and return journey in some areas; a few species breed in Ladakh also), summer migrants (arrives in the region during summer season and few of them act as passage migrants during journey), passage migrants (purely passage migrants seen only during onward and return journeys) and residents (stay in the region throughout the year, including few of them migrate locally owing to decreasing temperature in winter and food availability, within the region).

8.3 Results and Discussion

8.3.1 Species Composition

A total of 768 species of birds belonging to 21 orders and 95 families were enlisted in the present study including some rare/interesting records (Table 8.1); besides a few species remained doubtful. The most dominant family was Muscicapidae which represents 62 species, followed by family Accipitridae which represents 42 species and family Fringillidae and Anatidae with 32 and 31 species, respectively. The region encompasses a rich species diversity ranging from waterbirds to songbirds. Wetlands of Ladakh, Pong Dam, Asan Barrage and Corbett National Park are the main areas known to host a number of aquatic birds. Analysis of data revealed that about 54% (415 species) were passerine birds belonging to 51 families, comprised of various groups such as shrikes, minivets, babblers, drongos, crows, fantails, fly-catchers, larks, tits, bulbuls, warblers, laughingthrushes, robins, redstarts, wheat-ears, thrushes, starlings, flowerpeckers, sunbirds, pipits, buntings, finches and sparrows. The author recorded sightings of some uncommon/rare species such as Eurasian Linnet, wood snipe and forest wagtail from Himachal Pradesh during his field survey.

Review of literature revealed that studies on the avifauna of this region are limited and mostly confined to protected areas. In Jammu and Kashmir, over the decades Ladakh is the main destination for ornithologists. Avifauna of this area comprised of more than 300 species, most of which are migratory and about 100 species are known to breed there (Namgail and Yom-Tov 2009). Studies are scanty on the birds of Kashmir Valley and Jammu area, even though about 358 species of birds have been reported from the state (Khah et al. 2012). Price et al. (2003); Price and Jamdar (1990) enlisted 70 species of birds from the Overa-Aru Wildlife Sanctuary. One hundred forty-five species were recorded from Dachigam National Park (Katti 1989), and 78 species were recorded from Kishtwar National Park (Scott et al. 1988).

In Himachal Pradesh, Pong Dam and adjacent localities are well studied, and over 415 avian species of 65 families have been recorded from this area (Pandey 1989a, 1993; Editor-Director 2009; Kumar and Paliwal 2015). 555 species were enlisted from Kangra District (Besten 2004). One hundred eighty-three species of birds including 132 passerines were reported from Great Himalayan National Park (Gaston et al. 1994), and 106 species were reported from Majhatal Harsang Wildlife Sanctuary (Mishra 1996). One hundred three avian species were reported from Renuka Lake and adjoining areas (Editor-Director 2000). Kumar et al. (2014) enlisted 93 species from Kalatop Khajjiar Wildlife Sanctuary and adjacent localities based on their study and previously published work by Thakur et al. (2002). Zoological Survey of India reported the occurrence of 72 species from Pin Valley National Park (Editor-Director 2008). Sharma et al. (2009) recorded the occurrence of 210 species in Simbalbara Wildlife Sanctuary. Mahabal (1996) recorded the occurrence of 136 species from Una, Hamirpur and Bilaspur districts. In a recent study, avifauna (95 species) of Prashar Lake was documented by Singh et al. (2014).

Table 8.1 Checklist of the birds of northwest Himalaya

Sl. No.	Common name	Scientific name	Migratory status	IUCN (2014) status/remarks
	Order: Anseriformes			
	Family: Anatidae			
1	Lesser whistling duck	*Dendrocygna javanica*	SM	LC/scarce
2	Greater white-fronted goose	*Anser albifrons*	WM	LC/scarce
3	Lesser white-fronted goose	*Anser erythropus*	WM	VU/scarce
4	Greylag goose	*Anser anser*	WM/PM	LC
5	Bar-headed goose	*Anser indicus*	WM/BL	LC
6	Mute swan	*Cygnus olor*	WM/V	LC/scarce
7	Whooper swan	*Cygnus cygnus*	WM/V	LC/scarce
8	Comb duck	*Sarkidiornis melanotos*	R/V	LC/scarce
9	Ruddy shelduck	*Tadorna ferruginea*	WM/PM/BL	LC
10	Common shelduck	*Tadorna tadorna*	WM	LC/scarce
11	Cotton pygmy goose	*Nettapus coromandelianus*	SM/R	LC
12	Gadwall	*Anas strepera*	WM/PM	LC
13	Falcated duck	*Anas falcata*	WM	NT/scarce
14	Eurasian wigeon	*Anas penelope*	WM	LC
15	Mallard	*Anas platyrhynchos*	WM/PM/BL?	LC
16	Indian spot-billed duck	*Anas poecilorhyncha*	SM/R	LC
17	Northern shoveler	*Anas clypeata*	WM/PM/BL	LC
18	Northern pintail	*Anas acuta*	WM/PM/BL	LC
19	Garganey	*Anas querquedula*	WM/PM/BL	LC
20	Green-winged teal	*Anas crecca*	WM/PM	LC
21	Marbled teal	*Marmaronetta angustirostris*	WM	VU/scarce
22	Red-crested pochard	*Netta rufina*	WM/PM	LC
23	Common pochard	*Aythya ferina*	WM/PM	LC
24	Ferruginous duck	*Aythya nyroca*	WM/PM	NT
25	Tufted duck	*Aythya fuligula*	WM	LC
26	Greater scaup	*Aythya marila*	WM/V	LC
27	Long-tailed duck	*Clangula hyemalis*	WM/V	VU/scarce or accidental
28	Common goldeneye	*Bucephala clangula*	WM	LC/scarce
29	Smew	*Mergellus albellus*	WM	LC/scarce
30	Common merganser	*Mergus merganser*	WM/R in Ladakh	LC
31	White-headed duck	*Oxyura leucocephala*	WM	EN/scarce
	Order: Galliformes			
	Family: Phasianidae			
32	Chukar	*Alectoris chukar*	R	LC
33	Snow partridge	*Lerwa lerwa*	R	LC
34	Tibetan partridge	*Perdix hodgsoniae*	R	LC

(continued)

Table 8.1 (continued)

Sl. No.	Common name	Scientific name	Migratory status	IUCN (2014) status/remarks
35	Hill partridge	*Arborophila torqueola*	R	LC
36	Rufous-throated partridge	*Arborophila rufogularis*	R	LC
37	Tibetan snowcock	*Tetraogallus tibetanus*	R	LC
38	Himalayan snowcock	*Tetraogallus himalayensis*	R	LC
39	Grey francolin	*Francolinus pondicerianus*	R	LC
40	Black francolin	*Francolinus francolinus*	R	LC
41	Common quail	*Coturnix coturnix*	R	LC
42	Blue-breasted quail	*Coturnix chinensis*	R	LC
43	Jungle bush quail	*Perdicula asiatica*	R	LC
44	Western tragopan	*Tragopan melanocephalus*	R	VU/scarce
45	Satyr tragopan	*Tragopan satyra*	R	NT
46	Koklass pheasant	*Pucrasia macrolopha*	R	LC
47	Himalayan monal	*Lophophorus impejanus*	R	LC
48	Red junglefowl	*Gallus gallus*	R	LC
49	Red spurfowl	*Galloperdix spadicea*	R in south-east Uttarakhand	LC
50	Kalij pheasant	*Lophura leucomelanos*	R	LC
51	Cheer pheasant	*Catreus wallichii*	R	VU/scarce
52	Indian peafowl	*Pavo cristatus*	R	LC
	Order: Podicipediformes			
	Family: Podicipedidae			
53	Little grebe	*Tachybaptus ruficollis*	R/SM	LC
54	Horned grebe	*Podiceps auritus*	WM/V	LC/scarce
55	Red-necked grebe	*Podiceps grisegena*	WM/V	LC/scarce
56	Great crested grebe	*Podiceps cristatus*	WM/R/SM	LC
57	Eared grebe	*Podiceps nigricollis*	WM	LC/scarce
	Order: Phoenicopteriformes			
	Family: Phoenicopteridae			
58	Greater flamingo	*Phoenicopterus roseus*	WM in south-east Uttarakhand	LC/scarce
	Order: Ciconiiformes			
	Family: Ciconiidae			
59	Asian openbill	*Anastomus oscitans*	R/WM from Peninsular India	LC

(continued)

Table 8.1 (continued)

Sl. No.	Common name	Scientific name	Migratory status	IUCN (2014) status/remarks
60	Black stork	*Ciconia nigra*	WM/PM	LC
61	Woolly-necked stork	*Ciconia episcopus*	R/SM	VU
62	White stork	*Ciconia ciconia*	WM	LC
63	Black-necked stork	*Ephippiorhynchus asiaticus*	R	NT/scarce
64	Painted stork	*Mycteria leucocephala*	R	NT
65	Lesser adjutant	*Leptoptilos javanicus*	R/SM	VU
	Order: Suliformes			
	Family: Phalacrocoracidae			
66	Indian cormorant	*Phalacrocorax fuscicollis*	R	LC
67	Great cormorant	*Phalacrocorax carbo*	R/PM	LC
68	Little cormorant	*Phalacrocorax niger*	R	LC
	Family: Anhingidae			
69	Oriental darter	*Anhinga melanogaster*	R	NT
	Order: Pelecaniformes			
	Family: Pelecanidae			
70	Great white pelican	*Pelecanus onocrotalus*	WM	LC
71	Spot-billed pelican	*Pelecanus philippensis*	WM	NT
	Family: Ardeidae			
72	Great bittern	*Botaurus stellaris*	WM	LC
73	Yellow bittern	*Ixobrychus sinensis*	SM	LC
74	Little bittern	*Ixobrychus minutus*	SM	LC
75	Cinnamon bittern	*Ixobrychus cinnamomeus*	SM	LC
76	Black bittern	*Ixobrychus flavicollis*	SM	LC
77	Grey heron	*Ardea cinerea*	R/WM, locally from eastern India	LC
78	Purple heron	*Ardea purpurea*	R/WM locally	LC
79	Great egret	*Ardea alba*	R	LC
80	Intermediate egret	*Mesophoyx intermedia*	R	LC
81	Little egret	*Egretta garzetta*	R	LC
82	Cattle egret	*Bubulcus ibis*	R	LC
83	Indian pond heron	*Ardeola grayii*	R	LC
84	Striated heron	*Butorides striata*	R	LC/scarce
85	Black-crowned night heron	*Nycticorax nycticorax*	R/PM	LC

(continued)

Table 8.1 (continued)

Sl. No.	Common name	Scientific name	Migratory status	IUCN (2014) status/remarks
	Family: Threskiornithidae			
86	Glossy ibis	*Plegadis falcinellus*	WM/PM	LC/scarce
87	Red-naped ibis	*Pseudibis papillosa*	R	LC/scarce
88	Eurasian spoonbill	*Platalea leucorodia*	R/WM	LC/scarce
	Order: Accipitriformes			
	Family: Pandionidae			
89	Osprey	*Pandion haliaetus*	WM/PM	LC/scarce
	Family: Accipitridae			
90	Black-shouldered kite	*Elanus caeruleus*	R	LC
91	Lammergeier	*Gypaetus barbatus*	R	NT
92	Egyptian vulture	*Neophron percnopterus*	R	EN/scarce
93	Oriental honey buzzard	*Pernis ptilorhynchus*	R	LC
94	Red-headed vulture	*Sarcogyps calvus*	R	CR/scarce
95	Cinereous vulture	*Aegypius monachus*	WM	NT/scarce
96	White-rumped vulture	*Gyps bengalensis*	R	CR/scarce
97	Indian vulture	*Gyps indicus*	R	CR/scarce
98	Himalayan griffon	*Gyps himalayensis*	R	NT
99	Eurasian griffon	*Gyps fulvus*	WM	LC/scarce
100	Crested serpent eagle	*Spilornis cheela*	R	LC
101	Short-toed eagle	*Circaetus gallicus*	R	LC
102	Changeable hawk-eagle	*Nisaetus limnaeetus*	R	LC
103	Mountain hawk-eagle	*Nisaetus nipalensis*	R	LC
104	Rufous-bellied eagle	*Lophotriorchis kienerii*	R	LC
105	Black eagle	*Ictinaetus malayensis*	R	LC
106	Indian spotted eagle	*Clanga hastata*	R	VU
107	Greater spotted eagle	*Clanga clanga*	WM	VU/scarce
108	Booted eagle	*Hieraaetus pennatus*	R/PM	LC
109	Tawny eagle	*Aquila rapax*	R	LC
110	Steppe eagle	*Aquila nipalensis*	WM	LC
111	Imperial eagle	*Aquila heliaca*	PM	VU/scarce
112	Golden eagle	*Aquila chrysaetos*	R	LC
113	Bonelli's eagle	*Aquila fasciata*	R	LC
114	White-eyed buzzard	*Butastur teesa*	R/SM	LC
115	Eurasian marsh harrier	*Circus aeruginosus*	WM/PM	LC
116	Northern harrier	*Circus cyaneus*	WM/PM	LC
117	Pallid harrier	*Circus macrourus*	WM/PM	NT/scarce
118	Montagu's harrier	*Circus pygargus*	PM	LC
119	Shikra	*Accipiter badius*	R	LC
120	Besra	*Accipiter virgatus*	R	LC
121	Eurasian sparrowhawk	*Accipiter nisus*	R	LC
122	Northern goshawk	*Accipiter gentilis*	WM	LC

(continued)

Table 8.1 (continued)

Sl. No.	Common name	Scientific name	Migratory status	IUCN (2014) status/remarks
123	Red kite	*Milvus milvus*	WM	NT/scarce
124	Black kite	*Milvus migrans*	R/PM	LC
125	Pallas's fish eagle	*Haliaeetus leucoryphus*	R	VU/scarce
126	White-tailed eagle	*Haliaeetus albicilla*	WM	LC/scarce
127	Lesser fish eagle	*Ichthyophaga humilis*	R	NT/scarce
128	Grey-headed fish eagle	*Ichthyophaga ichthyaetus*	R	NT/scarce
129	Common buzzard	*Buteo buteo*	WM	LC
130	Long-legged buzzard	*Buteo rufinus*	WM/R	LC
131	Upland buzzard	*Buteo hemilasius*	WM	LC/scarce
	Order: Gruiformes			
	Family: Rallidae			
132	Corn crake	*Crex crex*	PM in Ladakh	LC/scarce or accidental
133	Water rail	*Rallus aquaticus*	WM	LC
134	Brown crake	*Amaurornis akool*	R	LC
135	White-breasted waterhen	*Amaurornis phoenicurus*	R	LC
136	Baillon's crake	*Porzana pusilla*	WM	LC
137	Spotted crake	*Porzana porzana*	WM/PM	LC
138	Ruddy-breasted crake	*Porzana fusca*	WM locally	LC
139	Slaty-legged crake	*Rallina eurizonoides*	SM	LC/scarce
140	Watercock	*Gallicrex cinerea*	SM	LC/scarce
141	Purple swamphen	*Porphyrio porphyrio*	R	LC
142	Eurasian moorhen	*Gallinula chloropus*	R/PM	LC
143	Eurasian coot	*Fulica atra*	R	LC
	Family: Gruidae			
144	Demoiselle crane	*Anthropoides virgo*	WM/PM	LC
145	Sarus crane	*Grus antigone*	R/WM from eastern India	VU
146	Common crane	*Grus grus*	WM/PM	LC
147	Black-necked crane	*Grus nigricollis*	R	VU/scarce
	Order: Charadriiformes			
	Family: Burhinidae			
148	Indian stone-curlew	*Burhinus indicus*	R/SM	LC
149	Great thick-knee	*Esacus recurvirostris*	R/SM	NT
	Family: Recurvirostridae			
150	Black-winged stilt	*Himantopus himantopus*	R/PM	LC
151	Pied avocet	*Recurvirostra avosetta*	WM	LC

(continued)

Table 8.1 (continued)

Sl. No.	Common name	Scientific name	Migratory status	IUCN (2014) status/remarks
	Family: Ibidorhynchidae			
152	Ibisbill	*Ibidorhyncha struthersii*	R	LC
	Family: Haematopodidae			
153	Eurasian oystercatcher	*Haematopus ostralegus*	WM/PM	LC
	Family: Charadriidae			
154	Black-bellied plover	*Pluvialis squatarola*	PM	LC
155	Pacific golden plover	*Pluvialis fulva*	WM/PM	LC/scarce
156	Northern lapwing	*Vanellus vanellus*	WM	LC
157	River lapwing	*Vanellus duvaucelii*	R	NT/scarce
158	Yellow-wattled lapwing	*Vanellus malabaricus*	R/WM from peninsular India	LC
159	Grey-headed lapwing	*Vanellus cinereus*	WM	LC
160	Red-wattled lapwing	*Vanellus indicus*	R	LC
161	Sociable lapwing	*Vanellus gregarius*	WM	CR/scarce
162	White-tailed lapwing	*Vanellus leucurus*	WM	LC
163	Lesser sand plover	*Charadrius mongolus*	SM/BL	LC
164	Greater sand plover	*Charadrius leschenaultii*	PM/SM	LC/scarce
165	Kentish plover	*Charadrius alexandrinus*	WM/PM	LC
166	Little ringed plover	*Charadrius dubius*	R/SM	LC
167	Long-billed plover	*Charadrius placidus*	WM in south-east Uttarakhand	LC/scarce
	Family: Rostratulidae			
168	Greater painted-snipe	*Rostratula benghalensis*	R	LC
	Family: Jacanidae			
169	Pheasant-tailed jacana	*Hydrophasianus chirurgus*	R/SM	LC
170	Bronze-winged jacana	*Metopidius indicus*	R	LC
	Family: Scolopacidae			
171	Terek sandpiper	*Xenus cinereus*	PM	LC
172	Common sandpiper	*Actitis hypoleucos*	SM	LC
173	Green sandpiper	*Tringa ochropus*	R/SM/WM	LC
174	Spotted redshank	*Tringa erythropus*	WM/PM	LC
175	Common greenshank	*Tringa nebularia*	WM/PM	LC
176	Marsh sandpiper	*Tringa stagnatilis*	WM/PM	LC
177	Wood sandpiper	*Tringa glareola*	PM/SM/R	LC
178	Common redshank	*Tringa totanus*	WM/SM	LC
179	Whimbrel	*Numenius phaeopus*	PM	LC/scarce
180	Eurasian curlew	*Numenius arquata*	PM	NT
181	Black-tailed godwit	*Limosa limosa*	WM	NT

(continued)

Table 8.1 (continued)

Sl. No.	Common name	Scientific name	Migratory status	IUCN (2014) status/remarks
182	Ruddy turnstone	*Arenaria interpres*	PM	LC
183	Ruff	*Calidris pugnax*	PM/WM	LC
184	Curlew sandpiper	*Calidris ferruginea*	PM	LC
185	Temminck's stint	*Calidris temminckii*	WM/PM	LC
186	Sanderling	*Calidris alba*	PM	LC/scarce
187	Dunlin	*Calidris alpina*	WM/PM	LC
188	Little stint	*Calidris minuta*	WM/PM	LC
189	Jack snipe	*Lymnocryptes minimus*	WM	LC/scarce
190	Solitary snipe	*Gallinago solitaria*	R/WM	LC
191	Wood snipe	*Gallinago nemoricola*	SM	VU/scarce
192	Common snipe	*Gallinago gallinago*	WM/PM	LC
193	Pin-tailed snipe	*Gallinago stenura*	PM	LC
194	Eurasian woodcock	*Scolopax rusticola*	SM	LC
195	Red-necked phalarope	*Phalaropus lobatus*	PM	LC
	Family: Turnicidae			
196	Small buttonquail	*Turnix sylvaticus*	SM	LC
197	Yellow-legged buttonquail	*Turnix tanki*	SM	LC/scarce
198	Barred buttonquail	*Turnix suscitator*	R	LC
	Family: Glareolidae			
199	Collared pratincole	*Glareola pratincola*	SM	LC/scarce
200	Small pratincole	*Glareola lactea*	SM	LC
	Family: Stercorariidae			
201	Parasitic jaeger	*Stercorarius parasiticus*	PM	LC/scarce
	Family: Laridae			
202	Slender-billed gull	*Chroicocephalus genei*	WM	LC
203	Black-headed gull	*Chroicocephalus ridibundus*	WM/PM	LC
204	Brown-headed gull	*Chroicocephalus brunnicephalus*	WM/BL	LC
205	Little gull	*Hydrocoloeus minutus*	PM	LC/scarce or accidental
206	Pallas's gull	*Ichthyaetus ichthyaetus*	WM/PM	LC
207	Mew gull	*Larus canus*	WM	LC
208	Little tern	*Sternula albifrons*	PM	LC
209	Gull-billed tern	*Gelochelidon nilotica*	PM	LC
210	White-winged tern	*Chlidonias leucopterus*	PM	LC
211	Whiskered tern	*Chlidonias hybrida*	PM	LC
212	Common tern	*Sterna hirundo*	WM/SM	LC

(continued)

Table 8.1 (continued)

Sl. No.	Common name	Scientific name	Migratory status	IUCN (2014) status/remarks
213	Arctic tern	*Sterna paradisaea*	SM/V	LC
214	River tern	*Sterna aurantia*	R	NT
215	Black-bellied tern	*Sterna acuticauda*	R	EN/scarce
216	Indian skimmer	*Rynchops albicollis*	SM/R/V	VU/scarce
	Order: Pterocliformes			
	Family: Pteroclidae			
217	Tibetan sandgrouse	*Syrrhaptes tibetanus*	R	LC
218	Black-bellied sandgrouse	*Pterocles orientalis*	WM	LC/scarce
	Order: Columbiformes			
	Family: Columbidae			
219	Rock pigeon	*Columba livia*	R	LC
220	Hill pigeon	*Columba rupestris*	R	LC
221	Snow pigeon	*Columba leuconota*	R	LC
222	Yellow-eyed pigeon	*Columba eversmanni*	WM	VU/scarce
223	Common wood pigeon	*Columba palumbus*	R/WM	LC
224	Speckled wood pigeon	*Columba hodgsonii*	R	LC
225	Ashy wood pigeon	*Columba pulchricollis*	R	LC
226	European turtle dove	*Streptopelia turtur*	PM	LC/scarce
227	Oriental turtle dove	*Streptopelia orientalis*	R/SM	LC
228	Eurasian collared dove	*Streptopelia decaocto*	R/SM	LC
229	Red collared dove	*Streptopelia tranquebarica*	SM	LC
230	Spotted dove	*Streptopelia chinensis*	R	LC
231	Laughing dove	*Streptopelia senegalensis*	R	LC
232	Emerald dove	*Chalcophaps indica*	R	LC
233	Orange-breasted pigeon	*Treron bicinctus*	R in south east Uttarakhand	LC
234	Yellow-footed pigeon	*Treron phoenicopterus*	R	LC
235	Pin-tailed pigeon	*Treron apicauda*	R	LC
236	Wedge-tailed pigeon	*Treron sphenurus*	R	LC
	Order: Cuculiformes			
	Family: Cuculidae			
237	Pied cuckoo	*Clamator jacobinus*	SM	LC
238	Chestnut-winged cuckoo	*Clamator coromandus*	SM in small area of south-eastern Uttarakhand	LC
239	Large hawk-cuckoo	*Hierococcyx sparverioides*	SM	LC
240	Common hawk-cuckoo	*Hierococcyx varius*	R	LC

(continued)

Table 8.1 (continued)

Sl. No.	Common name	Scientific name	Migratory status	IUCN (2014) status/remarks
241	Indian cuckoo	*Cuculus micropterus*	R	LC
242	Common cuckoo	*Cuculus canorus*	SM	LC
243	Himalayan cuckoo	*Cuculus saturatus*	SM	LC
244	Lesser cuckoo	*Cuculus poliocephalus*	SM	LC
245	Grey-bellied cuckoo	*Cacomantis passerinus*	R	LC
246	Asian emerald cuckoo	*Chrysococcyx maculatus*	SM	LC/scarce
247	Square-tailed drongo-cuckoo	*Surniculus lugubris*	SM	LC/scarce
248	Asian koel	*Eudynamys scolopaceus*	SM	LC
249	Green-billed malkoha	*Phaenicophaeus tristis*	R	LC/scarce
250	Sirkeer malkoha	*Phaenicophaeus leschenaultii*	R	LC/scarce
251	Greater coucal	*Centropus sinensis*	R	LC
252	Lesser coucal	*Centropus bengalensis*	R	LC/scarce
	Order: Strigiformes			
	Family: Tytonidae			
253	Eastern grass owl	*Tyto longimembris*	R	LC
254	Barn owl	*Tyto alba*	R	LC/scarce
	Family: Strigidae			
255	Mountain scops owl	*Otus spilocephalus*	R	LC
256	Collared scops owl	*Otus lettia*	R	LC
257	Oriental scops owl	*Otus sunia*	SM/WM	LC
258	Eurasian eagle-owl	*Bubo bubo*	R	LC/scarce
259	Indian eagle-owl	*Bubo bengalensis*	R	LC/scarce
260	Dusky eagle-owl	*Bubo coromandus*	R	LC/scarce
261	Forest eagle-owl	*Bubo nipalensis*	R	LC/scarce
262	Brown fish owl	*Ketupa zeylonensis*	R	LC
263	Tawny fish owl	*Ketupa flavipes*	R	LC
264	Collared owlet	*Glaucidium brodiei*	R	LC
265	Asian barred owlet	*Glaucidium cuculoides*	R	LC
266	Jungle owlet	*Glaucidium radiatum*	R	LC
267	Spotted owlet	*Athene brama*	R	LC
268	Little owl	*Athene noctua*	R	LC
269	Brown wood owl	*Strix leptogrammica*	R	LC
270	Tawny owl	*Strix aluco*	R	LC
271	Mottled wood owl	*Strix ocellata*	R	LC
272	Himalayan wood owl	*Strix nivicola*	R	LC

(continued)

Table 8.1 (continued)

Sl. No.	Common name	Scientific name	Migratory status	IUCN (2014) status/remarks
273	Long-eared owl	*Asio otus*	WM	LC
274	Short-eared owl	*Asio flammeus*	WM	LC
275	Boreal owl	*Aegolius funereus*	WM/V	LC/scarce
276	Brown boobook	*Ninox scutulata*	R	LC/scarce
	Order: Caprimulgiformes			
	Family: Caprimulgidae			
277	Grey nightjar	*Caprimulgus indicus*	R	LC
278	Eurasian nightjar	*Caprimulgus europaeus*	SM	LC
279	Large-tailed nightjar	*Caprimulgus macrurus*	R	LC
280	Savanna nightjar	*Caprimulgus affinis*	R/SM	LC
281	Indian nightjar	*Caprimulgus asiaticus*	R	LC
	Order: Apodiformes			
	Family: Apodidae			
282	White-rumped needletail	*Zoonavena sylvatica*	R in south-eastern Uttarakhand	LC
283	White-throated needletail	*Hirundapus caudacutus*	SM	LC
284	Himalayan swiftlet	*Aerodramus brevirostris*	R	LC
285	Alpine swift	*Apus melba*	PM	LC
286	Common swift	*Apus apus*	SM	LC
287	Pacific swift	*Apus pacificus*	SM	LC
288	House swift	*Apus nipalensis*	R	LC
289	Asian palm-swift	*Cypsiurus balasiensis*	R	LC
	Family: Hemiprocnidae			
290	Crested treeswift	*Hemiprocne coronata*	R	LC
	Order: Coraciiformes			
	Family: Alcedinidae			
291	Common kingfisher	*Alcedo atthis*	R/SM	LC
292	Stork-billed kingfisher	*Pelargopsis capensis*	R	LC
293	White-throated kingfisher	*Halcyon smyrnensis*	R	LC
294	Crested kingfisher	*Megaceryle lugubris*	R	LC
295	Pied kingfisher	*Ceryle rudis*	R	LC
	Family: Meropidae			
296	Blue-bearded bee-eater	*Nyctyornis athertoni*	R	LC
297	Green bee-eater	*Merops orientalis*	R	LC
298	Blue-cheeked bee-eater	*Merops persicus*	PM	LC
299	Blue-tailed bee-eater	*Merops philippinus*	SM	LC
300	European bee-eater	*Merops apiaster*	SM/PM	LC
301	Chestnut-headed bee-eater	*Merops leschenaulti*	R	LC

(continued)

Table 8.1 (continued)

Sl. No.	Common name	Scientific name	Migratory status	IUCN (2014) status/remarks
	Family: Coraciidae			
302	European roller	*Coracias garrulus*	SM/PM	NT
303	Indian roller	*Coracias benghalensis*	R	LC
304	Dollarbird	*Eurystomus orientalis*	R	LC/scarce
305	Red-headed trogon	*Harpactes erythrocephalus*	R in south-east Uttarakhand	LC/scarce
	Family: Upupidae			
306	Eurasian hoopoe	*Upupa epops*	R	LC
	Family: Bucerotidae			
307	Indian grey hornbill	*Ocyceros birostris*	R	LC
308	Oriental pied hornbill	*Anthracoceros albirostris*	R	LC
309	Great hornbill	*Buceros bicornis*	R	NT
	Order: Piciformes			
	Family: Megalaimidae			
310	Great barbet	*Megalaima virens*	R	LC
311	Brown-headed barbet	*Megalaima zeylanica*	R	LC
312	Lineated barbet	*Megalaima lineata*	R	LC
313	Blue-throated barbet	*Megalaima asiatica*	R	LC
314	Coppersmith barbet	*Megalaima haemacephala*	R	LC
	Family: Indicatoridae			
315	Yellow-rumped honeyguide	*Indicator xanthonotus*	R	NT/scarce
	Family: Picidae			
316	Eurasian wryneck	*Jynx torquilla*	WM/SM	LC
317	Speckled piculet	*Picumnus innominatus*	R	LC
318	White-browed piculet	*Sasia ochracea*	R/V	LC/scarce
319	Brown-capped woodpecker	*Dendrocopos nanus*	R	LC
320	Grey-capped woodpecker	*Dendrocopos canicapillus*	R	LC
321	Brown-fronted woodpecker	*Dendrocopos auriceps*	R	LC
322	Fulvous-breasted woodpecker	*Dendrocopos macei*	R	LC
323	Yellow-crowned woodpecker	*Dendrocopos mahrattensis*	R	LC
324	Rufous-bellied woodpecker	*Dendrocopos hyperythrus*	R	LC
325	Himalayan woodpecker	*Dendrocopos himalayensis*	R	LC
326	Rufous woodpecker	*Celeus brachyurus*	R	LC

(continued)

Table 8.1 (continued)

Sl. No.	Common name	Scientific name	Migratory status	IUCN (2014) status/remarks
327	Lesser yellownape	*Picus chlorolophus*	R	LC
328	Greater yellownape	*Picus flavinucha*	R	LC
329	Streak-throated woodpecker	*Picus xanthopygaeus*	R	LC
330	Scaly-bellied woodpecker	*Picus squamatus*	R	LC
331	Grey-faced woodpecker	*Picus canus*	R	LC
332	Himalayan flameback	*Dinopium shorii*	R	LC
333	Black-rumped flameback	*Dinopium benghalense*	R	LC
334	Greater flameback	*Chrysocolaptes guttacristatus*	R	LC
335	White-naped woodpecker	*Chrysocolaptes festivus*	R	LC
336	Bay woodpecker	*Blythipicus pyrrhotis*	R in south-eastern Uttarakhand	LC
337	Great slaty woodpecker	*Mulleripicus pulverulentus*	R	VU
	Order: Falconiformes			
	Family: Falconidae			
338	Collared falconet	*Microhierax caerulescens*	R	LC
339	Lesser kestrel	*Falco naumanni*	R/WM	LC/scarce
340	Common kestrel	*Falco tinnunculus*	R/SM	LC
341	Merlin	*Falco columbarius*	WM	LC
342	Eurasian hobby	*Falco subbuteo*	SM	LC
343	Oriental hobby	*Falco severus*	R	LC
344	Laggar falcon	*Falco jugger*	R	NT
345	Saker falcon	*Falco cherrug*	WM	EN
346	Peregrine falcon	*Falco peregrinus*	R	LC
347	Red-necked falcon	*Falco chicquera*	R	NT/scarce
348	Amur falcon	*Falco amurensis*	PM (fall migrant)	LC/scarce (occasional)
	Order: Psittaciformes			
	Family: Psittacidae			
349	Alexandrine parakeet	*Psittacula eupatria*	R	NT
350	Rose-ringed parakeet	*Psittacula krameri*	R	LC
351	Himalayan parakeet	*Psittacula himalayana*	R	LC
352	Plum-headed parakeet	*Psittacula cyanocephala*	R	LC
353	Red-breasted parakeet	*Psittacula alexandri*	R	NT
	Order: Passeriformes			
	Family: Eurylaimidae			
354	Long-tailed broadbill	*Psarisomus dalhousiae*	R	LC

(continued)

Table 8.1 (continued)

Sl. No.	Common name	Scientific name	Migratory status	IUCN (2014) status/remarks
	Family: Pittidae			
355	Hooded pitta	*Pitta sordida*	SM	LC/scarce
356	Indian pitta	*Pitta brachyura*	SM	LC
	Family: Vangidae			
357	Large woodshrike	*Tephrodornis gularis*	R in south-eastern Uttarakhand	LC
358	Common woodshrike	*Tephrodornis pondicerianus*	R	LC
359	Bar-winged flycatcher-shrike	*Hemipus picatus*	R	LC
	Family: Artamidae			
360	Ashy woodswallow	*Artamus fuscus*	R	LC
	Family: Aegithinidae			
361	Common iora	*Aegithina tiphia*	R	LC
	Family: Campephagidae			
362	Small minivet	*Pericrocotus cinnamomeus*	R	LC
363	Short-billed minivet	*Pericrocotus brevirostris*	R	LC/scarce
364	Long-tailed minivet	*Pericrocotus ethologus*	R/SM/WM	LC
365	Scarlet minivet	*Pericrocotus speciosus*	R	LC
366	Rosy minivet	*Pericrocotus roseus*	SM	LC
367	Large cuckooshrike	*Coracina macei*	R	LC
368	Black-winged cuckooshrike	*Lalage melaschistos*	R	LC
369	Black-headed cuckooshrike	*Lalage melanoptera*	SM	LC
	Family: Laniidae			
370	Red-backed shrike	*Lanius collurio*	PM	LC
371	Isabelline shrike	*Lanius isabellinus*	PM	LC
372	Brown shrike	*Lanius cristatus*	WM in south-eastern Uttarakhand	LC
373	Bay-backed shrike	*Lanius vittatus*	R	LC
374	Long-tailed shrike	*Lanius schach*	R/SM	LC
375	Grey-backed shrike	*Lanius tephronotus*	SM	LC
376	Northern shrike	*Lanius excubitor*	R/V	LC/scarce
377	Southern grey shrike	*Lanius meridionalis*	R	LC
	Family: Vireonidae			
378	Himalayan shrike-babbler	*Pteruthius ripleyi*[a]	R	LC
379	Green shrike-babbler	*Pteruthius xanthochlorus*	R	LC
380	White-bellied erpornis	*Erpornis zantholeuca*	R in south-eastern Uttarakhand	LC

(continued)

Table 8.1 (continued)

Sl. No.	Common name	Scientific name	Migratory status	IUCN (2014) status/remarks
	Family: Oriolidae			
381	European golden oriole	*Oriolus oriolus*	SM/V	LC/scarce
382	Indian golden oriole	*Oriolus kundoo*	SM	LC
383	Black-hooded oriole	*Oriolus xanthornus*	R	LC
384	Maroon oriole	*Oriolus traillii*	R	LC
	Family: Dicruridae			
385	Black drongo	*Dicrurus macrocercus*	R	LC
386	Ashy drongo	*Dicrurus leucophaeus*	SM	LC
387	White-bellied drongo	*Dicrurus caerulescens*	R	LC
388	Crow-billed drongo	*Dicrurus annectans*	SM	LC
389	Bronzed drongo	*Dicrurus aeneus*	R	LC
390	Lesser racket-tailed drongo	*Dicrurus remifer*	R	LC
391	Hair-crested drongo	*Dicrurus hottentottus*	R	LC
392	Greater racket-tailed drongo	*Dicrurus paradiseus*	R	LC
	Family: Rhipiduridae			
393	White-throated fantail	*Rhipidura albicollis*	R	LC
394	White-browed fantail	*Rhipidura aureola*	R	LC
	Family: Monarchidae			
395	Black-naped monarch	*Hypothymis azurea*	R	LC
396	Asian paradise flycatcher	*Terpsiphone paradisi*	SM	LC
	Family: Corvidae			
397	Eurasian jay	*Garrulus glandarius*	R	LC
398	Black-headed jay	*Garrulus lanceolatus*	R	LC
399	Yellow-billed blue magpie	*Urocissa flavirostris*	R	LC
400	Red-billed blue magpie	*Urocissa erythroryncha*	R	LC
401	Common green magpie	*Cissa chinensis*	R	LC
402	Rufous treepie	*Dendrocitta vagabunda*	R	LC
403	Grey treepie	*Dendrocitta formosae*	R	LC
404	Eurasian magpie	*Pica pica*	R	LC
405	Spotted nutcracker	*Nucifraga caryocatactes*	R	LC
406	Large-spotted nutcracker	*Nucifraga multipunctata*	R	LC
407	Red-billed chough	*Pyrrhocorax pyrrhocorax*	R	LC
408	Yellow-billed chough	*Pyrrhocorax graculus*	R	LC
409	Eurasian jackdaw	*Corvus monedula*	R	LC

(continued)

Table 8.1 (continued)

Sl. No.	Common name	Scientific name	Migratory status	IUCN (2014) status/remarks
410	House crow	*Corvus splendens*	R	LC
411	Rook	*Corvus frugilegus*	WM in Kashmir	LC
412	Carrion crow	*Corvus corone*	R	LC
413	Large-billed crow	*Corvus macrorhynchos*	R	LC
414	Common raven	*Corvus corax*	R	LC
	Family: Alaudidae			
415	Indian bushlark	*Mirafra erythroptera*	R	LC
416	Bengal bushlark	*Mirafra assamica*	R	LC
417	Singing bushlark	*Mirafra cantillans*	R	LC/scarce
418	Ashy-crowned sparrow-lark	*Eremopterix griseus*	R	LC
419	Bimaculated lark	*Melanocorypha bimaculata*	PM	LC
420	Tibetan lark	*Melanocorypha maxima*	SM	LC
421	Greater short-toed lark	*Calandrella brachydactyla*	PM	LC
422	Hume's lark	*Calandrella acutirostris*	SM	LC
423	Sand lark	*Calandrella raytal*	R	LC
424	Crested lark	*Galerida cristata*	R	LC
425	Sky lark	*Alauda arvensis*	WM (occasional)	LC/scarce
426	Oriental skylark	*Alauda gulgula*	SM	LC
427	Horned lark	*Eremophila alpestris*	R	LC
	Family: Hirundinidae			
428	Grey-throated martin	*Riparia chinensis*	R	LC
429	Pale sand martin	*Riparia diluta*	SM	LC
430	Eurasian crag martin	*Ptyonoprogne rupestris*	SM	LC
431	Dusky crag martin	*Ptyonoprogne concolor*	R	LC
432	Barn swallow	*Hirundo rustica*	SM/WM/PM	LC
433	Wire-tailed swallow	*Hirundo smithii*	SM/PM	LC
434	Red-rumped swallow	*Cecropis daurica*	SM/R	LC
435	Streak-throated swallow	*Petrochelidon fluvicola*	R	LC
436	Common house martin	*Delichon urbicum*	SM	LC
437	Asian house martin	*Delichon dasypus*	SM	LC
438	Nepal house martin	*Delichon nipalense*	R	LC
	Family: Stenostiridae			
439	Yellow-bellied fairy-fantail	*Chelidorhynx hypoxantha*	SM/WM (within Himalaya)	LC
440	Grey-headed canary-flycatcher	*Culicicapa ceylonensis*	SM	LC

(continued)

Table 8.1 (continued)

Sl. No.	Common name	Scientific name	Migratory status	IUCN (2014) status/remarks
	Family: Paridae			
441	Spot-winged tit	*Parus melanolophus*	R	LC
442	Black-breasted tit	*Periparus rufonuchalis*	R	LC
443	Rufous-vented tit	*Periparus rubidiventris*	R	LC
444	Grey-crested tit	*Lophophanes dichrous*	R	LC
445	Great tit	*Parus major*	R	LC
446	Green-backed tit	*Parus monticolus*	R	LC
447	Black-lored tit	*Parus xanthogenys*	R	LC
448	Yellow-browed tit	*Sylviparus modestus*	R	LC
449	Ground tit	*Pseudopodoces humilis*	R in trans-Himalayan area	LC/ scarce
	Family: Remizidae			
450	Fire-capped tit	*Cephalopyrus flammiceps*	SM	LC
	Family: Aegithalidae			
451	White-browed tit-warbler	*Leptopoecile sophiae*[a]	R	LC
452	White-cheeked tit	*Aegithalos leucogenys*	WM in western Kashmir	LC
453	Black-throated tit	*Aegithalos concinnus*	R	LC
454	White-throated tit	*Aegithalos niveogularis*	R	LC
	Family: Sittidae			
455	Indian nuthatch	*Sitta castanea*	R	LC
456	Chestnut-bellied nuthatch	*Sitta cinnamoventris*	R	LC
457	Kashmir nuthatch	*Sitta cashmirensis*	R	LC/scarce
458	White-tailed nuthatch	*Sitta himalayensis*	R	LC
459	White-cheeked nuthatch	*Sitta leucopsis*	R	LC
460	Velvet-fronted nuthatch	*Sitta frontalis*	R	LC
	Family: Tichodromidae			
461	Wallcreeper	*Tichodroma muraria*	R (in Ladakh)/ WM	LC
	Family: Certhiidae			
462	Hodgson's treecreeper	*Certhia hodgsoni*	R	LC
463	Bar-tailed treecreeper	*Certhia himalayana*	R	LC
464	Rusty-flanked treecreeper	*Certhia nipalensis*	R	LC
	Family: Troglodytidae			
465	Eurasian wren	*Troglodytes troglodytes*	R	LC
	Family: Cinclidae			
466	White-throated dipper	*Cinclus cinclus*	R	LC
467	Brown dipper	*Cinclus pallasii*	R	LC

(continued)

Table 8.1 (continued)

Sl. No.	Common name	Scientific name	Migratory status	IUCN (2014) status/remarks
	Family: Pycnonotidae			
468	Black-crested bulbul	*Pycnonotus flaviventris*	R	LC
469	Red-vented bulbul	*Pycnonotus cafer*	R	LC
470	Red-whiskered bulbul	*Pycnonotus jocosus*	R	LC
471	Himalayan bulbul	*Pycnonotus leucogenys*	R	LC
472	Himalayan black bulbul	*Hypsipetes leucocephalus*	R	LC
473	Ashy bulbul	*Hemixos flavala*	R	LC
474	Mountain bulbul	*Ixos mcclellandii*	R	LC
	Family: Regulidae			
475	Goldcrest	*Regulus regulus*	R	LC
	Family: Pnoepygidae			
476	Scaly-breasted cupwing or scaly-breasted wren-babbler	*Pnoepyga albiventer*	R	LC
477	Immaculate cupwing or Nepal wren-babbler	*Pnoepyga immaculata*	R	LC
478	Pygmy cupwing or pygmy wren-babbler	*Pnoepyga pusilla*	R	LC
	Family: Cettiidae			
479	Grey-bellied tesia	*Tesia cyaniventer*	R	LC
480	Chestnut-crowned bush warbler	*Cettia major*	R	LC
481	Grey-sided bush warbler	*Cettia brunnifrons*	SM/WM	LC
482	Pale-footed bush warbler	*Cettia pallidipes*	SM	LC/scarce
483	Chestnut-headed tesia	*Cettia castaneocoronata*	R	LC
484	Cetti's warbler	*Cettia cetti*	R	LC
485	Black-faced warbler	*Abroscopus schisticeps*	R	LC/scarce
486	Brown-flanked bush warbler	*Horornis fortipes*	SM	LC
487	Hume's bush warbler	*Horornis brunnescens*	R	LC
488	Aberrant bush warbler	*Horornis flavolivaceus*	R	LC
	Family: Phylloscopidae			
489	Common chiffchaff	*Phylloscopus collybita*	WM	LC
490	Mountain chiffchaff	*Phylloscopus sindianus*	SM	LC
491	Plain leaf warbler	*Phylloscopus neglectus*	SM	LC/scarce
492	Dusky warbler	*Phylloscopus fuscatus*	SM/PM	LC/scarce

(continued)

Table 8.1 (continued)

Sl. No.	Common name	Scientific name	Migratory status	IUCN (2014) status/remarks
493	Smoky warbler	*Phylloscopus fuligiventer*	WM	LC/scarce
494	Tickell's leaf warbler	*Phylloscopus affinis*	SM	LC
495	Sulphur-bellied warbler	*Phylloscopus griseolus*	SM	LC
496	Buff-barred warbler	*Phylloscopus pulcher*	SM/WM	LC
497	Ashy-throated warbler	*Phylloscopus maculipennis*	R	LC
498	Pale-rumped warbler	*Phylloscopus chloronotus*	SM/WM	LC
499	Hume's warbler	*Phylloscopus humei*	SM/WM	LC
500	Bright-green warbler	*Phylloscopus nitidus*	SM/PM	LC
501	Greenish warbler	*Phylloscopus trochiloides*	SM	LC
502	Large-billed leaf warbler	*Phylloscopus magnirostris*	SM	LC
503	Tytler's leaf warbler	*Phylloscopus tytleri*	SM	NT/endemic (country/ region) and scarce
504	Western crowned leaf warbler	*Phylloscopus occipitalis*	SM/PM	LC
505	Blyth's leaf warbler	*Phylloscopus reguloides*	SM	LC
506	Grey-hooded warbler	*Phylloscopus xanthoschistos*	R	LC
507	Chestnut-crowned warbler	*Seicercus castaniceps*	SM/WM in eastern Uttarakhand	LC
508	Golden-spectacled warbler	*Seicercus burkii*	SM/R/WM	LC
509	Whistler's warbler	*Seicercus whistleri*	SM/WM	LC
510	Grey-cheeked warbler	*Seicercus poliogenys*	R	LC/scarce
	Family: Acrocephalidae			
511	Booted warbler	*Iduna caligata*	PM	LC
512	Sykes's warbler	*Iduna rama*	PM	LC
513	Black-browed reed warbler	*Acrocephalus bistrigiceps*	PM	LC/scarce
514	Moustached warbler	*Acrocephalus melanopogon*	WM	LC
515	Sedge warbler	*Acrocephalus schoenobaenus*	PM	LC/scarce
516	Paddyfield warbler	*Acrocephalus agricola*	PM	LC
517	Blunt-winged warbler	*Acrocephalus concinens*	SM	LC

(continued)

Table 8.1 (continued)

Sl. No.	Common name	Scientific name	Migratory status	IUCN (2014) status/remarks
518	Blyth's reed warbler	*Acrocephalus dumetorum*	PM	LC
519	Great reed warbler	*Acrocephalus arundinaceus*	PM	LC/scarce
520	Clamorous reed warbler	*Acrocephalus stentoreus*	SM/V	LC
	Family: Locustellidae			
521	Striated grassbird	*Megalurus palustris*	R in south-eastern Uttarakhand	LC/occasional
522	Long-billed bush warbler	*Locustella major*	SM	NT/scarce
523	Common grasshopper warbler	*Locustella naevia*	PM	LC
524	West Himalayan bush warbler	*Locustella kashmirensis*[a]	R?	Status uncertain
525	Spotted bush warbler	*Locustella thoracica*	SM	LC
526	Bristled grassbird	*Chaetornis striata*	R in south-eastern Uttarakhand	VU/scarce
	Family: Cisticolidae			
527	Zitting cisticola	*Cisticola juncidis*	R	LC
528	Golden-headed cisticola	*Cisticola exilis*	R in south-eastern Uttarakhand	LC
529	Common tailorbird	*Orthotomus sutorius*	R	LC
530	Rufous-vented prinia	*Prinia burnesii*	R	NT
531	Striated prinia	*Prinia crinigera*	R	LC
532	Grey-crowned prinia	*Prinia cinereocapilla*	R in south-eastern Uttarakhand	VU
533	Rufous-fronted prinia	*Prinia buchanani*	R	LC
534	Grey-breasted prinia	*Prinia hodgsonii*	R	LC
535	Graceful prinia	*Prinia gracilis*	R	LC
536	Jungle prinia	*Prinia sylvatica*	R	LC
537	Yellow-bellied prinia	*Prinia flaviventris*	R	LC
538	Ashy prinia	*Prinia socialis*	R	LC
539	Plain prinia	*Prinia inornata*	R	LC
	Family: Sylviidae			
540	Garden warbler	*Sylvia borin*	PM	LC/scarce
541	Barred warbler	*Sylvia nisoria*	PM	LC
542	Hume's whitethroat	*Sylvia althaea*	SM/WM	LC
543	Lesser whitethroat	*Sylvia curruca*	WM/PM	LC
544	Eastern Orphean warbler	*Sylvia crassirostris*	PM	LC
545	Greater whitethroat	*Sylvia communis*	PM	LC
546	Asian desert warbler	*Sylvia nana*	WM/V in foothills	LC

(continued)

Table 8.1 (continued)

Sl. No.	Common name	Scientific name	Migratory status	IUCN (2014) status/remarks
	Family: Paradoxornithidae			
547	Yellow-eyed babbler	*Chrysomma sinense*	R	LC
548	White-browed fulvetta	*Fulvetta vinipectus*	R	LC
549	Great parrotbill	*Conostoma oemodium*	R	LC/scarce
550	Black-throated parrotbill	*Suthora nipalensis*	R	LC
	Family: Zosteropidae			
551	Whiskered yuhina	*Yuhina flavicollis*	R	LC
552	Stripe-throated yuhina	*Yuhina gularis*	R	LC
553	Black-chinned yuhina	*Yuhina nigrimenta*	R	LC
554	Oriental white-eye	*Zosterops palpebrosus*	R	LC
	Family: Timaliidae			
555	Chestnut-capped babbler	*Timalia pileata*	R in south-eastern Uttarakhand	LC
556	Pin-striped tit-babbler	*Mixornis gularis*	R in south-eastern Uttarakhand	LC
557	Tawny-bellied babbler	*Dumetia hyperythra*	R	LC
558	Black-chinned babbler	*Cyanoderma pyrrhops*	R	LC
559	Streak-breasted scimitar babbler	*Pomatorhinus ruficollis*	R	LC
560	White-browed scimitar babbler	*Pomatorhinus schisticeps*	R	LC
561	Rusty-cheeked scimitar babbler	*Megapomatorhinus erythrogenys*	R	LC
	Family: Pellorneidae			
562	Puff-throated babbler	*Pellorneum ruficeps*	R	LC
	Family: Leiothrichidae			
563	Striated laughingthrush	*Grammatoptila striata*	R	LC
564	Himalayan cutia	*Cutia nipalensis*	R	LC
565	Common babbler	*Turdoides caudata*	R	LC
566	Striated babbler	*Turdoides earlei*	R	LC
567	Large grey babbler	*Turdoides malcolmi*	R	LC
568	Jungle babbler	*Turdoides striata*	R	LC
569	White-crested laughingthrush	*Garrulax leucolophus*	R	LC
570	Rufous-chinned laughingthrush	*Ianthocincla rufogularis*	R	LC
571	Spotted laughingthrush	*Ianthocincla ocellata*	R	LC
572	White-throated laughingthrush	*Ianthocincla albogularis*	R	LC
573	Streaked laughingthrush	*Trochalopteron lineatum*	R	LC

(continued)

Table 8.1 (continued)

Sl. No.	Common name	Scientific name	Migratory status	IUCN (2014) status/remarks
574	Variegated laughingthrush	*Trochalopteron variegatum*	R	LC
575	Chestnut-crowned laughingthrush	*Trochalopteron erythrocephalum*	R	LC
576	Rufous sibia	*Heterophasia capistrata*	R	LC
577	Silver-eared mesia	*Leiothrix argentauris*	R in south-east Uttarakhand	LC
578	Red-billed leiothrix	*Leiothrix lutea*	R	LC
579	Blue-winged minla	*Actinodura cyanouroptera*	R	LC
580	Chestnut-tailed minla	*Actinodura strigula*	R	LC
	Family: Muscicapidae			
581	Spotted flycatcher	*Muscicapa striata*	SM	LC
582	Dark-sided flycatcher	*Muscicapa sibirica*	SM	LC
583	Asian brown flycatcher	*Muscicapa latirostris*	SM	LC
584	Rusty-tailed flycatcher	*Muscicapa ruficauda*	SM	LC
585	Indian robin	*Copsychus fulicatus*	R	LC
586	Oriental magpie-robin	*Copsychus saularis*	R	LC
587	White-rumped shama	*Copsychus malabaricus*	R in south-eastern Uttarakhand	LC
588	Blue-throated flycatcher	*Cyornis rubeculoides*	SM	LC
589	Tickell's blue flycatcher	*Cyornis tickelliae*	R	LC
590	Small niltava	*Niltava macgrigoriae*	R	LC
591	Rufous-bellied niltava	*Niltava sundara*	SM/WM/R	LC
592	Verditer flycatcher	*Eumyias thalassinus*	SM	LC
593	Gould's shortwing	*Brachypteryx stellata*	SM	LC
594	White-browed shortwing	*Brachypteryx montana*	R/V	LC/scarce
595	Lesser shortwing	*Brachypteryx leucophrys*	R in southern Uttarakhand	LC/scarce
596	Indian blue robin	*Luscinia brunnea*	SM	LC
597	Bluethroat	*Luscinia svecica*	SM/PM	LC
598	Siberian rubythroat	*Luscinia calliope*	WM in south-east Uttarakhand	LC/scarce
599	White-tailed rubythroat	*Calliope pectoralis*	SM/WM	LC
600	White-bellied redstart or Hodgson's blue robin	*Hodgsonius phoenicuroides*	SM	LC/scarce
601	Blue whistling thrush	*Myophonus caeruleus*	R	LC
602	Little forktail	*Enicurus scouleri*	R	LC/scarce
603	Spotted forktail	*Enicurus maculatus*	R	LC
604	Black-backed forktail	*Enicurus immaculatus*	R	LC
605	Slaty-backed forktail	*Enicurus schistaceus*	R	LC

(continued)

Table 8.1 (continued)

Sl. No.	Common name	Scientific name	Migratory status	IUCN (2014) status/remarks
606	Grandala	*Grandala coelicolor*	R/SM	LC
607	Red-flanked bluetail	*Tarsiger cyanurus*	R	LC
608	Himalayan red-flanked bush-robin	*Tarsiger rufilatus*[a]	R	Under review
609	White-browed bush-robin	*Tarsiger indicus*	R	LC
610	Golden bush-robin	*Tarsiger chrysaeus*	R	LC
611	Little pied flycatcher	*Ficedula westermanni*	SM	LC
612	Kashmir flycatcher	*Ficedula subrubra*	SM	VU/endemic (country/region)
613	Ultramarine flycatcher	*Ficedula superciliaris*	SM	LC
614	Rufous-gorgeted flycatcher	*Ficedula strophiata*	R	LC
615	Snowy-browed flycatcher	*Ficedula hyperythra*	SM	LC
616	Red-breasted flycatcher	*Ficedula parva*	PM/WM	LC
617	Slaty-blue flycatcher	*Ficedula tricolor*	SM	LC
618	Blue-fronted redstart	*Phoenicurus frontalis*	SM/WM	LC
619	Plumbeous redstart	*Phoenicurus fuliginosus*	R	LC
620	Rufous-backed redstart	*Phoenicurus erythronotus*	WM	LC
621	White-capped redstart	*Phoenicurus leucocephalus*	R/SM	LC
622	Blue-capped redstart	*Phoenicurus caeruleocephala*	R	LC
623	Hodgson's redstart	*Phoenicurus hodgsoni*	WM	LC/scarce
624	White-winged redstart	*Phoenicurus erythrogastrus*	R/SM/WM	LC
625	Black redstart	*Phoenicurus ochruros*	SM/PM	LC
626	Chestnut-bellied rock thrush	*Monticola rufiventris*	R	LC
627	Blue-capped rock thrush	*Monticola cinclorhynchus*	SM	LC
628	Rufous-tailed rock thrush	*Monticola saxatilis*	PM	LC
629	Blue rock thrush	*Monticola solitarius*	SM	LC
630	White-browed bush chat	*Saxicola macrorhynchus*	R	VU/scarce or accidental
631	White-throated bush chat	*Saxicola insignis*	WM/V	VU/scarce
632	Siberian stonechat	*Saxicola maurus*	SM/PM	LC
633	White-tailed stonechat	*Saxicola leucurus*	R	LC
634	Pied bush chat	*Saxicola caprata*	SM/R	LC
635	Grey bush chat	*Saxicola ferreus*	R	LC

(continued)

Table 8.1 (continued)

Sl. No.	Common name	Scientific name	Migratory status	IUCN (2014) status/remarks
636	Indian chat	*Cercomela fusca*	R	LC
637	Hume's wheatear	*Oenanthe albonigra*	R	LC/scarce
638	Variable wheatear	*Oenanthe picata*	SM	LC/scarce
639	Pied wheatear	*Oenanthe pleschanka*	SM	LC
640	Red-tailed wheatear	*Oenanthe xanthoprymna*	SM	LC/scarce
641	Desert wheatear	*Oenanthe deserti*	SM	LC
642	Isabelline wheatear	*Oenanthe isabellina*	WM/SM	LC
	Family: Turdidae			
643	Orange-headed thrush	*Geokichla citrina*	SM	LC
644	Long-tailed thrush	*Zoothera dixoni*	R	LC
645	Plain-backed thrush	*Zoothera mollissima*	R	LC
646	Long-billed thrush	*Zoothera monticola*	SM	LC
647	Scaly thrush	*Zoothera dauma*	SM	LC
648	Dark-sided thrush	*Zoothera marginata*	SM in south-east Uttarakhand	LC/scarce
649	Tickell's thrush	*Turdus unicolor*	SM/PM	LC
650	White-collared blackbird	*Turdus albocinctus*	R	LC
651	Grey-winged blackbird	*Turdus boulboul*	R	LC
652	Tibetan blackbird	*Turdus maximus*	R	LC
653	Chestnut thrush	*Turdus rubrocanus*	R	LC
654	Black-throated thrush	*Turdus atrogularis*	WM	LC
655	Red-throated thrush	*Turdus ruficollis*	WM	LC
656	Dusky thrush	*Turdus eunomus*	PM	LC/scarce
657	Fieldfare	*Turdus pilaris*	SM	LC/scarce
658	Song thrush	*Turdus philomelos*	WM	LC
659	Mistle thrush	*Turdus viscivorus*	R	LC
660	Purple cochoa	*Cochoa purpurea*	R	LC/scarce
661	Green cochoa	*Cochoa viridis*	R/V	LC/scarce
	Family: Sturnidae			
662	Spot-winged starling	*Saroglossa spiloptera*	SM	LC
663	Common hill myna	*Gracula religiosa*	R	LC/scarce
664	Jungle myna	*Acridotheres fuscus*	R	LC
665	Bank myna	*Acridotheres ginginianus*	R	LC
666	Common myna	*Acridotheres tristis*	R	LC
667	Asian pied starling	*Gracupica contra*	R	LC
668	Chestnut-tailed starling	*Sturnia malabarica*	SM	LC
669	Brahminy starling	*Temenuchus pagodarum*	R	LC
670	Rosy starling	*Pastor roseus*	PM	LC
671	European starling	*Sturnus vulgaris*	WM	LC

(continued)

Table 8.1 (continued)

Sl. No.	Common name	Scientific name	Migratory status	IUCN (2014) status/remarks
	Family: Chloropseidae			
672	Golden-fronted leafbird	*Chloropsis aurifrons*	R	LC
673	Orange-bellied leafbird	*Chloropsis hardwickii*	R	LC
	Family: Dicaeidae			
674	Thick-billed flowerpecker	*Dicaeum agile*	R	LC
675	Yellow-bellied flowerpecker	*Dicaeum melanoxanthum*	SM	LC/scarce
676	Pale-billed flowerpecker	*Dicaeum erythrorhynchos*	R	LC
677	Fire-breasted flowerpecker	*Dicaeum ignipectus*	R	LC
	Family: Nectariniidae			
678	Purple sunbird	*Cinnyris asiaticus*	SM/R	LC
679	Fire-tailed sunbird	*Aethopyga ignicauda*	R/SM	LC/scarce
680	Black-throated sunbird	*Aethopyga saturata*	R	LC/scarce
681	Gould's sunbird	*Aethopyga gouldiae*	R	LC
682	Green-tailed sunbird	*Aethopyga nipalensis*	R	LC
683	Crimson sunbird	*Aethopyga siparaja*	R	LC
	Family: Prunellidae			
684	Alpine accentor	*Prunella collaris*	R	LC
685	Altai accentor	*Prunella himalayana*	WM	LC
686	Robin accentor	*Prunella rubeculoides*	R	LC
687	Rufous-breasted accentor	*Prunella strophiata*	R/SM	LC
688	Brown accentor	*Prunella fulvescens*	R	LC
689	Black-throated accentor	*Prunella atrogularis*	WM	LC
	Family: Motacillidae			
690	Western yellow wagtail	*Motacilla flava*	WM/PM	LC
691	Citrine wagtail	*Motacilla citreola*	SM	LC
692	Grey wagtail	*Motacilla cinerea*	SM	LC
693	White wagtail	*Motacilla alba*	SM/PM	LC
694	White-browed wagtail	*Motacilla maderaspatensis*	R	LC
695	Richard's pipit	*Anthus richardi*	WM	LC
696	Oriental pipit	*Anthus rufulus*	R	LC
697	Long-billed pipit	*Anthus similis*	SM	LC
698	Blyth's pipit	*Anthus godlewskii*	PM	LC/scarce
699	Tawny pipit	*Anthus campestris*	WM	LC
700	Upland pipit	*Anthus sylvanus*	R	LC
701	Rosy pipit	*Anthus roseatus*	SM	LC
702	Tree pipit	*Anthus trivialis*	PM/SM	LC
703	Olive-backed pipit	*Anthus hodgsoni*	SM/WM	LC
704	Red-throated pipit	*Anthus cervinus*	PM	LC

(continued)

Table 8.1 (continued)

Sl. No.	Common name	Scientific name	Migratory status	IUCN (2014) status/remarks
705	Water pipit	*Anthus spinoletta*	WM/PM	LC
706	Buff-bellied pipit	*Anthus rubescens*	WM	LC
707	Forest wagtail	*Dendronanthus indicus*	PM	LC/scarce
	Family: Emberizidae			
708	Crested bunting	*Melophus lathami*	SM	LC
709	Yellowhammer	*Emberiza citrinella*	V/WM	LC/scarce
710	Pine bunting	*Emberiza leucocephalos*	WM	LC
711	Rock bunting	*Emberiza cia*	SM/R	LC
712	Grey-hooded bunting	*Emberiza buchanani*	WM	LC
713	Ortolan bunting	*Emberiza hortulana*	PM	LC/scarce
714	Chestnut-breasted bunting	*Emberiza stewarti*	SM/WM	LC
715	Chestnut-eared bunting	*Emberiza fucata*	R	LC
716	Little bunting	*Emberiza pusilla*	WM/PM	LC
717	Yellow-breasted bunting	*Emberiza aureola*	WM	EN/scarce
718	Chestnut bunting	*Emberiza rutila*	PM	LC/scarce
719	Black-headed bunting	*Emberiza melanocephala*	PM	LC
720	Red-headed bunting	*Emberiza bruniceps*	PM	LC
721	Reed bunting	*Emberiza schoeniclus*	WM	LC/scarce
	Family: Fringillidae			
722	Common chaffinch	*Fringilla coelebs*	WM	LC
723	Brambling	*Fringilla montifringilla*	WM	LC
724	Plain mountain finch	*Leucosticte nemoricola*	R/WM	LC
725	Black-headed mountain finch	*Leucosticte brandti*	R	LC
726	Spectacled finch	*Callacanthis burtoni*	R	LC
727	Mongolian finch	*Bucanetes mongolicus*	R/WM	LC
728	Brown bullfinch	*Pyrrhula nipalensis*	R	LC
729	Orange bullfinch	*Pyrrhula aurantiaca*	R	LC
730	Red-headed bullfinch	*Pyrrhula erythrocephala*	R/WM	LC
731	Dark-breasted rosefinch	*Carpodacus nipalensis*	SM	LC
732	Common rosefinch	*Carpodacus erythrinus*	SM/PM/WM	LC
733	Himalayan beautiful rosefinch	*Carpodacus pulcherrimus*	SM	LC
734	Pink-browed rosefinch	*Carpodacus rodochroa*	SM/WM	LC

(continued)

Table 8.1 (continued)

Sl. No.	Common name	Scientific name	Migratory status	IUCN (2014) status/remarks
735	Vinaceous rosefinch	*Carpodacus vinaceus*	WM	LC/scarce
736	Spot-winged rosefinch	*Carpodacus rhodopeplus*	R	LC/scarce
737	Himalayan white-browed rosefinch	*Carpodacus thura*	R	LC
738	Blyth's rosefinch	*Carpodacus grandis*	R/WM	LC
739	Streaked rosefinch	*Carpodacus rubicilloides*	R	LC
740	Spotted great rosefinch	*Carpodacus severtzovi*	R	LC
741	Red-fronted rosefinch	*Carpodacus puniceus*	R	LC
742	Scarlet finch	*Haematospiza sipahi*	SM/WM	LC/scarce
743	Yellow-breasted greenfinch	*Chloris spinoides*	SM/WM	LC
744	Red crossbill	*Loxia curvirostra*	R/V	LC
745	Eurasian siskin	*Spinus spinus*	WM/V	LC/scarce
746	European goldfinch	*Carduelis carduelis*	R/WM	LC
747	Twite	*Carduelis flavirostris*	R	LC
748	Eurasian linnet	*Carduelis cannabina*	WM	LC
749	Fire-fronted serin	*Serinus pusillus*	R/WM	LC
750	Black-and-yellow grosbeak	*Mycerobas icterioides*	R	LC
751	Collared grosbeak	*Mycerobas affinis*	R	LC
752	Spot-winged grosbeak	*Mycerobas melanozanthos*	SM	LC
753	White-winged grosbeak	*Mycerobas carnipes*	R	LC
	Family: Passeridae			
754	House sparrow	*Passer domesticus*	R/SM	LC
755	Spanish sparrow	*Passer hispaniolensis*	WM	LC/scarce
756	Russet sparrow	*Passer rutilans*	R	LC
757	Eurasian tree sparrow	*Passer montanus*	R in eastern Uttarakhand	LC
758	Chestnut-shouldered petronia	*Petronia xanthocollis*	SM	LC
759	Rock petronia	*Petronia petronia*	WM	LC/scarce
760	Black-winged snowfinch	*Montifringilla adamsi*	R	LC
761	Blanford's snowfinch	*Montifringilla blanfordi*	R	LC
	Family: Ploceidae			
762	Streaked weaver	*Ploceus manyar*	R	LC
763	Baya weaver	*Ploceus philippinus*	R	LC
764	Finn's weaver	*Ploceus megarhynchus*	R in south-east Uttarakhand	VU/ scarce

(continued)

Table 8.1 (continued)

Sl. No.	Common name	Scientific name	Migratory status	IUCN (2014) status/remarks
	Family: Estrildidae			
765	Red avadavat	*Amandava amandava*	R	LC/scarce
766	Indian silverbill	*Euodice malabarica*	R	LC
767	White-rumped munia	*Lonchura striata*	R in south-east Uttarakhand	LC
768	Nutmeg mannikin	*Lonchura punctulata*	R	LC

WM Winter migrants, *SM* summer migrants, *PM* passage migrants, *R* residents, *BL* breeds in Ladakh, *V* vagrants, *LC* least concern, *NT* near threatened, *VU* vulnerable, *EN* endangered, *CR* critically endangered
[a]Taxonomic status uncertain

Studies showed that in Uttarakhand, lower Himalaya and Shiwalik hills are quite rich in avian diversity. 549 species have been reported from Corbett Tiger Reserve Landscape (Dhakate et al. 2008), while 312 avian species (including 89 migrants and 53 altitudinal migrants) were reported from Rajaji National Park (Pandey et al. 1994). Sultana and Khan (2000) recorded 185 species (151 residents, 8 migrants and 26 residents/migrants) in oak forests of Almora. Sankaran (1995) reported the occurrence of 112 species from Nanda Devi National Park. Over 550 species have been recorded from Dehradun District (Singh 2006). Recently, an updated checklist of the birds (comprising 686 species) of Uttarakhand has been compiled by Mohan and Sondhi (2014).

8.3.2 Migratory Species

Western Himalaya is an important destination for migratory waterbirds and stopover for a number of passage migrants, owing to its geographical position and supporting habitats. The region is mountainous with low-range Shiwalik hills to snow-covered Trans-Himalayan peaks, drained by a number of rain-/snow-fed perennial rivers and streams. Analysis of the data revealed that about 48% species of avifauna of this region are migratory. They have different patterns of migration. Some species (16.02%) are long-distance winter migrants in this region. These birds mainly breed in north-central part of Palearctic region and winter in India subcontinent. This category of migrants mainly comprised of water fowls, raptors, shore birds and some passerine birds. A small population of some of these species (such as bar-headed goose, ruddy shelduck, northern pintail, garganey and northern shoveler) have been reported to breed in Ladakh region. Some species (6.51%) are passage migrants, which are known to spend few days in the area during fall and spring. During summer (breeding season), some species (14.58%) migrate from other parts of the country mainly from southern India, eastern India and Gangetic plains to breed in western Himalaya. These are usually treated as summer migrants. The rest of the species (483 species) are resident in the area. However, among them about 10.55% (81 species) are known to migrate locally/partially mainly higher Himalaya to lower Himalaya. Some of these species are altitudinal migrants, generally move just few hundred metres in valley due to snow fall in winter. The status of local migration of some of these species is uncertain and/or undocumented. For example, red-vented bulbul is treated as resident throughout its distribution range. However, I observed that in Solan area during the extreme cold, it moves locally in nearby warmer areas such as in valley, while himalayan bulbul remains in the area throughout the winter. Another species of bulbul, the himalayan black bulbul, moves locally according to availability of food, rather than temperature.

Migratory birds of the region are documented by several workers. In Jammu and Kashmir, the birds of Dal Lake were documented by Gousia (2014), and a total of 76 species were reported, out of which 26 species were summer visitors, 20 species were winter visitors, 9 species were local migrants and 21 species were residents (Gousia 2014). The birds of Ladakh area are also documented by some workers, despite its remoteness and limited accessibility (Osmaston 1925; Holmes 1983; Mallon 1987; Gole 1992; Pfister 2001; Namgail 2005). The marshes around the high altitude lakes (such as Tso Moriri, Tso Kar and Pangong Tso) in eastern Ladakh are known to provide abundant food and shelter to the large number of migratory birds. Pfister (2004) produced a comprehensive book, which encompasses updated information on avifauna of Ladakh. Over 100 species of birds have been reported to breed in Ladakh, including some rare or occasional breeders and about 60 migrants (Namgail and Yom-Tov 2009).

Some wetlands/water reservoirs such as Pong Dam, Govind Sagar, Renuka wetland and Chandertal and a number of rain-fed/drain-fed small depressions/lakes and some Trans-Himalayan lakes located in Himachal Pradesh are important destination for migratory waterbirds. Over 85 species of migratory aquatic birds belonging to 6 orders and 16 families have been reported from Himachal Pradesh (Kumar 2015). Pong Dam (32°01′ N; 76°05′ E) is one of the most important wetlands of this region as it attracts a large number of migratory birds from the plains of India and Central Asian countries. It was created in 1975 as a water storage reservoir impounded across Beas River primarily for irrigation and hydroelectric power generation (Editor-Director 2009). It is one of the 25 international wetland sites recognized in India by the Ramsar Convention (Kumar 2011). The occurrence of more than 412 avian species of 65 families has been reported (Pandey 1989a, 1993; Editor-Director 2009; Kumar and Paliwal 2015). Govind Sagar reservoir is also known to host most of the migratory bird species (both divers and waders) of Himachal Pradesh. However, it is not well documented. Over 100 avian species have been reported from Renuka wetland also (Editor-Director 2000), but the representation by migratory waterbirds was almost negligible (Kumar 2015).

In Uttarakhand, Asan Barrage, Bhimgoda reservoir and water bodies in Corbett National Park are the main destinations for migratory birds. On the basis of studies conducted in the past (by Kumar and Bhatt 2000; Tak and Sati 2003; Bhattacharjee and Bargali 2012; Kaushik and Gupta 2013), about 175 wetland birds (134 species of waterbird and 41 species of wetland-dependent bird) were enlisted in a recent review by Bhatt et al. (2015). This study showed that Corbett Landscape, Asan Barrage and Bhimgoda Barrage were represented by 114, 76 and 55 species of waterbird, respectively (Bhatt et al. 2015).

8.3.3 Globally Threatened Species and Endemic Species

The region supports the occurrence of a number of uncommon, rare and threatened species. Analysis of the data revealed that out of 768 species, 61 (7.94%) belong to threatened categories of IUCN (2014). Four species (namely, red-headed vulture, white-rumped vulture, Indian vulture and sociable lapwing) were 'critically endangered', while five (namely, white-headed duck, Egyptian culture, black-bellied tern, yellow-breasted bunting and saker falcon) were 'endangered'. Twenty-three species belong to 'vulnerable' category followed by 29 species in 'near-threatened' category. Apart from this many species were rare or scarce in the region (Table 8.1).

This area is a major constituent of the 'Important Bird Area' and 'Western Himalayas Endemic Bird Area' declared by Birdlife International (2015a). Some areas/protected areas are declared as important bird areas (Table 8.2). Some important and rare restricted-range birds breed in west Himalayan temperate forests. Eleven species belonging to pheasants, tits, leaf-warblers, flycatchers, nuthatches and finches have reported to inhibit/breed in this region (Stattersfield et al. 1998; Islam and Rahmani 2004).

1. Western Tragopan (*Tragopan melanocephalus*): It is a beautiful, magnificent and rare species of pheasant. Males (68–73 cm) are larger than females (60 cm) in size and have an orange to red collar, red facial skin and white-spotted black belly. It has a discontinuous distribution in the western Himalayas, occurring from Indus-Kohistan district of north Pakistan, east through Kashmir and Himachal Pradesh to Uttarakhand in northwest India. It inhabits in temperate coniferous and deciduous forests ranging from 2400 to 3600 m ASL. During winter, it locally migrates from lower altitude (between 1750 and 3000 m ASL) to grassy or shrubby areas (Kazmierczak and Perlo 2000; Rasmussen and Anderton 2005).
2. Cheer Pheasant (*Catreus wallichii*): It is a large-sized (male, 90–118 cm; female, 61–76 cm), grey-brown, long, buff bar-tailed uncommon species of pheasant, having a long crest and red facial skin. Males possess largely plain pale-greyish upper neck and clear, dark barring on upper plumage, while females are smaller in size, slightly duller in colour and more heavily marked. It occurs in the western Himalaya from north Pakistan through Kashmir into Himachal Pradesh and Uttarakhand, India, and east to central Nepal. It has a patchy distribution in precipitous, rocky terrain dominated by scrub, tall grass and scattered clumps of trees, usually located between 1445 and 3050 m ASL. It prefers combination of low shrubs with grass growing through spring and summer. It is also recorded in regenerating coniferous and broadleaved forests, as well as juniper and rhododendron on grassy slopes (Kazmierczak and Perlo 2000; Rasmussen and Anderton 2005).

3. Himalayan Quail (*Ophrysia superciliosa*): It is a medium-sized (25 cm), extremely elusive and rare (most probably extinct!) species of quail (Rasmussen and Anderton 2005). Males have greyish plumage, with black face and throat and white forehead and narrow supercilium with red bill and legs, while females have dark-marked brown upper parts, buffish head sides and underparts and contrasting dark mask and dark streaks on breast to vent. It is known from a small pocket in Uttarakhand, India, where about a dozen specimens were collected near Mussoorie and Nainital prior to 1877 (Ali and Ripley 1983). During recent decades, surveys undertaken to understand the population status indicate no records. On the basis of old information, it was known to inhabit in the flocks of 6–12 birds, in long grass and scrub on steep hillsides, particularly south-facing slope crests, between 1650 and 2400 m ASL (Rasmussen and Anderton 2005).
4. White-Cheeked Tit (*Aegithalos leucogenys*): It is a small-sized (11 cm) typically long-tailed tit having white cheek patch and blackish bib. Wings and upper plumage are dull grey blue with buff belly, while forehead and centre of crown are cinnamon drab. It is reported to be common in Afghanistan, locally common in Pakistan and scarce or even rare in Kashmir (BirdLife International 2012). Its natural habitat is open dry scrubby forest of Holly oak (*Quercus ilex*), juniper (*Juniperus*) and pine forests.
5. White-Throated Tit (*Aegithalos niveogularis*): It is a small-sized (11·5 cm) typically long-tailed tit. It has a dull buff-brown plumage, white fore-crown, half-collar forehead, broad white crown stripe and grading to cinnamon-brown throat. It is found in India, Nepal, and Pakistan (Rasmussen and Anderton 2005). Its natural habitat is subtropical or tropical moist montane forests.
6. Brooks's Leaf Warbler (*Phylloscopus subviridis*): It is a small-sized (9–10 cm), insectivorous, poorly known, olive-green leaf warbler with paler yellowish rump, well-defined supercilium, two wing bars, dark olive lateral crown stripe and yellowish coronal strip. It breeds in coniferous forest, mainly spruce and silver fir vegetation in north-east Afghanistan and northwest Pakistan (Rasmussen and Anderton 2005). During winter, it migrates to north-central Pakistan and east Punjab to West Uttar Pradesh in northwest India.
7. Tytler's Leaf Warbler (*Phylloscopus tytleri*): It is a medium-sized (10–12 cm), poorly known species of leaf warbler having olive-greyish upper plumage and whitish under plumage, with a long and slender bill, short tail, a prominent supercilium and no wing bar. It is found in Afghanistan, Pakistan, Nepal and India (Rasmussen and Anderton 2005). It passes through the western Himalayas to southern India in winter, particularly in the Western Ghats and the Nilgiris. Its natural habitat is subtropical or tropical moist montane forests. During the breeding season, it inhabits in coniferous forest, subalpine dwarf willows and birches. In winter, it prefers shola forest in the Western Ghats.
8. Kashmir Nuthatch (*Sitta cashmirensis*): It is a medium-sized (14 cm) nuthatch species, typical in shape and behaviour. Male has dull blue crown and upper parts, including upper-wing coverts and tertials. Lower belly and under plumage are rufous buff with under marked vent. Female is slightly duller than male. It is distributed in the eastern Afghanistan and Pakistan, in northwestern India

Table 8.2 List of important bird areas of the region (BirdLife International 2015a)

IBA code	Name of the site	State
IN002	Dachigam National Park	Jammu and Kashmir
IN003	Dehra Gali (DKG) forest	Jammu and Kashmir
IN004	Gulmarg Wildlife Sanctuary	Jammu and Kashmir
IN010	Kishtwar National Park	Jammu and Kashmir
IN011	Lachipora Wildlife Sanctuary	Jammu and Kashmir
IN012	Limber Valley Wildlife Sanctuary	Jammu and Kashmir
IN014	Overa-Aru Wildlife Sanctuary	Jammu and Kashmir
IN022	Bandli Wildlife Sanctuary	Himachal Pradesh
IN023	Chail Wildlife Sanctuary	Himachal Pradesh
IN025	Daranghati Wildlife Sanctuary	Himachal Pradesh
IN026	Dhauladhar Wildlife Sanctuary and McLeod Ganj	Himachal Pradesh
IN027	Gamgul Siahbehi Wildlife Sanctuary	Himachal Pradesh
IN029	Great Himalayan National Park	Himachal Pradesh
IN030	Kais Wildlife Sanctuary	Himachal Pradesh
IN031	Kalatop Khajjiar Wildlife Sanctuary	Himachal Pradesh
IN032	Kanwar Wildlife Sanctuary	Himachal Pradesh
IN034	Kugti Wildlife Sanctuary	Himachal Pradesh
IN036	Majathal Wildlife Sanctuary	Himachal Pradesh
IN037	Manali Wildlife Sanctuary	Himachal Pradesh
IN041	Rupi Bhaba Wildlife Sanctuary	Himachal Pradesh
IN042	Sangla (Raksham Chitkul) Wildlife Sanctuary	Himachal Pradesh
IN044	Sechu Tuan Nala Wildlife Sanctuary	Himachal Pradesh
IN045	Shikari Devi Wildlife Sanctuary	Himachal Pradesh
IN046	Shimla Water Catchment Wildlife Sanctuary	Himachal Pradesh
IN047	Talra Wildlife Sanctuary	Himachal Pradesh
IN048	Tirthan Wildlife Sanctuary	Himachal Pradesh
IN099	Askot Wildlife Sanctuary and Goriganga Basin	Uttarakhand
IN100	Binog Sanctuary – Bhadraj – Jharipani	Uttarakhand
IN103	Govind National Park and Wildlife Sanctuary, Sandra, Kotinad and Singtur ranges (tons forest division)	Uttarakhand
IN104	Kedarnath Musk Deer Sanctuary and surrounding Reserve Forests	Uttarakhand
IN105	Nanda Devi Biosphere Reserve	Uttarakhand
IN109	Upper Pindar Catchment in East Almora Forest Division	Uttarakhand
IN111	Gangotri National Park	Uttarakhand

and in Nepal (Rasmussen and Anderton 2005; Birdlife International 2015b). Its natural habitats are boreal forests and temperate forests.

9. Kashmir Flycatcher (*Ficedula subrubra*): It is a small (13 cm) flycatcher with black-bordered, orange-red throat, breast and flanks. Females and first-winter birds have dark base to bill and paler, slightly browner upper parts (Rasmussen and Anderton 2005). It is endemic to the Indian subcontinent and breeds in the Neelum Valley and Kaz-i-nag Range in Pakistan and Kashmir and the Pir Panjal Range in India. During winter, it migrates to some areas in south-

ern India and Sri Lanka. It is a strict pairing bird known to form winter territories and exhibit site fidelity.

10. Spectacled Finch (*Callacanthis burtoni*): It is a medium-sized, brown-coloured, sexually dimorphic bird. Male has light brownish underparts and upper parts, reddish forehead and broad supercilium black wings (with white spots) and tail with white tips, yellowish conical-shaped bill and pinkish yellow legs (Ali and Ripley 1983). In female, the brownish colour of plumage is replaced with buff-grey and reddish colour of forehead and supercilium with buff. It is found in temperate northern regions of the Indian subcontinent, ranging across Afghanistan, India, Nepal and Pakistan.
11. Orange Bullfinch (*Pyrrhula aurantiaca*): It is relatively small (4 cm) bullfinch with short, curved bill and slightly notched black tail. Male has a bright orange (yellowish in subadults) plumage with black mask from forehead to lores, cheek and chin. Female has grey nape, ochre-brown mantle and yellower belly (Ali and Ripley 1983). It is found in east to north Kashmir in India and Chitral and Gilgit in Pakistan. During winter it comes to Himachal Pradesh. Its natural habitat is temperate forests.

8.3.4 Major Conservation Issues and Implications

The major threat to avian diversity is loss, degradation and fragmentation of habitats. In this region forests are being destroyed for unorganized development, agriculture, fuel wood and timber. However, over the years, forest department is implementing social forestry and afforestation schemes to increase forest cover and to cope with the biodiversity loss.

Among natural disasters, forest fire (or I should say man made!) is a prominent factor for the destruction of flora and fauna in western Himalayas existing between 1000 and 1800 m, dominated by pine forests. Owing to repeated fires over the decades, mixed forests of oak and pine are converted into monoculture pine forest. Such fires have made the situation more favourable for pines to grow because pines are relatively strong to face fires. The uncontrolled fires in such areas help in spreading pine forest at the cost of indigenous oak forest, which is very serious threat to the ecological balance in this region. Forest fires are adversely affect the avifauna. Birds are known to lose their lives due to increased temperature and flames. Eggs of birds and insects are destroyed due to fire impact. The birds can save themselves by flight, but their nests/eggs/nestlings are usually destroyed.

In terms of anthropogenic pressures, Jammu region is worth mentioning. It is known for over exploitation of timber, non-timber forest products (especially wild medicinal plants), livestock grazing, unplanned developmental activities and deforestation (Sharma 2008). As a result, natural forests and grasslands have degraded and top soil has eroded from most of the mountain slopes. The soil erosion is particularly severe in the Shiwalik hills (Sharma 2008). Situation is almost similar in foothills of Uttarakhand and Himachal Pradesh, also.

8.4 Conclusion

The present article consolidated the information on the avifauna of northwest Himalayas. This region is comprised of various fragile ecosystems vulnerable to climate change. Analysis of the data revealed the occurrence of a sizable number of avian species in this region. Moreover, this area hosts a number of winter migrants, summer migrants and passage migrants. Yet, information is scanty and restricted to the short-term/opportunistic studies mostly conducted in some protected areas. Population of most species is declining fast and it resulted into biodiversity loss. There is need for long-term systematic studies on the avifauna of the region, their conservation status and issues, so that preventive measures could be adopted to save and sustain the avian diversity of the region.

Acknowledgements I am grateful to the Director, Zoological Survey of India, Kolkata, for his kind encouragements and support. Inputs by Dr. Rahul Paliwal, Zoological Survey of India, Solan, are highly appreciated. Valuable support and active cooperation at various levels from staff of Zoological Survey of India, Solan, and Forest Department of Himachal Pradesh, are also gratefully acknowledged.

References

Ali S, Ripley SD (1983) Compact handbook of the birds of India and Pakistan. Oxford University Press, Bombay

den Besten JW (2004) Birds of Kangra. Moonpeak Publishers, Dharamsala, p 173

Bhargav BK, Uniyal VP, Kittur S, Sivakumar K (2007) Bird records from Simbalbara Wildlife Sanctuary, Himachal Pradesh. Indian For 133(10):1411–1418

Bhatt D, Sethi VK, Sharma S, Kumar A, Saini V, Singh A (2015) Waterbirds of selected wetlands of Uttarakhand. In: Gopi GV, Hussain SA (eds) 2014, Waterbirds of India, ENVIS Bulletin: wildlife & protected areas, vol 16. Wildlife Institute of India, Dehradun, pp 140–159, 367

Bhattacharjee A, Bargali HS (2012) Status and trend of waterbirds in Wetlands around Corbett. The Corbett Foundation, India. Technical report, Uttarakhand State Forest Department

Bhattacharya T, Sathyakumar S (2007) An ornithological survey of Chenab Valley, Chamoli District, Uttaranchal, including notes on pheasants. Indian Birds 3(4):138–145

BirdLife International (2012) *Aegithalos leucogenys*. IUCN red list of threatened species. Version 2013.2. International Union for Conservation of Nature. Retrieved 26 November 2013

BirdLife International (2015a) Endemic bird area factsheet: western Himalayas. Downloaded from http://www.birdlife.org on 18 May 2015

BirdLife International (2015b) Species factsheet: *Sitta cashmirensis*. Downloaded from http://www.birdlife.org on 18 May 2015

Clements JF, Schulenberg TS, Iliff MJ, Roberson D, Fredericks TA, Sullivan BL, Wood CL (2014) The eBird/clements checklist of birds of the world: version 6.9. Downloaded from http://www.birds.cornell.edu/clements checklist/download/

Dar GH (2008) Special habitats and threatened plants of Kashmir Himalaya. In: Rawat GS (ed) Special habitats and threatened plants of India. ENVIS Bulletin: wildlife and protected areas, vol 11(1). Wildlife Institute of India, Dehradun, pp 29–36, 239

Dhakate PM, Patil TA, Bhartari R (2008) Wetland birds of Corbett Tiger Reserve Landscape. In: Sengupta M, Dalwani R (eds) Proceedings of Taal 2007: the 12th world lake conference, pp 1974–1982

Dodsworth PTL (1910) Notes relating to the distribution, habits and nidification of *Certhia himalayana* Vigors (The Himalayan Tree-creeper) in and around Simla and the adjacent ranges. J Bombay Nat Hist Soc 20(2):463–467

Dodsworth PTL (1911) Occurrence of *Hemilophus pulverulentus* Temm, the Great Slaty Woodpecker in the neighborhood of Simla, N.W. Himalayas. J Bombay Nat Hist Soc 21(1):263

Dodsworth PTL (1912a) Occurrence of the Common Peafowl, *Pavo cristatus* Linnaeus in the neighbourhood of Simla, N.W. Himalayas. J Bombay Nat Hist Soc 21(3):1082–1083

Dodsworth PTL (1912b) Question whether *Gyps fulvus* Gmelin, the Griffon, occurs in the Himalayan districts of the Punjab. J Bombay Nat Hist Soc 21(4):1331–1332

Dodsworth PTL (1913a) XXXI-notes on the vultures found in the neighbourhood of Simla and adjacent ranges of the Himalayas. Ibis 55(4):534–544

Dodsworth PTL (1913b) Occurrence of the red-tailed chat, *Saxicola chrysopygia* De Filippi in the vicinity of Simla. J Bombay Nat Hist Soc 22(1):196

Dodsworth PTL (1913c) Occurrence of the emerald dove, *Chalcophaps indica* (Linn.) in the Simla District. J Bombay Nat Hist Soc 22(2):398

Dodsworth PTL (1914) Occurrence of the white-browed Bush-Robin, *Ianthia indica* (Vieill.) in the North-West Himalayas. J Bombay Nat Hist Soc 22(4):795–796

Editor-Director (2000) Fauna of Renuka Wetland: Wetland ecosystem series 2, pp. vi+187 (Published- Director, ZSI, Kolkata)

Editor-Director (2008) Fauna of Pin Valley National Park. Conservation Area Series, 34. pp 1–147 (Published by Director, Zool. Surv. India, Kolkata)

Editor-Director (2009) Faunal diversity of Pong Dam and its catchment area. Wetland Ecosystem Series, 12, pp 1–138 (Published by the Director, Zool. Surv. India, Kolkata)

Frome NF (1946) Birds noted in the Mahasu-Narkanda-Baghi area of the Simla Hills. J Bombay Nat Hist Soc 46(2):308–316

Gaston AJ, Singh J (1980) The status of the cheer pheasant *Catreus wallichii* in the Chail wildlife sanctuary, Himachal Pradesh. World Peasant Assoc J 5:68–73

Gaston AJ, Garson PJ, Pandey S (1994) Birds recorded in Great Himalayan National Park, Himachal Pradesh, India. Forktail 9:45–57

Gill F, Donsker D (2014) IOC world bird list (v 4.2). http://www.worldbirdnames.org

Gole P (1992) On the tract of wintering black-necked cranes in India. J Ecol Soc 6:7–22

Gousia S (2014) Avifauna and its conservation in Dal Lake. Int J Curr Res 6(2):4911–4915

Grimmett R, Inskipp C, Inskipp T (2003) Pocket guide to the birds of the Indian subcontinent. Oxford University Press, New Delhi

Holmes PR (1983) Avifauna of the Suru River Valley. Forktail 2:21–41

Islam MZ, Rahmani AR (2004) Important bird areas in India: priority sites for conservation. Indian Bird Conservation Network: Bombay Natural History Society and BirdLife International (UK). p xviii+1133

ISRF (2013) India State of Forest Report, Forest Survey of India, Ministry of Environment & Forests, Government of India, Dehradun, p xxvi+252

IUCN (2014) The IUCN red list of threatened species. Version 2014.3. http://www.iucnredlist.org. Downloaded on 12 December 2014

Jones AE (1947a) The birds of the Simla and adjacent hills. Part I. J Bombay Nat Hist Soc 47(1):117–125

Jones AE (1947b) The birds of the Simla and adjacent hills. Part II. J Bombay Nat Hist Soc 47(2):219–249

Jones AE (1948) The birds of the Simla and adjacent hills. Part III. J Bombay Nat Hist Soc 47(3):409–432

Katti MV (1989) Bird communities of lower Dachigam Valley, Kashmir. M.Sc. thesis, Saurashtra University, Rajkot, p 58

Kaushik TK, Gupta RC (2013) Understanding and analysing the coordinates of diversity of wetland birds of Asan Barrage near Paonta Sahib, northern India. Our Nat 11:192–200

Kazmierczak K, van Perlo B (2000) A field guide to the birds of India. Om Book Service, New Delhi, p 352

Khah SK, Rao RJ, Wani KA (2012) Studies on bird diversity of Overa-Aru wildlife sanctuary of Jammu and Kashmir, India. J Threat Taxa 4(13):3228–3232
Koelz W (1937) IV.-Notes on the birds of Spiti, a Himalayan Province of the Punjab. Ibis 79(1):86–104
Kumar A (2011) Pong Dam Wetlands and zoological survey of India. Curr Sci 85(10):1398
Kumar A (2015) Migratory waterbirds of Himachal Pradesh: status and conservation. In: Gopi GV, Hussain SA (eds) 2014, Waterbirds of India, ENVIS Bulletin: wildlife & protected areas, vol 16. Wildlife Institute of India, Dehradun, pp 258–267, , 367
Kumar A, Bhatt D (2000) Status of migratory avifauna of a subtropical wetland in Ganga Valley, Haridwar, India. Ann For 8(1):17–24
Kumar A, Paliwal R (2015) Galloanserae and aquatic Neoaves of Pong Dam Wetland, Himachal Pradesh: status and conservation issues, 307–321. M. Rawat et al., Aquatic ecosystem: biodiversity, ecology and conservation, https://doi.org/10.1007/978-81-322-2178-4_18, © Springer India, pp. xi+333
Kumar A, Paliwal R, Saikia U (2014) Avifauna of Kalatop-Khajjiar Wildlife Sanctuary and adjacent areas, Himachal Pradesh, India. Bird Popul 13:36–48
Mahabal A (1992a) Natural distribution of some bird species in Chamba district, Himachal Pradesh. Newsl Birdwatchers 32(1&2):17–25
Mahabal A (1992b) Avifauna of Chamba district (Himachal Pradesh) with emphasis on their altitudinal distribution. Pavo 30(1&2):17–25
Mahabal A (1996) Bird survey in Siwalik Himalaya in Himachal Pradseh. Pavo 34(1&2):7–16
Mahabal A (2000a) Birds of Talra wildlife sanctuary in lower western Himalaya, H.P. with notes on their status and altitudinal movements. Zoo's Print J 15(10):334–338
Mahabal A (2000b) Avifauna. In: Fauna of Renuka wetland: wetland ecosystem series 2. Zoological Survey of India, Kolkata, pp 169–176
Mahabal A (2005) Aves. In: Fauna of Western Himalaya (part2). Zoological Survey of India, Kolkata, pp 275–339
Mahabal A, Mukherjee R (1991) Birds of Mandi district (Himachal Pradesh). Newsl Birdwatchers 31(1&2):8–9
Mahabal A, Sharma TR (1992) Distribution pattern of birds in Kangra Valley (Himachal Pradesh). Himal J Environ Zool 6(2):85–96
Mahabal A, Sharma TR (1993) Birds of Naina Devi wildlife sanctuary in lower Himalayas. Newsl Birdwatchers 33(3):43–44
Mallon DP (1987) The winter birds of Ladakh. Forktail 3:27–41
Mishra C (1996) Pheasants and other birds of Majhatal Harsang wildlife sanctuary, Himachal Pradesh, India. Forktail 12:1–8
Mohan D, Sondhi S (2014) An updated checklist of the birds of Uttarakhand. Uttarakhand Forest Department, Dehradun
Namgail T (2005) Winter birds of the Gya-Miru wildlife sanctuary, Ladakh, Jammu and Kashmir, India. Indian Birds 1:26–28
Namgail T, Yom-Tov Y (2009) Elevational range and timing of breeding in the birds of Ladakh: the effects of body mass, status and diet. J Ornithol 150(2):505–510
Narang ML (1989) Birds of Sangla Valley. Newsl Birdwatchers 29(5&6):8
Narwade SS, Jathar GA, Rahmani AR (2006) Bibliography of the birds of North India. Buceros 11(1):1–144
Osmaston BB (1925) The birds of Ladakh. Ibis 67(3):663–719
Pandey S (1989a) The birds of Pong Dam Lake bird sanctuary. Tigerpaper 16(2):20–26
Pandey S (1989b) Some observations on the birds of Pin Valley National Park. Newsl Birdwatchers 29(1&2):9
Pandey S (1993) Changes in waterbird diversity due to the conservation of Pong Dam reservoir, Himachal Pradesh, India. Biol Conserv 66(2):125–130
Pandey S, Joshua J, Rai ND, Mohan D, Rawat GS, Sankar K, Katti MV, Khati DVS, Johnsingh AJT (1994) Birds of Rajaji National Park, India. Forktail 10:105–113
Pfister O (2001) Birds recorded during visits to Ladakh, India from 1994 to 1997. Forktail 17:81–90

Pfister O (2004) Birds and mammals of Ladakh. Oxford University Press, Oxford

Price T, Jamdar N (1990) The breeding birds of Overa wildlife sanctuary, Kashmir. J Bombay Nat Hist Soc 87:1–15

Price T, Zee J, Jamdar K, Jamdar N (2003) Bird species diversity along the Himalaya: a comparison of Himachal Pradesh with Kashmir. J Bombay Nat Hist Soc 100(2&3):394–410

Rasmussen PC, Anderton JC (2005) Birds of South Asia. The Ripley Guide, vol 1. Smithsonian Institution and Lynx Edicions, Washington, DC/Barcelona, p 378

Rawat GS (2008) Special habitats and threatened plants of Ladakh. In: Rawat GS (ed) 2008 Special habitats and threatened plants of India. ENVIS Bulletin: wildlife and protected areas, vol 11(1). Wildlife Institute of India, Dehradun, pp 1–7, 239

Rodgers WA, Panwar HS (1988) Planning a wildlife protection area network in India, vol 1. Report. Wildlife Institute of India, Dehra Dun

Sankaran R (1995) Ornithological survey of Nanda Devi National Park, India. Forktail 10:115–128

Scott G, Rowcliffe M, Stoneman J, Watts S (1988) University of Newcastle Kashmir expedition 1988. Final Report, Unpublished, p 40

Schickhoff U (2005) The upper timberline in the Himalayas, Hindu Kush and Karakorum: a review of geographical and ecological aspects. In: Broll G, Keplin B (eds) Mountain ecosystems: studies in treeline ecology. Springer, Berlin, pp 275–354

Sharma OP (2008) Threatened plants of Jammu Region, North-West Himalaya and strategies for the conservation. In: Rawat GS (ed) Special habitats and threatened plants of India. ENVIS Bulletin: wildlife and protected areas, vol 11(1). Wildlife Institute of India, Dehradun, pp 37–40, 239

Sharma TR, Mahabal A (1997) Seasonal changes in bird species in two altitudinal locations of Solan District, Himachal Pradesh. Rec Zool Surv India 96(1&4):151–165

Sharma DK, Paliwal R, Saikia U (2009) Avifauna. In: Faunal diversity of Simbalbara WLS. Conservation Area Series, 41. Zoological Survey of India, Kolkata, pp 81–101

Singh AP (2006) New and significant records from Dehra Dun valley and adjacent hills, lower Garhwal Himalayas, India: May 2002–March 2006. Forktail 22:160–163

Singh G, Dutt Rai I (2008) Threatened plants of Kedarnath wildlife sanctuary, Western Himalaya. In: Rawat GS (ed) Special habitats and threatened plants of India. ENVIS Bulletin: wildlife and protected areas, vol 11(1). Wildlife Institute of India, Dehradun, pp 51–54, 239

Singh J, Thakur ML, Banyal HS (2014) Avifauna of Prashar Lake and its surrounding area in Mandi District (Himachal Pradesh), India. Asian J Biol Sci 7:47–56

Stattersfield AJ, Crosby MJ, Long A, Wege DC (1998) Endemic bird areas of the world: priorities for biodiversity conservation. BirdLife International, Cambridge

Stoliczka F (1868) Ornithological observations in the Sutlej valley, N. W. Himalaya. J Asiat Soc Bengal 37(1):1–70

Sultana A, Khan JA (2000) Birds of oak forests in the Kumaon Himalaya, Uttar Pradesh, India. Forktail 16:131–146

Suyal BO (1992) Birds of Sarahan Bushar, Shimla district. Newsl Birdwatchers 32(9&10):14–15

Tak PC, Kumar G (1987) Wildlife of Nanda Devi National Park-an update. Indian J For 10(13):184–190

Tak PC, Paliwal R (2008) Aves. In: Fauna of Pin Valley National Park. Conservation Area Series, 34. Zoological Survey of India, Kolkata, pp 97–136

Tak PC, Sati JP (2003) AVES. In: Fauna of Asan wetland, wetland ecosystem series. Zoological Survey of India, 5. pp 31–38

Thakur ML, Mattu VK (2011) Avifauna of Kaza area of Spiti (Himachal Pradesh), India. Int J Sci Nat 2(3):483–487

Thakur ML, Paliwal R, Tak PC, Mehta HS, Mattu VK (2002) Birds of Kalatop-Khajjiar wildlife sanctuary, Chamba district, Himachal Pradesh, India. Cheetal 41(3&4):29–36

Thakur ML, Mattu VK, Paliwal R, Mehta HS, Thakur V (2008) Birds of Shahnahar reservoir, Kangra, Himachal Pradesh. Ann For 16(1):129–151

Theobald W (1862) Notes on a trip from Simla to the Spiti valley and Chomoriri (Tshomoriri) Lake during the months of July, August and September, 1861. J Asiat Soc Bengal 31:480–527
Tytler RC (1868) Notes on the birds observed during a march from Simla to Mussoorie. Ibis 10(2):190–203
Whistler H (1923) A note on the birds of Spiti. Ibis 65(4):611–629
Whistler H (1926a) XXII.-The birds of Kangra district, Punjab. Ibis 68(3):521–581
Whistler H (1926b) A note on the birds of Kullu. J Bombay Nat Hist Soc 31:458–485

Chiropteran Faunal Diversity in the Tropical Forest Ecosystem of Agasthyamalai Biosphere Reserve

9

Juliet Vanitharani

Abstract

In this chapter, we presented the status and distribution of chiropteran fauna of Agasthyamalai Biosphere Reserve. It is one of the biosphere reserves recognized among the world's hotspots treasuring the least disturbed forest area. Of the 119 bat species reported from India, Agasthyamalai Biosphere Reserve supports 46 bat species. Detailed suitable conservation and management suggestions are discussed.

Keywords

Biosphere reserve · Chiroptera · Distribution · Diversity

9.1 Introduction

The night is really alive with the only winged mammal, the bat. Bats are different from other mammals. Because their hands are winged, they hang upside down, they see through their ears, and their activities are essentially restricted to be nocturnal. As they speed through the air at twilight, they cry loudly but are heard only by the other fellows of their own kind. If the nights are silent without them, it is sure the insects devoured by them will proliferate and gain upper hand in all forest ecosystem, thereby losing the ecological balance. Bats are mammals instantly recognized yet poorly known. They yet remain creatures of mystery and subject of more prejudice and lots of misinformation than any other group of animals. Their incredible diversity, status and their ecosystem services renowned bats as important

J. Vanitharani (✉)
Bat Research Laboratory, Zoology Department and Research Centre, Sarah Tucker College (Autonomous), Tirunelveli, Tamil Nadu, India
e-mail: jvanitharani@gmail.com

C. Sivaperuman, K. Venkataraman (eds.), *Indian Hotspots*,
https://doi.org/10.1007/978-981-10-6605-4_9

component of vertebrate communities; they play key roles in forest dynamics and regeneration. They are the only mammals who mastered true, sustained flight much before man's own lineage began. India being a tropical country, the ecosystem immensely depends on this group of mammals for land (vegetation) management, pest management, biodiversity maintenance and also to the general well-being of mankind and plant and animal kingdom.

'Hand wing' forms a basis for classifying bats as separate order of mammals. The order Chiroptera (Greek: Cheiros, hand; Pteros, wing) is a species-rich mammalian order with 1240 species. Karl Koopman, a bat specialist from the American Museum of Natural History in New York, classified the bats based on their dietary variation as fruit-eating megabats or megachiropterans and the highly specialized insect-eating, echolocating microbats or michrochiropterans. About 70% of bat species are insectivores, and the rest are frugivores.

9.2 Status of Bats in India

In spite of bats playing a highly beneficial role, they have a negative public image and legislative support. Legislations in India are still based on false economic principles and political salvation without sufficient attention to genuine scientific information from active field biologists. Indian Wildlife Protection Act, 1972, Schedule V, treats fruit bats as vermin. According to IUCN, the bat species number is declining all over the world, drastically in India. If appropriate measures are not taken to conserve the bat species, their diversity could be lost forever. Conservation and management action plans alone can bring positive and effective changes in Indian bat diversity protection.

9.3 Status of Bats in Agasthyamalai Biosphere Reserve

The bats in general exploit tremendous variety of food resources and achieve remarkable abundance and diversity in certain habitats. One such landscape with varied habitat is Kalakkad Mundanthurai Tiger Reserve. The climatic conditions, water resources and plant diversity of Agasthyamalai Biosphere Reserve are actually a boon to support a variety of bat species. Agasthyamalai Biosphere Reserve is recognized as one among the world's hotspots treasuring the least disturbed forest area, which acts as a best environment with diversified flora provided with food resources for foraging and a wide range of roosting habitats with caves and crevices on the riversides scattered at different elevations (Vanitharani 2006, 2007a, b). Out of the available 119 bat species of India, Agasthyamalai Biosphere Reserve hosts 46 bat species. The bat species distribution started from the foothills up to the peaks of Agathiyar hill range; but the species assemblage varies in accordance to the elevation and forest type (Vanitharani et al. 2013a, b, 2014). Majority of these assemblages are the cave dwellers. In foothill tropical plains, most of the bat species prefer abandoned anthropogenic structures. The foraging strategy and their external

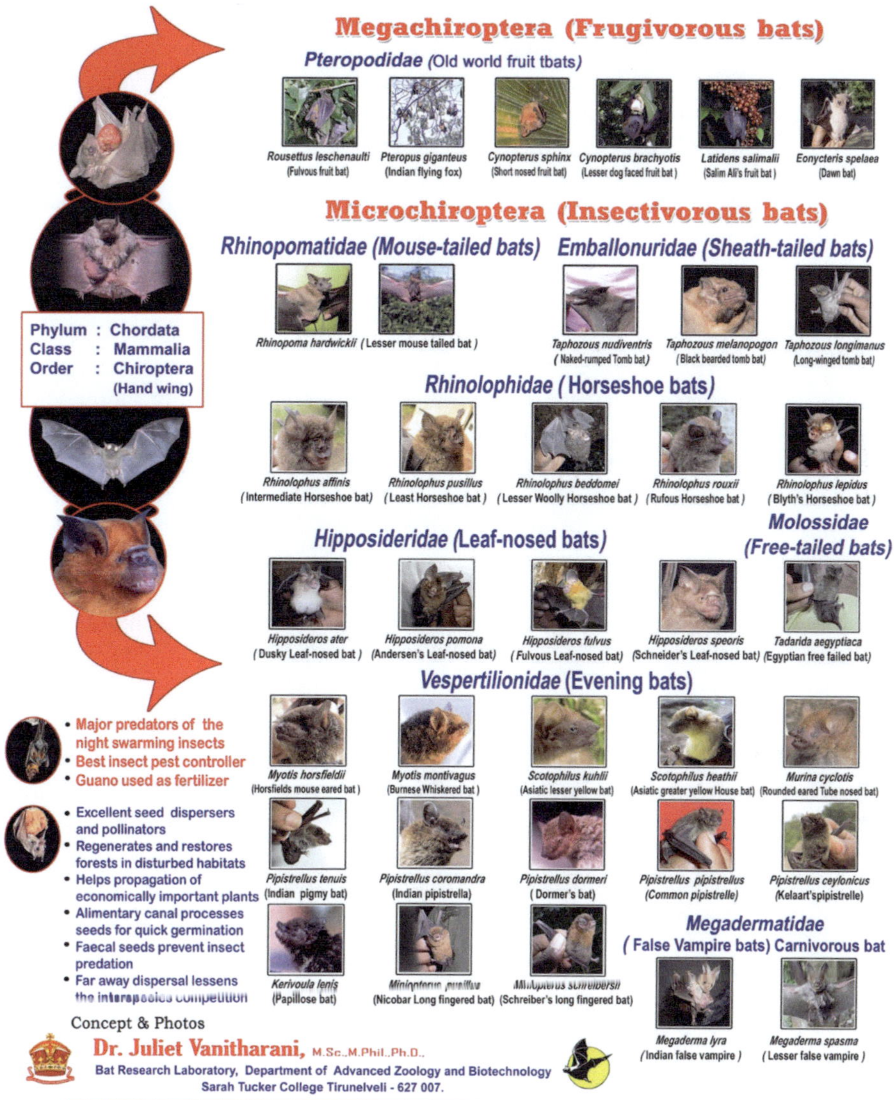

Plate 9.1 Bat species representation in Agasthiyamali Biosphere Reserve

morphology especially cranial and wing enabled them to occupy almost every niche of the forest ecosystem and play a key role in forest maintenance and also to render ecosystem services. Their presence, diversity and abundance give indications about the ecosystem health and sustainability (Vanitharani 2014) (Plate 9.1).

Without knowing the ethology of species, conservation measures cannot be made. In situ conservation in Agasthyamalai Biosphere Reserve is an excellent storehouse to protect bat species diversity. Among the noteworthy representatives of

Agasthyamalai Biosphere Reserve, the microchiropteran bats, *Rhinolophus beddomei* and *Kerivoula lenis*, reached the Near Threatened category. The three microchiropteran representatives endemic to South Asia present here are *Rhinolophus beddomei*, which is Near Threatened, and *Hipposideros speoris* and *Pipistrellus dormeri*, which are Least Concerned. The megachiropteran *Latidens salimalii* is not only endemic to southern Western Ghats but is declared endangered by IUCN (Vanitharani 1998, 2003; Vanitharani et al. 2003a, b; Addline Esther Pushparani et al. 2004).

9.4 Fruit Bats of Agasthyamalai Biosphere Reserve

The megachiropteran fruit bats are represented by only one family Pteropodidae (Table 9.1). There are about six fruit bat species widespread starting from foothill to evergreen mountaintops of Kalakkad Mundanthurai Tiger Reserve (KMTR). Three common bat species, namely, *Rousettus leschenaulti*, *Cynopterus sphinx* and *Pteropus giganteus* inhabit in the scrub jungle of the foothills <200 m AMSL. *Cynopterus brachyotis* hosts in the moist deciduous lowland forest and the semievergreen forest <1000 m AMSL, and *Cynopterus brachyotis*, *Latidens salimalii* and *Eonycteris spelaea* inhabit in the tropical wet evergreen forest >1000 m AMSL.

Table 9.1 Describes the noteworthy bat species representation in Agasthyamalai Biosphere Reserve with IUCN status

	Elevation (feet)						
Order/Family/Species	A	B	C	D	E	IUCN status	Roosting pattern
Suborder – Megachiroptera							
Family: Pteropodidae							
1. *Rousettus leschenaulti*	+	+	+	+	+	LC	Cave
2. *Pteropus giganteus*	+	–	–	–	–	LC	Open Tree
3. *Cynopterus sphinx*	+	–	–	–	–	LC	Tent maker – foliage
4. *Cynopterus brachyotis*	+	+	+	+	+	LC	Tent maker – foliage
5. *Latidens salimalii*	–	–	+	+	+	EN	Cave
6. *Eonycteris spelaea*	–	–	–	–	+	LC	Cave
Suborder – Microchiroptera							
Family: Rhinopomatidae							
1. *Rhinopoma hardwickii*	+	+	–	–	–	LC	Cave
2. *Rhinopoma muscatellum*	–	+	–	–	–	LC	Cave
Family: Emballonuridae	+	+	–	–	–	LC	Cave
3. *Taphozous melanopogon*	+	+	–	–	–	LC	Cave
4. *Taphozous longimanus*	+	–	–	–	–	LC	Cave
5. *Taphozous kachhensis*							
Family: Megadermatidae							
6. *Megaderma lyra*	+	–	–	–	–	LC	Abudened buildings
7. *Megaderma spacma*	+	+	+	+	–	LC	Tree hole

(continued)

Table 9.1 (continued)

Order/Family/Species	Elevation (feet) A	B	C	D	E	IUCN status	Roosting pattern
Family: Rhinolophidae							
8. Rhinolophus rouxii	–	+	+	+	+	LC	Cave
9. Rhinolophus lepidus	–	+	+	+	+	LC	Cave
10. Rhinolophus beddomei	–	+	+	+	+	NT	Cave
11. Rhinolophus pusillus	–	+	+	–	–	LC	Cave
Family: Hipposideridae	+	–	–	–	–	LC	Abudened buildings
12. Hipposideros ater	+	–	–	–	–	LC	Cave
13. Hipposideros fulvus	+	+	–	–	–	LC	Cave
14. Hipposideros speoris	–	+	+	+	+	NT	Cave
15. Hipposideros Pomona	+	+	–	–	–	LC	Care
16. Hipposideros durgadasi	+	+	–	–	–		
Family: Molossidae							
17. Tadarida aegyptiaca	–	+	+	+	+	NT	Crevices
Family: Vespertilionidae							
18. Myotis montivagus	–	–	+	+	+	LC	Cave
19. Myotis horsfieldii	+	–	–	–	–	LC	Cave
20. Scotophilus heathii	+	+	+	–	–	LC	Tree bark, crown
21. Scotophilus kuhlii	+	+	+	–	–	LC	Tree bark, crown
22. Pipistrellus tenuis	+	–	–	–	–	LC	Crevices
23. Pipistrellus dormeri	+	+	–	–	–	LC	Crevices
24. Pipistrellus ceylonicus	+	+	–	–	–	LC	Crevices
25. Pipistrellus coromandra	+	+	–	–	–	LC	Crevices
26. Pipstrellus pipistrellus	–	+	–	–	–	LC	Crevices
27. Miniopterus schreibersii	–	–	+	+	+	NT	Foliage
28. Miniopterus pusillus	–	–	+	+	+	NT	Foliage
29. Murina cyclotis	–	–	+	–	–	NT	Foliage
30. Kerivoulalenis	+	–	–	–	–	DD	Foliage

LC Least Concern, *NT* Near Threatened, *DD* Data Deficient, *EN* Endangered
Abbreviations used to denote the elevation. A; foothills to 1000 ft, B: 1000 to 2000 ft, C: 2000 3000 ft, D: 3000–4000 ft, E: above 4000 ft

9.5 Morphology of Fruit Bats

The fruit bats are morphologically (wing, skull and dental) adapted to forage within the cluttered forest cover. The chiropterophilic plants also bear fruits and flowers on open defoliated branches. The muzzle and jaws are strongly built. Absence of uropatagium (tail membrane) in fruit bats gives freedom of the hind limb to crawl over vegetation, When they select fruits and nectar from the wide open bat preferred flowers. The eyes of these bats are comparatively large and capable of sight even in dull night illumination. Their sense of smell is well developed. They have good memory and can locate the fruiting trees in their foraging area. The ears are simple. There is a large claw on the first digit and a smaller one on the second. These claws aid to grab fruits and to cling and crawl on the branches while feeding. The fruit bat *Rousettus* alone can echolocate and uses tongue clicking methods to produce ultrasonic calls. They use this mechanism to approach dark roosting locations in the caves.

9.6 Seed Dispersal Role of Fruit Bats

Fruit bats contribute the maximum as much as 95% of seed dispersal when compared to other propagators (Fig. 9.1). Fruit bats depend mainly on plant resources throughout the year. Their diet is primarily composed of fruits, flowers and flower products such as nectar, pollen, etc. They are catholic in food selection. Each fruit bat species has its own core plant species for dietary selection. These flying mammals alone can carry larger fruits even equivalent to 1/3 of its own body weight to longer distance. Fruit bats fly long distances nightly to locate food (Vanitharani's unpublished radiotelemetry data), as most of the core plant species are seasonal and located at different patches in their foraging area. They visit more than one species of plant in a single night to get balanced diet and have vast foraging area which provides the plant species not only the faraway dispersal of seeds but also lessened interspecies competition.

Fruit bats never feed while they are on wings. They are very sensitive to disturbance; therefore they select secluded feeding roosts to avoid nocturnal predators, for food processing and consumption at leisure. The fruits handled and processed (bat-bitten fruits by saliva, faecal seeds by intestinal juices) by fruit bats show quick germination, retrieve seed dormancy and have resistance against insect and microbial attacks (Vanitharani 2011; Vanitharani and Pandian 2012; Vanitharani 2014).

Fig. 9.1 Fruit bat feeding

Cataloguing of these chemicals is very helpful in the propagation of many rare, endangered, endemic plants and implementing conservation management action plans for forest restoration and the survival of the dependent diversity of species. Very often the feeding roost is full of bat-treated seeds and seedlings of various bat-dependent endemic trees. The recovery can help forest managers to replenish forest cover in the affected protected areas with native tree assemblage.

9.7 Pollination Role of Fruit Bats

Bat pollination is a phenomenon restricted to the tropics and subtropics (Fig. 9.2). Observations indicate that in addition to fruits, the bats feed on pollen and nectar for their basic mineral requirement and energy resource. Flower pollinators by bats exhibit a number of characteristic adaptations collectively described as "syndrome of chiropterophily". Major characteristics of bat-pollinated flowers exhibit nocturnal anthesis which correlates with bat activity, often limited to single night; shape and sturdiness (often brushlike flowers or bells, which sometimes are only the size of a head mask for a bat); production of larger quantity of nectar and pollen; an intense typical scent with strong odour and unpleasantly stale, inconspicuous colours (white or green or brown or brownish red); and a freely exposed position on the plant specifically to attract the fruit bats. Even though *Eonycteris spelaea* the only nectar feeder of Agasthyamalai Biosphere Reserve benefits pollination, other fruit bats also visit chiropterophilic flowers to get balanced diet and to meet the high energy requirement to be on wing. Fruit bats with long snout and bristly tongue were able to lap up nectar and pollen and pollinate flowers. The sticky scent gland socked facial hair of bats also act as a pollen basket to gather pollen and aid cross-pollination. Mostly the bat-pollinated flowers are almost exclusive to bats because

Fig. 9.2 Pollination role of fruit bats

no other visitors are capable of transferring such appreciable quantity of pollen for the propagation. For those plant species, the bat pollination is indispensable.

The earlier project completion reports of the author to State Forestry Research Institute of Tamil Nadu Forest Department on plant-animal interaction (pollinators and seed dispersers) and their impacts in the recovery of native flora and biodiversity in some parts of Western Ghats in Tamil Nadu suggest that the megachiropterans help a lot in the propagation and recovery of many native species (Vanitharani 2007a, b, 2011, 2012, 2013a, b, c, 2014; Vanitharani and Jeyapraba 2011).

9.8 Special Note on the Endemic and Endangered Fruit Bat Species *Latidens salimalii*

Agasthyamalai Biosphere Reserve shelters *Latidens salimalii*, the endemic endangered fruit bat species (Fig. 9.3). The High Wavy Mountains of southern Western Ghats remained the only recorded distribution of *Latidens* until 1999, when its presence was recorded for the first time in 2000 by the author and her research team. They made the first report about the distribution of *Latidens salimalii* in Courtallam hills (foraging area) and about its cave roosts in Nagapothigai (Vanitharani et al. 2003a, b, 2004; Vanitharani 2005, 2006, 2007a, b). Till 2000, the bat was described 'unknown' by IUCN, but their diurnal roost was located on the riversides of the rainforest interiors of Pothigai, Servalar and Manimuthar hills of Agasthyamalai Biosphere Reserve; then after their ethology, foraging and breeding ecology were

Fig. 9.3 An endemic fruit bat species *Latidens salimalii*

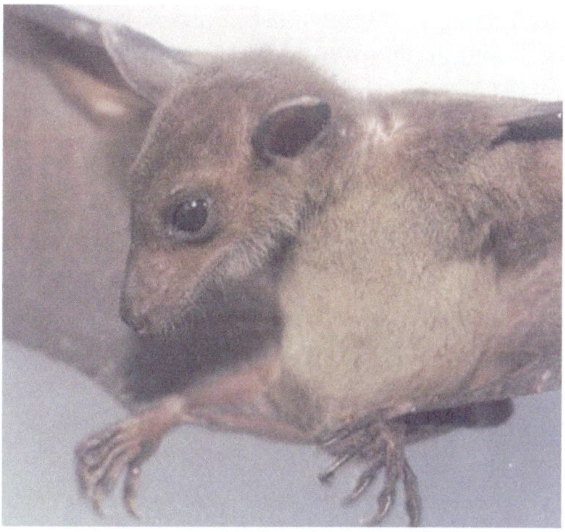

Fig. 9.4 Nectarivorous bat species *Eonycteris spelaea*

studied in Agasthyamalai Biosphere Reserve with the financial support from Rufford Small Grants, UK; University Grant's Commission, New Delhi; and Ministry of Environment and Forest (Government of India).

9.9 Other Red-Listed Fruit Bats

Eonycteris spelaea is the only nectarivorus bat species available above 1400 m elevation (Vanitharani 2006) (Fig. 9.4). Few bats were caught in the mist net during the bat survey at Injikuli area of Pothigai hills. The author still couldn't locate the cave roost of the bat, and the population assessment is difficult. So far no report has been made about their existence in other parts of southern Western Ghats, Palani and Anaimalai hills. While considering the distribution status of this bat species within Tamil Nadu, it is a rare species.

9.10 Insect-Eating Bats of Agasthyamalai Biosphere Reserve

India being a tropical country is home to a diversified insect fauna and the bats. The michrochiropterans have very small eyes, which characterizes them as insectivores. They are the only predators of the night swarming insects and keep the insect population under check. They help to bring sustainable balanced ecosystem. They roost in large numbers in caves, crevices, underground tunnels and abandoned old buildings.

In general, these winged hunters can swallow insects equal to one half of their own weight (Vanitharani and Jeyapraba 2011; Selva Ponmalar and Vanitharani 2014); in addition they are morphologically adapted with the skull (feeding apparatus) and wing (flight apparatus) to prey upon a variety of insects to get their balanced diet (Selva Ponmalar and Vanitharani 2014). The dietary selection and composition of the insectivorous bats mainly include the lepidopteran, dipteran, orthopteran, hemipteran and coleopteran groups of insects. Incidentally they are the insect groups where the predominant forest and crop insect pests belong. The dietary composition of each species is selectively species specific. With the aid of considerable diversity in wing morphology and flight style, they show clear partitioning in the food source selection, and they avoided direct competition among species. Interestingly KMTR harbours two carnivorous bats. They are commonly known as false vampires *Megaderma lyra* distributed <200 m and *Megaderma spasma* >500 m elevation. They can catch large insects like locusts, grasshoppers, beetles, moths, spiders and small vertebrates such as lizards, birds, rodents and fishes. Carnivorous bats as a rodent controller act as a good friend of farmers.

Just one microbat can catch hundreds of insects in an hour, and a large colony in turn catches tons of insects including common pests such as beetles, stem borers, bugs, caterpillars, moths, mosquitoes, etc. These efficient biocontrol agents are a kind of 'bug police', who fly around and catch insects using echolocation, a superbly developed navigational system that allows them to picture the environment through sound as much as vision. Echolocation is the single characteristic that easily separates bats from other mammals. This possession of natural radar system inspires scientists. These pest managers of KMTR maintain the ecological integrity of the KMTR forest ecosystem.

9.11 Commonly Asked Questions About Bats

9.11.1 What Is Echolocation?

Insectivorous bats emit pulses of ultrasonic sound (frequencies beyond hearing capability – over about 15 kh), produced through the larynx or voice box, and are emitted through the mouth in some bats and through the nose in others. They analyse information from the returning echoes to "see" the environment around them. Variations in their elaborate leaf nose help to produce species-specific ultrasonic sound waves. By using echolocation calls, they navigate and capture prey (Fig. 9.5).

9.11.2 How Fast Can Bats Fly?

Bats' actual flight speeds in the wild are difficult to measure, but some calculations based on morphology and also flight studies in captive studies predicted the flight speeds fall in the range of 30–50 km per hour. The fastest flying bats are with high aspect ratios and high wing loading; they normally fly high above the vegetation and are not as manoeuvrable as slower flying bats.

Fig. 9.5 Insectivorous bat

9.11.3 How Long Do Bats Live?

Compared with most small mammals, bats are amazingly long-lived. Some individuals may live for 30 or more years. The maximum age varies greatly from species to species and is heavily influenced by geography and lifestyle. In tropical species, it has been estimated to live upward of 10 years. Individual flying foxes have lived for over 25 years in zoos.

9.11.4 Why Do Bats Hang Upside Down?

Bats before taking off their flight drop a metre or so to gain the necessary momentum and also to propel them forward. Although some bats are able to take off from the ground, most are not. Therefore, hanging upside down enables them to fly much more quickly and efficiently than they could be from an upright posture.

9.11.5 How Do Bats Reproduce?

Tropical species, freed from the constraints of hibernation and migration, have more variation in reproductive patterns. Some species have restricted breeding seasons, with only a single young produced per year. This is primarily tied to the food availability and the cyclical insect abundance. Courtship displays usually involve wing flapping, sometimes vocalizations and mutual grooming. By shaking their wings, the males presumably waft a pheromonal substance towards the females, and they

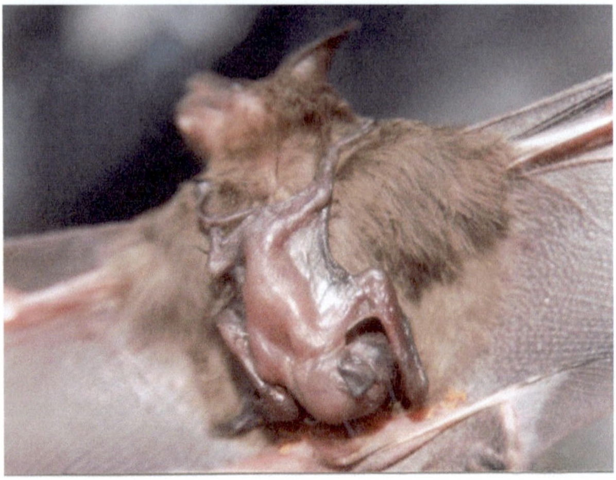

Fig. 9.6 Newborn bat with mother

vocalize at the same time. In all species studied to date, males play no role in rearing young: their involvement in reproduction essentially ends with copulation. They show sexual segregation during the reproductive period. The females show parental care and carry their young ones in an inverted position. The false ducts of the pubic region of the females help the infants for their attachment to the mother.

9.11.6 How and Where Do Mother Bats Give Birth?

Females give birth in the roost site, and young bats are born naked and helpless. The young bat assists in its own birth by grasping its mother's fur with its claws and pulling its upper body free from the birth canal. The newborn moves quickly down to its mother's ventral surface, holding to the hair of her belly until it finds and attaches to an axillary nipple (Fig. 9.6). The nipples, tucked away in the armpits, are the real mammary gland. Once the young bat attaches to the nipple, dislodging it becomes quite difficult.

9.11.7 How Do Mother Bats Find Their Young?

Mother prefers to leave their babies behind when they go out to forage, but they may return several times during the night to feed their youngsters, especially when they are very young. Otherwise, the young spend their nights hanging in cluster 'nursery colonies' and their days safely tucked into their moms' armpits attached to the nipple. When the mothers return to the colony, they pick up their babies based on their spatial memory of the general area where their young were left. The mothers use a combination of vocalizations and olfactory cues to locate their babies and quickly allow them to attach to their nipples. Young bats grow quickly. Once they are capable of flight, the bats quickly become independent of their mothers.

9.12 Suggestions for Bat Conservation

India is not an exception to believe and trust a lot of myths about bats. Bats are persecuted as they attack people, drink blood and transmit disease like rabies. They are depicted as agents of evil, indicators of death, portrayed symbol of bad luck and in auspicious bad omen. The only remedy is to create informative awareness among different ages and status of society. Even bat biologists admit that they know very little about the Indian bat species. In India bats are of the least studied mammalian group. The scientific studies only can make management recommendations. Mammal surveys in protected areas (PA) in India have a very strong bias towards larger and conspicuous animals. Bats should be included in their management action plans and also should be listed in their inventories. It is high time to establish the status of bats in India and give recommendations to the government body to make viable legislations and policies to protect Indian bats.

References

Addline Esther Pushparani D, Vanitharani J, Rajkumar S, Sundararaj T (2004) Government law and temple bat management. Muthuchelian K Biodiversity resources management and sustainable use. Centre for Biodiversity and Forest Studies School of Energy, Environment and Natural Resources, Madurai Kamaraj University, Madurai, 450 p

Basil GS, Vanitharani J, Antonysamy A, Jayapriya K (2014a) An analytical report of perceptual, spectrum and cepstral feature-based bat classification in Kalakad Mundanthurai tiger reserve with K-NN and Bayes classifiers. Int J Emerging Technol Adv Eng. Website: www.ijetae.com ISSN 2250-2459, ISO 9001:2008 Certified Journal, Volume 4, Issue 3, March 2014

Basil GS, Vanitharani J, Jayapriya K (2014b) An extensive review of methods of identification of bat species through acoustics. Int J Comput Appl Technol Res 3(4):186. 2014 www.ijcat.com ISSN 0952-8091

Jeyapraba L, Vanitharani J (2004) Bat species diversity in India. In: Muthuchelian K (ed) Biodiversity resources management and sustainable use. Centre for Biodiversity and Forest Studies School of Energy, Environment and Natural Resources, Madurai Kamaraj University. Madurai, 450 p

Juliet Vanitharani, L. Jeyapraba, Nikky Thomas and Gladriene Sheena Basil 2014 Inventory Message about Bat Species of Kalakad Mundanthurai Tiger Reserve (Tamil Nadu, India) through Global Positioning System (GPS). J Theor Exp Biol. (ISSN 0972–9720), 10 (1 and 2): 01–20, 2013. Elias Academic Publishers. www.jteb.webs.com

Selva Ponmalar S, Vanitharani J (2014) Insect pest management by the horse shoe bats (*Rhinolophus* species) in the forest ecosystem of Kalakad Mundanthurai Tiger Reserve, India. Scrutiny SIRJ-BES 1(4–5). (2014) ISSN 2348–5787

Vanitharani J (1998) Studies on the changes of wing morphology and body weight in few species of South Indian bats, Ph.D. dissertation, Madurai Kamaraj University, Madurai, Tamil Nadu, India

Vanitharani J (2003) The emerging trend in the bio-diversity of bats in Tamil Nadu. J Map Sci 2(2):14–27

Vanitharani J, Jeyapraba L, Annamalai R(2003a) New record of distribution and roosting in Salim Ali's fruit bat Latidens salimalii Thonglongya 1972. In Annamalai R, Narayanan M, Vanitharani J (eds) Proceedings of the 28th conference on Ethology Society India. UGC &DST sponsored Publication by Department of Zoology Sarah Tucker College Tirunelveli abstract 5P9 book with research papers P179 In: 60–62, 2003

Vanitharani JA, PJJ R, Bates DL, Harrison MJ, Pearch A (2003b) A taxonomic reassessment of Kerivoula lenis Thomas, 1916 (Chiroptera: Vespertilionidae) including a first record from peninsular India. Acta Chirop 5(1):49–60. www.miz.waw.pl/periodicals/acta_chiroptrologica

Vanitharani J (2004) Conservation strategies and action plan for bats of Tamil Nadu. In: Annamalai R (ed) Tamil Nadu biodiversity strategy and action plan- chordate diversity. Forest Department of Tamil Nadu, Chennai

Vanitharani J (2004b) Bat diversity management in India In: Muthuchelian K (ed) Biodiversity resources management and sustainable use. Centre for Biodiversity and Forest Studies School of Energy, Environment and Natural Resources, Madurai Kamaraj University. Madurai, . 450 p

Vanitharani J (2005) Special report about *Latiden's salimalii* of Kalakad Mundanthurai Tiger Reserve. Bat Net Newsl CCINSA 6(1):7

Vanitharani J (2006) Noteworthy representatives of bat species in Agasthiayamalai Bosphere reserve, Tamil Nadu, India. J Theor Exp Biol 2(2):47–59

Vanitharani J (2007a) Conservation of Latiden's salimalii and other red-listed bats in Tamil Nadu. Bat Net Newsl CCINSA 8(1–2.) www.zooreach.org

Vanitharani J (2007b) Status of bats in Tamil Nadu and integration of bat conservation in management work plans. J Theor Exp Biol 3(4):175–188

Vanitharani J (2011) Bat-Plant Chemical Signaling a Boon for Conservation of Endangered Fauna and Flora of Southern Western Ghats, India. J Adv Biotechnol 10(10):61–66. Pub: ISSN 0973–0109

Vanitharani J, Jeya Praba L (2011) The impact of Bat diversity in the conservation of Kalakad Mundanthurai Tiger Reserve Biodiversity. J Adv Biotechnol 10(10):57–60. Pub: ISSN 0973–0109

Vanitharani J (2012) Impact of anthropogenic perturbation through science and developments on the complex networks of bat diversity and their interactions in Agasthiyamalai biosphere reserve. In: Rahunathan C, Sivaperuman C, Venkataraman K (eds) Recent advances in biodiversity of India Zoological Survey of India, Kolkata, 529 p

Vanitharani J (2014) Sustainable management of forest through ecosystem services of bats. Scrutiny Int Res J Biol Environ Sci SIRJ-BES 1(6) (2014) ISSN 2348–5787

Vanitharani J, Jeyapraba L, Annamalai R, Bates PJJ, Harrison DL (2003a) New record of distribution and roosting in Salim Ali's fruit bat (*Latidens salimalii* Thonglongya 1972). In: Annamalai R, Narayanan M. and Vanitharani J. (eds) Proceedings of 28th conference of Ethological Society of India

Vanitharani J, Rajendran A, Bates PJJ, Harrison DL, Pearch MJ (2003b) A taxonomic reassessment of *Kerivoula lenis* Thomas, 1916 (Chiroptera: Vespertilionidae) including a first record from peninsular India. Acta Chirop 5(1):49–60

Vanitharani J, Malathi USU, Arul Sundari K (2005a) New records of bats from Kalakad Mundathurai Tiger reserve. Bat Net Newsl CCINSA 6:13–14

Vanitharani J, Pearch MJ, Jeyapraba L, Annamalai R (2005b) A review of the distribution of *Latidens salimalii* (Chiroptera: Pteropodidae) with new records from the Western Ghats. India J Lutra 47(1):21–32

Vanitharani J, Singaravelan N, Marimuthu G (2013a) Population, ecology and conservation of Salim Ali's fruit bat (*Latidens Salimalii*). In: Rare animals of India. Bentham Science Publishers, Sharjah, pp 156–174

Vanitharani J, Thomas N, Straka T, Selva Ponmalar S, Mercy C, Basil GS (2013b) Expedition with bat detector at Mundanthurai plateau of KMTR, India. J Theo Exp Biol 9(3 and 4):141–149

Vanitharani J, Jeyapraba L, Thomas N, Basil GS (2013c) Inventory message about bat species of Kalakad Mundanthurai Tiger Reserve (Tamil Nadu, India) through global positioning system (GPS). J Theo Exp Biol 10(1& 2):01–20

Diversity of Bat Fauna in States of Northeast and Around Western Ghats of India with Reference to Its Conservation

10

Vishakha Korad

Abstract

Diversity of bat fauna was studied in the states of Northeast India and states around Western Ghats, as both regions are recognized global hot spots for their biological values. The diversity of bat fauna from the tropical and semitropical forests of hilly area of the two regions is compiled on the basis of available literature mentioning their taxonomical details, distribution in the regions, and IUCN status. Thirty-six genera and ninety-two bat species reported from these regions comprise about 79% of the total bat species diversity reported from India. Sixty-eight bat species are reported from states of Northeast India and 58 bat species from the states around Western Ghats. Fifty percent of bat species reported from Northeast region of India is common to those reported from states around Western Ghats. The bat fauna of these two regions is dominated by evening bats of the family Vespertilionidae, representing 45 species, which comprise about 49% of bat species reported from the two regions. For conservation of the bat species in the area, endemism of bat species, species diversity richness, and evenness in the regions are taken into consideration. The states of Western Ghats have more number of endemic bat species, while states of Northeast India have higher value for diversity indices indicating higher diversity and evenness in distribution of bat species in the area. Thus both regions are to be considered seriously for the conservation of bat fauna and their habitats.

Keywords

Bat · Distribution · Endemism of Northeast India · Western Ghats

V. Korad (✉)
Department of Zoology, Fergusson College, Pune, Maharashtra State, India
e-mail: vishakhakorad@yahoo.com

C. Sivaperuman, K. Venkataraman (eds.), *Indian Hotspots*,
https://doi.org/10.1007/978-981-10-6605-4_10

10.1 Introduction

Among mammals of the world, bats make up 25% of the total number (Mickleburgh et al. 2002). Bats have lived for the past 60 million years as a successful group of mammals both in terms of species diversity and the area of the earth inhabited by them. However the bats are not reported from the polar region, extreme deserts, and a few isolated oceanic islands (Ronald and Nowak 1994).

The order Chiroptera is further classified as suborders (1) Megachiroptera (fruit bats) and (2) Microchiroptera (mostly insectivorous bats) on the basis of feeding habits, mode of perception, orientation, and morphological adaptations. The members of the suborder Megachiroptera belong to the single family Pteropodidae, which includes Old World fruit bats of tropical regions of Africa, Asia, and Indo-Australia (Hill and Smith 1984). They have large eyes, the sense of smell is well developed, the ears are simple in form, and muzzle and jaws are strongly built. The tail is short or absent, and the dentition is reduced in general. The fruit bats feed almost exclusively on plants and consume the soft, pulpy, and juicy parts including leaves, flowers, pollens, fruits, and often seeds. The megachiropteran bats feed upon at least 145 genera of fruit plants of 30 families, which are widely distributed throughout the world (Marshall 1985).

The fruit bats are very important pollinators and seed dispersers in tropical forests throughout the world (Marshall 1983, 1985; Fleming et al. 1987; Pierson and Rainey 1992) and have shared a long evolutionary history with angiosperms. Fujita and Tuttle (1991) reported that at least 443 products useful to mankind derive from 143 plant species that rely to some extent on bats for pollination or seed dispersal.

The suborder Microchiroptera includes bats of medium to small size and depends on power of echolocation for their orientation and feeding. In general, these bats have small to remarkably big ears and have genus-specific nose leaf complex aiding in echolocation. It is the well-diversified group feeding mostly on insects; but some are phytophagus, carnivorous, piscivorous, or sanguivorous.

About 103 species of microchiropteran bats found in India are all insectivorous. The insectivorous bats are reported to consume insects in large volumes up to 100% of their body weight per night (Eckrich and Neuweiler 1988; Davison and Zubaid 1992). They play a major role in regulating the insect population and thus become very important components of the food web of the forest as well as of agroecosystem. These bats consume insects of the order Lepidoptera, Diptera, Orthoptera, Hemiptera, Homoptera, Trichoptera, and Coleoptera (Kunz 1988; Swift et al. 1985; Whitekar 1993, 1995; Whitekar et al. 1997, 1999). Most of the Indian agricultural crop and forest pests belong to aforementioned orders of class Insecta, and thus the insectivorous microchiropteran bats play a vital role in the biological control of them. In the urban area, mosquitoes comprise the important and prime dietary item for these bats.

The bats, though mammals, are poor in thermoregulation. They are very sensitive to climatic conditions, particularly temperature and humidity of the atmosphere. With few exceptions, most of the bat species prefer cool and dark places, which maintain more or less static relative humidity and temperature as their diurnal biotopes.

Roost site fidelity is generally high in those genera of bats that roost communally (Marshall 1983). And those bat genera roosting singly or in small groups show less site fidelity but may use the same perch for considerable period (Marshall 1983). Bats are often observed shifting their roosting sites, as they are sensitive to the disturbance to their habitats. Therefore, abundance of the bat fauna and their species diversity indicate the "well-being" condition of the ecosystem.

The study of diversity of bat fauna in the states of Northeast India and states in periphery of Western Ghats of India is very important from various aspects. Both the states of Northeast and Western Ghats of India are the regions, which are recognized as biodiversity hot spots (Myers et al. 2000). The present study aims at reporting of Chiroptera species from states of Northeast India and states in periphery of Western Ghats of India. The endemism and IUCN status of these species will be considered and discussed from the conservation point of view.

10.2 Methods

10.2.1 Study Area

Northeast states of India represent the easternmost region of India connected to East India via a narrow corridor squeezed between Nepal and Bangladesh. It comprises the contiguous Seven Sister States (Arunachal Pradesh, Assam, Manipur, Meghalaya, Mizoram, Nagaland, and Tripura), plus the Himalayan state of Sikkim (Maps 10.1 and 10.2). The Northeast India (22°–30° N and 89°–97° E) spreads over 2,62,379 km^2 and represents the transition zone between the Indian, Indo-Malayan, and Indo-Chinese biogeographic regions and a meeting place of the Himalayan Mountains and Peninsular India.

Western Ghats of India runs 1600 km from just north of Mumbai and south through the states of Maharashtra, Goa, Karnataka, and Kerala and then extends little eastern side bordering the state of Tamil Nadu (Maps 10.1 and 10.3). This mountain range runs almost parallel to the western coast of Peninsular India and separates the Deccan Plateau from the narrow coastal plains. The major hill range in the north is referred as Sahyadri, and in the south it is known as Sahya Parvatam. The Nilgiris, an offshoot of the Western Ghats, is located northwest of the state of Tamil Nadu.

10.2.2 Compilation of Data

Data regarding the presence and distribution of bat species from states of Northeast and around Western Ghats of India is compiled from the available published literature (Corbet and Hill 1992; Bates and Harrison 1997; Nameer et al. 2001; Nameer 2008; Pradhan 2008; Molur et al. 2002; Wilson and Reeder 2005; Korad et al. 2007; Talmale and Pradhan 2009, Pradhan and Talmale 2012, 2013; Korad 2014).

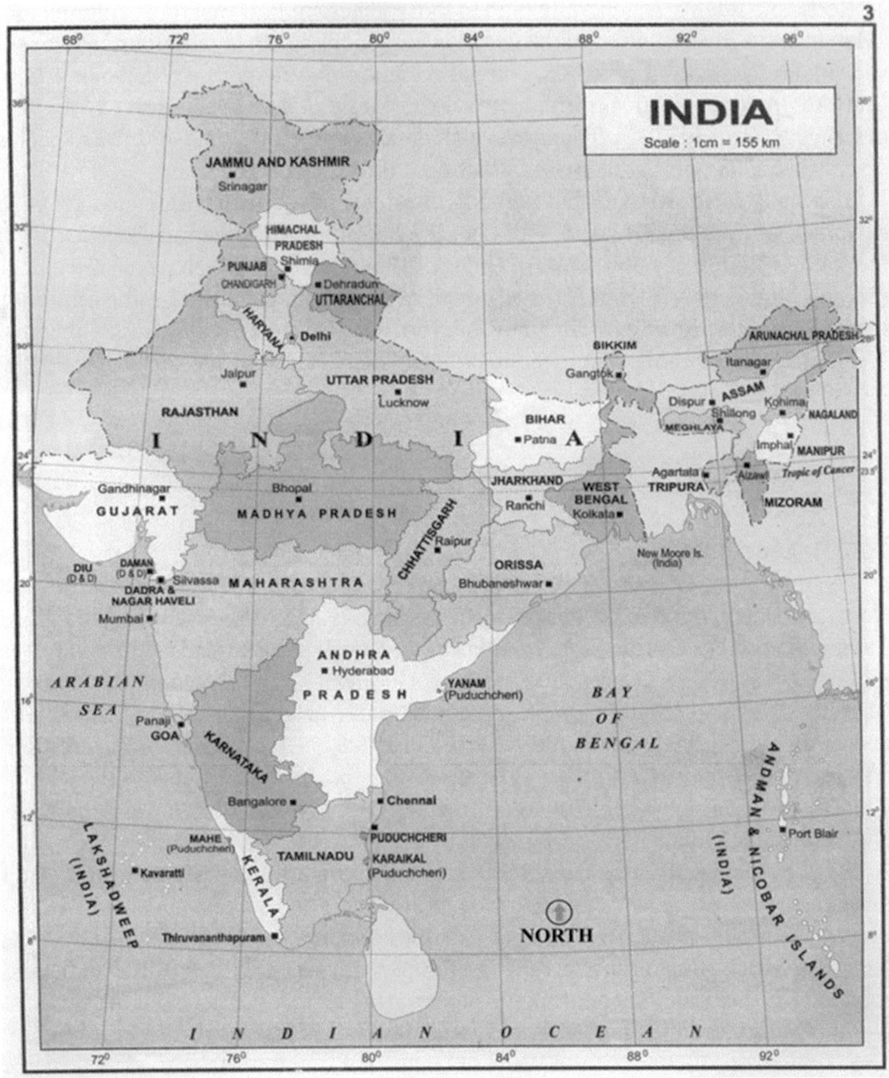

Map 10.1 Map of India showing states and general layout

10.2.3 Statistical Analysis of Data

The data is presented in tabular and graphical forms. The data is further analyzed by using Shannon-Weaver Index and compared for understanding of the species diversity richness and evenness in the two regions.

The bat species reported from previous literature are systemically listed in Table 10.1. This table includes three parts: (I) bat species common to both the regions (states of Northeast India and states around Western Ghats), (II) bat species reported additionally from states of Northeast India, and (III) additional list of bat

Map 10.2 States of Northeast India referred for the present study (Arunachal Pradesh, Assam, Nagaland, Manipur, Mizoram, Tripura, Meghalaya, and Sikkim)

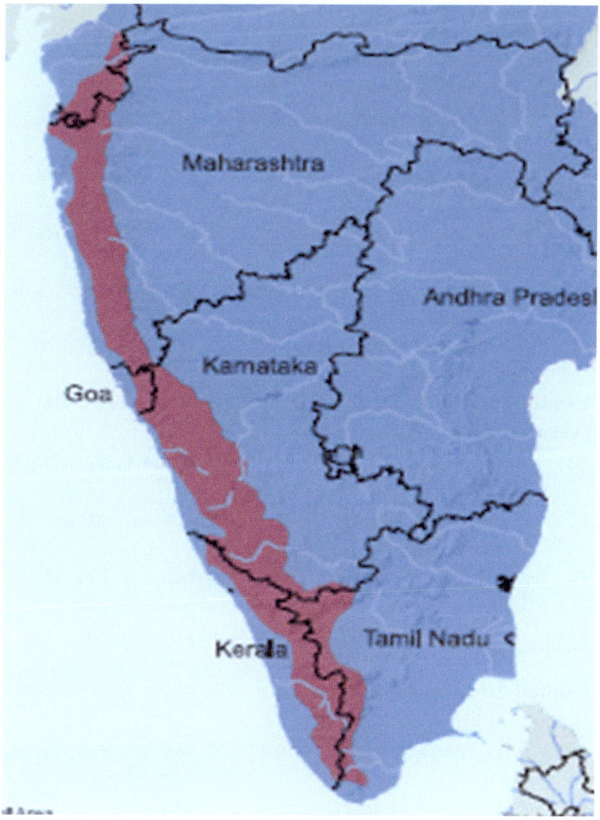

Map 10.3 States around Western Ghats of India (State of Maharashtra, Goa, Karnataka, Kerala, and Tamil Nadu)

Table 10.1 Bat species diversity reported from states of Northeast and states around Western Ghats of India

Sl. No.	Name of species	Presence in states of India	IUCN status in South Asia and India
I. Bat species common to states of Northeast and states around Western Ghats of India			
A. Suborder: Megachiroptera			
Family: Pteropodidae Gray, 1821 (Old World fruit bats)			
Genus: *Rousettus* Gray, 1821			
1	*Rousettus leschenaulti* (Desmarest, 1820)	All states	LC
Genus: *Pteropus* Brisson, 1762			
2	*Pteropus giganteus* Brunnich, 1782	All states	LC
Genus: *Cynopterus* Cuvier, F., 1824			
3	*Cynopterus sphinx* Vahl, 1797	All states except Sikkim	LC
Genus: *Eonycteris* Dobson, 1873			
4	*Eonycteris spelaea* (Dobson, 1871)	KS, TN, Asm, Sik, Megh, NE	LC
B. Suborder: Microchiroptera			
Family: Rhinolophidae Gray, 1825 (horseshoe bats)			
Genus: *Rhinolophus* Lacepede, 1799			
5	*Rhinolophus rouxii* Temminck, 1835	MS, Goa, KS, Ker, TN, Sik, Arn, NE	NT
6	*Rhinolophus sinicus* (Andersen, 1905)	KS, Ker, Sik, Arn, NE	LC
7	*Rhinolophus pusillus* Temminck, 1834	KS, Ker, Sik, Arn, NE	LC
8	*Rhinolophus lepidus* Blyth, 1844	MS, KS, Ker, TN, Asm, Megh	LC
9	*Rhinolophus luctus* Temminck, 1835	MS, Asm, Sik, NE	NT
Family: Hipposideridae Lydekker, 1891 (leaf-nosed bats)			
Genus: *Hipposideros* Gray, 1831			
10	*Hipposideros ater* Templeton, 1848	MS, KS, Ker, TN, Megh	LC
11	*Hipposideros cineraceus* Blyth 1853	TN, Arn, Megh	NT
12	*Hipposideros pomona* Andersen, 1918	MS, KS, Ker, TN, Asm, Sik, Arn, NE	LC
13	*Hipposideros lankadiva* Kelaart, 1850	MS, KS, NE	LC#
Family: Megadermatidae H. Allen, 1864 (false vampire bats)			
Genus: *Megaderma* E. Geoffroy, 1810			
14	*Megaderma lyra* E. Geoffroy, 1810	MS, KS, Ker, TN, Asm, Arn, Megh	LC
15	*Megaderma spasma* Linneus, 1758	MS, Goa, KS, Ker, TN, Asm, NE	LC

(continued)

Table 10.1 (continued)

Sl. No.	Name of species	Presence in states of India	IUCN status in South Asia and India
Family: Emballonuridae Gervais, 1855 (sheath-tailed bats)			
Subfamily Taphozoinae Jerdon, 1867			
Genus: *Saccolaimus* Temminck, 1838			
16	*Saccolaimus saccolaimus* Temminck, 1838	MS, KS, Ker, TN, Megh	LC
Genus: *Taphozous* E. Geoffroy, 1818			
17	*Taphozous longimanus* Hardwicke,1825	MS, KS, Ker, TN, NE	LC
18	*Taphozous nudiventris* Cretzschmer, 1830	MS, KS, TN, Sik	LC
Family: Molossidae Gervais, 1856 (free-tailed bats)			
Subfamily Molossinae Gervais, 1856			
Genus: *Chaerephon* Dobson, 1874			
19	*Chaerephon plicata* (Buchana, 1800)	MS, Goa, TN, Megh	LC
Genus: *Otomops* Thomas, 1913			
20	*Otomops wroughtoni* (Thomas, 1913)	KS, Megh	CR
Family: Vespertilionidae Gray, 1821 (evening bats)			
Subfamily Vespertilioninae Gray, 1821			
Tribe Nycticeiini Gervais, 1855			
Genus: *Scotophilus* Leach, 1821			
21	*Scotophilus heathii* Horsfield, 1831	MS, KS, Ker, TN, Asm, Megh	LC
22	*Scotophilus kuhlii* Leach, 1821	MS, KS, Ker, TN, NE	LC
Tribe Pipistrellini Tate, 1942			
Genus: *Pipistrellus* Kaup, 1829			
23	*Pipistrellus pipistrellus* (Schreber, 1774)	MS, TN, Asm	LC
24	*Pipistrellus javanicus* (Gray, 1838)	MS, Asm, Sik, Arn, NE	LC
25	*Pipistrellus coromandra* (Gray, 1838)	MS, Goa, KS, TN, Asm, Sik, Arn, Megh, NE	LC
26	*Pipistrellus tenuis* (Temminck, 1840)	MS, KS, Ker, TN, all states of NE except Arn	LC
27	*Pipistrellus kuhlii* (Kuhl, 1819)	MS, KS, Ker, TN, Asm, Megh	LC
Tribe Vespertilionini Gray, 1821			
Genus: *Hypsugo* Kolenati, 1856			
28	*Hypsugo savii* (Bonaparte, 1837	MS, Megh	VU
Genus: *Tylonycteris* Peters, 1872			
29	*Tylonycteris pachypus* (Temminck, 1840)	KS, Ker, Sik, Megh, NE	NT

(continued)

Table 10.1 (continued)

Sl. No.	Name of species	Presence in states of India	IUCN status in South Asia and India
Subfamily Myotinae Tate, 1942			
Genus: *Myotis* Kaup, 1829			
30	*Myotis formosus* (Hodgson, 1835)	MS, Asm, Sik, Megh	LC
Genus: *Harpiocephalus* Gray, 1842			
31	*Harpiocephalus harpia* (Temminck, 1840)	Ker, TN, Asm, Sik, Megh, NE	NT
Genus: *Murina* Gray, 1842			
32	*Murina cyclotis* Dobson, 1872	TN, Sik, Megh	LC
Subfamily Kerivoulinae Miller, 1907			
Genus: *Kerivoula* Gray, 1842			
33	*Kerivoula picta* (Pallas, 1767)	MS, Goa, KS, TN, Asm, Sik	LC
34	*Kerivoula hardwickii* (Horsefield, 1824)	KS, Asm, Megh, NE	LC
II. Additional list of bat species reported from states of North-East of India			
A. Suborder: Megachiroptera			
Family: Pteropodidae Gray, 1821 (Old World fruit bats)			
Genus: *Megaerops* Peters, 1865			
35	*Megaerops niphanae* Yenbutra & Felten, 1983	Arn	NT
Genus: *Sphaerias* Miller, 1906			
36	*Sphaerias blanfordi* (Thomas, 1891)	Arn	NT
Genus: *Macroglossus* F. Cuvier, 1824			
37	*Macroglossus sobrinus* (K. Andersen, 1911)	Sik, Arn, Megh, NE	NT
B. Suborder: Microchiroptera			
Family: Rhinolophidae Gray, 1825 (horseshoe bats)			
Genus: *Rhinolophus* Lacepede, 1799			
38	*Rhinolophus ferrumequinum* (Schreber, 1774)	Sik, Arn, NE	VU#
39	*Rhinolophus affinis* Horsfield,1823	Arn, NE	LC
40	*Rhinolophus shortridgei* K. Andersen, 1918	NE	NE
41	*Rhinolophus subbadius* Blyth, 1844	Arn, Megh	VU
42	*Rhinolophus macrotis* Blyth, 1844	Arn, Megh	NT
43	*Rhinolophus pearsonii* Horsfield, 1851	Sik, Megh	LC
44	*Rhinolophus yunanensis* Dobson, 1872	Arn, NE	VU

(continued)

Table 10.1 (continued)

Sl. No.	Name of species	Presence in states of India	IUCN status in South Asia and India
Family: Hipposideridae Lydekker, 1891 (leaf-nosed bats)			
Genus: *Coelops* Blyth, 1848			
45	*Coelops frithi* Blyth, 1848	NE	NT
Genus: *Hipposideros* Gray, 1831			
46	*Hipposideros armiger* Hodgson, 1835	Sik, NE	LC
Family: Vespertilionidae Gray, 1821 (evening bats)			
Subfamily Vespertilioninae Gray, 1821			
Tribe Eptesicini Volleth and Heller, 1994			
Genus: *Arielulus* Hill and Harrison, 1987			
47	*Arielulus circumdatus* (Temminck, 1840)	Megh	LC
Genus: *Eptesicus* Rafinesque, 1820			
48	*Eptesicus serotinus* (Schreber, 1774)	Asm, NE	NT
49	*Eptesicus pachyotis* (Dobson, 1871)	Megh	DD
Tribe Nycticeiini Gervais, 1855			
Genus: *Scotomanes* Dobson, 1875			
50	*Scotomanes ornatus* (Blyth, 1851)	Sik, Arn, Megh, NE	LC
Tribe Pipistrellini Tate, 1942			
Genus: *Nyctalus* Bowditch, 1825			
51	*Nyctalus noctula* (Schreber, 1774)	Sik, NE	LC
Genus: *Pipistrellus* Kaup, 1829			
52	*Pipistrellus paterculus* Thomas, 1915	Asm, NE	LC
53	*Pipistrellus abramus* (Temminck, 1838)	Arn	DD
Tribe Plecotini Gray, 1866			
Genus: *Barbastella* Gray, 1821			
54	*Barbastella leucomelas* (Cretzschmar, 1830/31)	Sik, Megh	NT
Tribe Vespertilionini Gray, 1821			
Genus: *Ia* Thomas, 1902			
55	*Ia io* Thomas, 1902	Megh	EN
Genus: *Tylonycteris* Peters, 1872			
56	*Tylonycteris robustula* Thomas, 1915	NE	NE
Subfamily Myotinae Tate, 1942			
Genus: *Myotis* Kaup, 1829			
57	*Myotis sicarius* Thomas, 1915	Sik	VU#
58	*Myotis mystacinus* (Kuhl, 1819)	Sik, Megh	VU

(continued)

Table 10.1 (continued)

Sl. No.	Name of species	Presence in states of India	IUCN status in South Asia and India
59	*Myotis muricola* (Gray, 1846)	Asm, Sik, Megh	LC
60	*Myotis siligorensis* (Horsfield, 1855)	Sik, Megh	NT
61	*Myotis annectans (Dobson, 1871*)	NE	VU
62	*Myotis longipes (Dobson, 1873*)	Megh	NT
63	*Myotis laniger Peters, 1871*	Megh	EN
Genus: *Miniopterus* Bonaparte, 1837			
64	*Miniopterus magnater* Sanborn, 1931	NE	NE
Genus: *Harpiocephalus* Gray, 1842			
65	*Harpiocephalus mordax* Thomas,1923	NE	DD
Genus: *Murina* Gray, 1842			
66	*Murina aurata* Milne-Edwards, 1872	Sik, Megh	NT
67	*Murina tubinaris* (Scully, 1881)	Sik, Arn, Megh, NE	NT
68	*Murina huttonii* (Peters, 1872)	Asm	LC
III. Additional list of bat species reported from states around Western Ghats of India			
Suborder: Megachiroptera			
Family: Pteropodidae Gray, 1821 (Old World fruit bats)			
Genus: *Cynopterus* Cuvier, F., 1824			
69	*Cynopterus brachiotis* (Muller, 1838)	MS, Goa, KS, TN	LC
Genus: *Latidens* Thonglongya, 1972			
70	*Latidens salimalii* Thonglongya,1972	TN	EN#
Suborder: Microchiroptera			
Family: Rhinolophidae Gray, 1825 (horseshoe bats)			
Genus: *Rhinolophus* Lacepede, 1799			
71	*Rhinolophus cognatus* Andesen, 1906	MS	VU#
72	*Rhinolophus beddomei* Andersen, 1905	MS, KS, Ker, TN	NT#
Family: Hipposideridae Lydekker, 1891 (leaf-nosed bats)			
Genus: *Hipposideros* Gray, 1831			
73	*Hipposideros fulvus* Gray, 1838	MS, KS, Ker, TN	LC
74	*Hipposideros hypophyllus* Kock & Bhat, 1994	KS	EN#
75	*Hipposideros galeritus* Cantor, 1846	MS, KS	NT
76	*Hipposideros speoris* (Schneider, 1800)	MS, KS, Ker, TN	LC#

(continued)

Table 10.1 (continued)

Sl. No.	Name of species	Presence in states of India	IUCN status in South Asia and India
Family: Rhinopomatidae Bonaparte, 1838 (mouse-tailed bats)			
Genus: *Rhinopoma* E. Geoffroy, 1818			
77	*Rhinopoma microphyllum* (Brinich, 1782)	MS, TN	LC
78	*Rhinopoma hardwickii* Gray, 1831	MS, KS, TN	LC
79	*Rhinopoma muscatellum* Thomas, 1903	TN	NT
Family: Emballonuridae Gervais, 1855 (sheath-tailed bats)			
Subfamily Taphozoinae Jerdon, 1867			
Genus: *Taphozous* E. Geoffroy, 1818			
80	*Taphozous melanopogon* Temminck, 1841	MS, Goa, KS, Ker, TN	LC
81	*Taphozous perforatus* E. Geoffroy, 1818	MS	LC#
82	*Taphozous theobaldi* Dobson, 1872	KS	VU
Family: Molossidae Gervais, 1856 (free-tailed bats)			
Subfamily Molossinae Gervais, 1856			
Genus: *Tadarida* Rafinesque, 1814			
83	*Tadarida aegyptiaca* (E. Geoffroy, 1818)	MS, KS, Ker, TN	LC
Family: Vespertilionidae Gray, 1821 (evening bats)			
Subfamily Vespertilioninae Gray, 1821			
Tribe Eptesicini Volleth and Heller, 1994			
Genus: *Hesperoptenus* Peters, 1868			
84	*Hesperoptenus tickelli* (Blyth, 1851)	MS, Goa, KS, TN	LC
Tribe Pipistrellini Tate, 1942			
Genus: *Pipistrellus* Kaup, 1829			
85	*Pipistrellus ceylonicus* (Kelaart, 1852)	MS, KS, Ker, TN	LC
Genus: *Scotozous* Dobson, 1875			
86	*Scotozous dormeri* (Dobson, 1875)	MS, Goa, KS, Ker, TN	LC#
Tribe Vespertilionini Gray, 1821			
Genus: *Falsistrellus* Troughton, 1943			
87	*Falsistrellus affinis* Dobson, 1871	MS, Ker, TN	NT
Subfamily Myotinae Tate, 1942			
Genus: *Myotis* Kaup, 1829			
88	*Myotis montivagus* (Dobson, 1874)	MS, KS, Ker, TN	VU
89	*Myotis horsfieldii* (Temminck, 1840)	MS, KS, Ker, TN	LC

(continued)

Table 10.1 (continued)

Sl. No.	Name of species	Presence in states of India	IUCN status in South Asia and India
Genus: *Miniopterus* Bonaparte, 1837			
90	*Miniopterus schreibersii* (Kuhl, 1819)	MS, TN	LC
91	*Miniopterus pusillus* Dobson, 1876	KS, TN	VU
Subfamily Kerivoulinae Miller, 1907			
Genus: *Kerivoula* Gray, 1842			
92	*Kerivoula lenis* Thomas, 1916	TN	NT

Name of states: *Arn* Arunachal Pradesh, *Asm* Assam, *Goa*, *Ker* Kerala, *KS* Karnataka State, *Megh* Meghalaya, *MS* Maharashtra State, *NE* states of Northeast India, *TN* Tamil Nadu
IUCN status: *LC* least concern, *NT* near threatened, *VU* vulnerable, *EN* endangered, *CR* critically endangered, *DD* data deficient, *NE* not evaluated, # endemic to South Asia and India

Table 10.2 Distribution of chiropteran fauna in Northeast (NE) and Western Ghats (WG) regions of India

Sl. No.	Family	No. of genera reported	No. of species common to two regions	No. of species reported from NE region	No. of species reported from WG region
1	Pteropodidae	8	4	7	6
2	Rhinolophidae	1	5	12	7
3	Hipposideridae	2	4	6	8
4	Megadermatidae	1	2	2	2
5	Rhinopomatidae	1	0	0	3
6	Emballonuridae	2	3	3	6
7	Molossidae	3	2	2	3
8	Vespertilionidae	18	14	36	23
	Total	**36**	**34**	**68**	**58**

species reported from states around Western Ghats. This table also mentions the distribution of bat species in the study area and its IUCN status. The distribution of bat species in these two regions is summarized in Table 10.2. The number of species per bat family reported from study area is presented in Fig. 10.1. IUCN status of bat species reported from states of Northeast and states around Western Ghats of India is summarized graphically in Fig. 10.2. IUCN status of bat species reported from these two regions is reported separately in Figs. 10.3 and 10.4 for comparison. Table 10.3 mentions the endemic bat species reported from states of Northeast India, and those from states around Western Ghats is mentioned in Table 10.4. For evaluation of the bat species diversity richness, and evenness in the two regions,; the data is analyzed by Shannon-Weaver Index. The results are presented in Fig. 10.5.

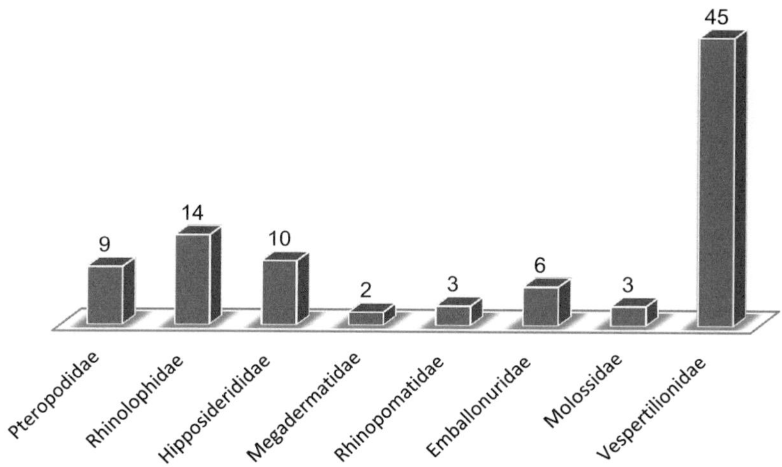

Fig. 10.1 Bat species reported from Northeast and Western Ghats regions of India

Fig. 10.2 IUCN status of bat species reported from states of Northeast and around Western Ghats of India

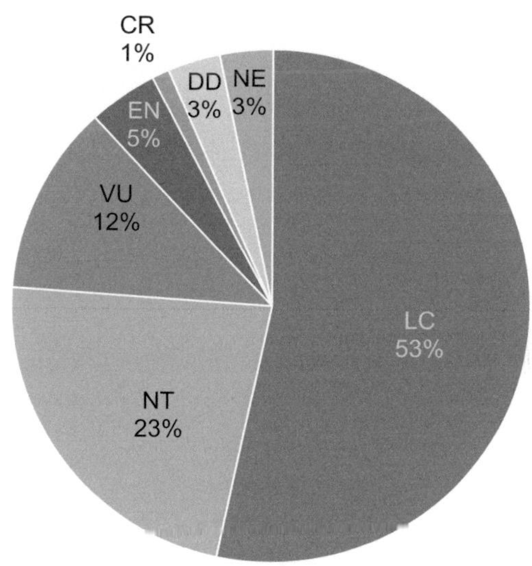

Fig. 10.3 IUCN status of bat species reported from states of Northeast India

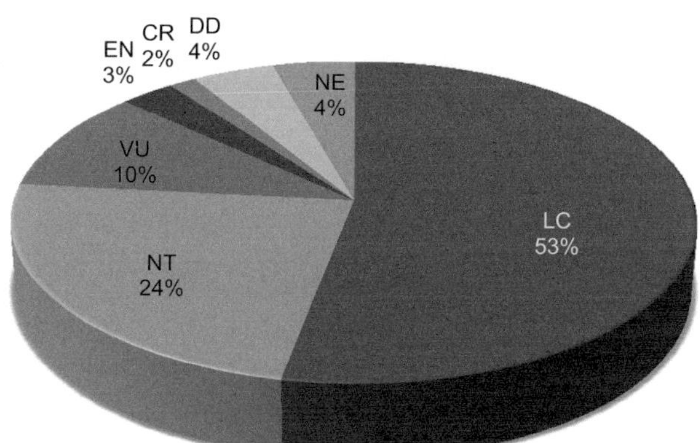

Western Ghats can be seen as a belt running from the north (Maharashtra) to the south (on the border of Kerala and Tamil Nadu) on the western border of the Peninsular India.

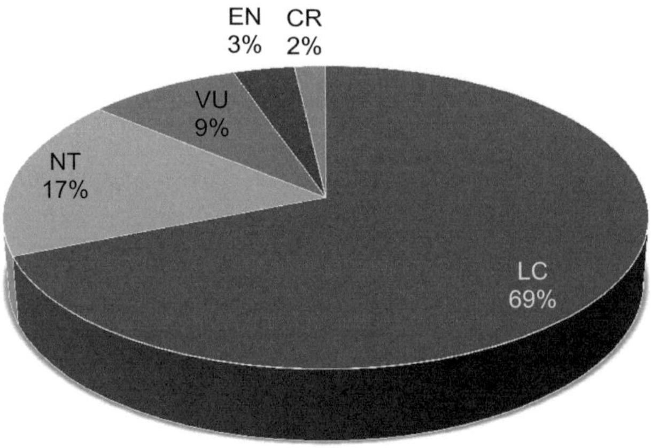

Fig. 10.4 IUCN status of bat species reported from states around Western Ghats of India. *LC* least concern, *NT* near threatened *VU* vulnerable, *EN* endangered, *CR* critically endangered, *DD* data deficient, *NE* not evaluated, # endemic to South Asia and India

Table 10.3 Diversity and IUCN status of endemic (to South Asia) bat species of Northeast India

Sl. No.	Bat species	IUCN status
1	*Rhinolophus ferrumequinum* (Schreber,1774)	VU
2	*Hipposideros lankadiva* Kelaart, 1850	LC
3	*Myotis sicarius* Thomas, 1915	VU

LC least concern, *VU* vulnerable

Table 10.4 Endemic bat species reported from states around Western Ghats of India

Sl. No.	Bat species	IUCN status
1	*Latidens salimalii* Thonglongya, 1972	EN
2	*Taphozous perforatus* Geoffroy, 1818	LC
3	*Hipposideros speoris* (Schneider, 1800)	LC
4	*Hipposideros hypophyllus* Kock & Bhatt, 1994	EN
5	*Hipposideros lankadiva* Kelaart, 1850	LC
6	*Scotozous dormeri* (Dobson, 1875)	LC
7	*Rhinolophus beddomei* Andesen, 1905	NT
8	*Rhinolophus cognatus* Andersen, 1906	VU

IUCN status: *LC* least concern, *NT* near threatened, *VU* vulnerable, *EN* endangered

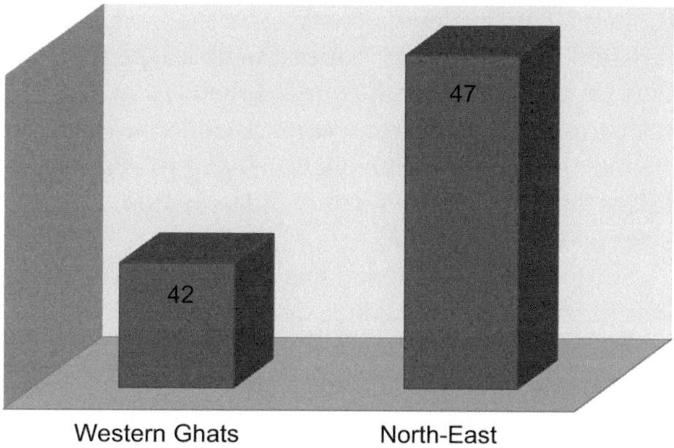

Fig. 10.5 Comparison of bat species diversity by Shannon-Weaver Index in states of Northeast and around Western Ghats of India

10.3 Results and Discussion

Northeast India is at the confluence of Indo-Malayan, Indo-Chinese, and Indian biogeographical realms. Northeast India forms one of the major regions, particularly its tropical rain forests. The tropical semievergreen and moist deciduous forests in the lowlands of this region extend south and west into the subcontinent. The hilly region is known for the presence of deep crevices and caves, which serve as the natural abode for many species of wild fauna of the region. The area is one of the richest in biological values, high in endemism and holds a large number of rare species that are now under serious threat. On the other hand, the biome of Western Ghats is formed of tropical and semitropical moist broad-leaved forests. The Nilgiris in the south of Western Ghats is known for dense grassy table lands, dense evergreen forests, and Sholas. Similar Sholas are also present in the area of Anaimalai Hills, Palni Hills, Kudremukh, and other south Indian ranges. They provide the main shelter to wild elephants, gaur, and other large animals. The forests of the Western Ghats and the south Indian hill ranges have a richer fauna than the remaining areas of the peninsular region and known as home to numerous endemic species.

Thus for the study of species diversity of bat fauna, states of Northeast India and states around Western Ghats are found to be the most appropriate regions as both these regions are globally recognized biological hot spots. Comparing the area occupied by these two regions, the states around Western Ghats occupy about 2.4 times bigger area (633,264 km^2) than that of states of Northeast India (262,379 km^2). The bat fauna is the largest aggregation of mammals in India and is represented by 8 families, 39 genera, 117 species, and 100 subspecies (Wilson and Reeder 2005; Talmale and Pradhan 2009).

The suborder Megachiroptera of the order Chiroptera is represented by the sole family Pteropodidae (Old World fruit bats) including eight genera and nine species. The suborder Microchiroptera is represented by 7 families, 28 genera, and 83 species. Thus total 36 genera and 92 species of the order Chiroptera reported from states of Northeast and states around Western Ghats of India comprise about 79% of the total bat species diversity reported from the country. Of the 92 bat species reported from states of Northeast India and states around Western Ghats, 34 bat species are common to both regions (Table 10.2). The states of Northeast India harbor 68 bat species, and 58 bat species are reported from the states around Western Ghats. As mentioned before, though the area of Northeast region of India is about 2.4 times lesser than that of the Western Ghats region, this region seems to be richer in bat species diversity.

Secondly it is also found that 50% of bat species diversity reported from states of Northeast India is common to those of Western Ghats regions, while states of Western Ghats share about 59% similarity regarding bat species with those of states of Northeast region. Further it is found that of 36 genera were reported from the two regions; the presence of about ten genera, i.e., *Megaerops* Peters 1865, *Sphaerias* Miller 1906, *Macroglossus* F. Cuvier 1824 (all belong to the family Pteropodidae of the suborder Megachiroptera), *Coelops* Blyth 1848 (belongs to the suborder Microchiroptera and the family Hipposideridae), *Barbastella* Gray 1821, *Scotomanes* Dobson 1875, *Eptesicus* Rafinesque 1820, *Ia* Thomas 1902, *Arielulus* Hill and Harrison 1987, and *Nyctalus* Bowditch 1825 (all belong to the family Vespertilionidae of the suborder Microchiroptera), is the characteristic of the Northeast region of India.

As mentioned in Fig. 10.1, the two regions of India are dominated by the species of the family Vespertilionidae, which is represented by 45 bat species, which comprises about 49% of bat species reported from the two regions. These hilly areas among forests are natural homes for the evening bats, which prefer mostly the crevices and narrow gaps in the rocks. The second large aggregation is of the family Rhinolophidae (a monogeneric group as mentioned by Corbet and Hill 1992) and is represented by 14 species. These species are known to depend exclusively on forest for roosting and foraging. The third large group is of Old World fruit bats of the family Pteropodidae in these regions and is represented by eight genera and nine species (Table 10.2, Fig. 10.1), which depend on wild plants of the forest.

As per the IUCN status (Fig. 10.2), about 53% of bat speces reported from the regions of Northeast India and Western Ghats are categorized as *least concern*, and 23% of bat species is considered as *near threatened*. However, 18% of bat species are threatened in one way or other (12% as *vulnerable*, 5% as *endangered*, and 1% *as critically endangered*), while 3% of bat species are reported as *data deficient* and 3% as *not evaluated*.

In states of Northeast India, the percentage of nonthreatened categories of bat species is almost similar (bat species of LC category-53%, and NT category-24%). Among the threatened taxa, 10% bat species come under vulnerable category; 3% bat species is categorized as endangered and 2% under critically endangered (Fig. 10.3). Three bat species, namely, *Rhinolophus shortridgei* Andersen, *Tylonycteris robustula* Thomas, and *Miniopterus magnater* Sanborn, reported from states of Northeast India are categorized as not evaluated and three bat species, viz., *Eptesicus pachyotis* (Dobson), *Pipistrellus abramus* (Temminck), and *Harpiocephalus mordax* Temminck, as data deficient. Thus about 8% of bat species reported from states of Northeast India need special attention and intensive field survey pertaining to the distribution and roosting habits of these species.

The bat species diversity in the states around Western Ghats seems to be in better condition, as 69% of bat fauna belong to least concern (Fig. 10.4) and 17% bat species is reported to be near threatened. Remaining 14% of bat species reported from this region are threatened (9% bat species is considered as vulnerable, 3% as endangered, and 2% as critically endangered). The characteristic of the bat fauna of this region is that none of the bat species is data deficient or comes under the category not evaluated. This indicates that the bat fauna in the states around Western Ghats region is relatively well surveyed.

It is further observed that the diversity of Old World fruit bats is remarkable in the states of Northeast India, representing seven species under seven genera. But of these, three species, *Megaerops niphanae* Yenbutra & Felten, *Sphaerias blanfordi* (Thomas), and *Macroglossus sobrinus* (Andersen), are categorized as near threatened. This might be due to uncontrolled disturbing anthropogenic activities in the natural forests including deforestation.

In the Western Ghats region, the Old World fruit bats are represented by six species. Among these species, Salim Ali's fruit bat (*Latidens salimalii* Thonglongya) is endemic to South Asia and categorized as endangered, due mainly to the restricted extent and area as well as continuing decline in the quality of habitat.

Wroughton's free-tailed bat *Otomops wroughtoni* (Thomas 1913) is reported from Meghalaya state of Northeast as well as from Karnataka state of Western India. This is the only bat species of these two regions, which is categorized as critically endangered, due to restricted extent and continuing decline in the quality of habitat.

Further, the insectivorous bat of the genus *Rhinolophus* is represented by 12 species from states of Northeast India. This number of species is remarkably high, and all are truly forest species, indicating high sustainability of the forests of this region. Similarly the species diversity of water bats of the genus *Myotis* is also high (eight species) in this region and needs special attention to conserve natural water bodies. The species diversity of the family Vespertilionidae in the hilly area of states of Northeast India is remarkable, represented by 36 species, comprising about 56% of total bat species diversity of this region.

As far as endemism of bat fauna is concerned, only three species (Table 10.3) reported from states of Northeast India (*Rhinolophus ferrumequinum* (Schreber), *Hipposideros lankadiva* Kelaart, and *Myotis sicarius* Thomas) are endemic to South Asia. *Rhinolophus ferrumequinum* and *Myotis sicarius* are categorized as vulnerable endemic species to South Asia, but *Myotis sicarius* is categorized as endemic endangered species in India, while *Hipposideros lankadiva*, the species common to both regions, is the endemic species in South Asia and categorized as least concern.

The endemism of bat fauna is far greater in states of Western Ghats (Table 10.4) than that from states of Northeast India. The states around Western Ghats harbor eight bat species, which are endemic to South Asia. Of these four species belong to least concern category, and one species (*Rhinolophus beddomei*) is categorized as near threatened. Thus about 63% of endemic species reported from states around Western Ghats belong to nonthreatened categories. Among the threatened endemic taxa of this region, *Rhinolophus cognatus* is categorized as vulnerable;, and *Latidens salimalii* and *Hipposideros hypophyllus* are categorized as endangered due to their restricted extent and continual decline in habitats

The bat species diversity richness and evenness are evaluated using Shannon-Weaver Index. The values are used for comparison of the bat fauna in two regions. The results indicate that the value of index is greater for the states of Northeast India than that of states around Western Ghats. This is probably due to the greater number of bat species reported from comparatively smaller area of states of Northeast India. This result further tempts to suggest that the states of Northeast India are to be surveyed systematically and conserved for the well-being of the wild fauna of this region, which is peculiar due mainly to the exclusiveness of bat genera and topography, climatic conditions and vegetation of the region as well.

The Western Ghats region is relatively well surveyed for bat fauna. While surveying the forest area of Western Ghats it is found that the rate of decline in habitats and foraging grounds of the fauna is reported to be high. Decline in atmospheric relative humidity due to deforestation in this region seems to be additional restriction factor for the bat fauna. For the conservation of the wild fauna of bats, inclusion of native varieties of broad-leaved plants under the schemes of afforestation is essential. These plants not only provide shelter and food to the wild fauna of the forest but also help to maintain the atmospheric relative humidity in the area, which promotes life to both the flora and fauna of the forests.

Acknowledgments The author is thankful to the Principal of Fergusson College, Pune, and Head of Department of Zoology, Fergusson College, Pune, for providing facilities and encouragement.

References

Bates PJJ, Harrison DL (1997) Bats of the Indian subcontinent. Harrison Zoological Museum Publication, England

Corbet GB, Hill JE (1992). The mammals of the Indomalayan region. Nat Hist Mus/OUP, Oxford, 488 p

Davison GWH, Zubaid A (1992) Food habits of the lesser false vampire, *Megaderma spasma,* from Kuala Lompat, Peninsular Malaysia. Zeitschrift Saugetierk 57(5):310–332

Eckrich M, Neuweiler G (1988) Food habits of the sympatric insectivorous bat Rhinolophus rouxi and Hipposideros lankadiva from Sri Lanka. J Zool (Lond) 215:729–737

Fleming TH, Breitwisch R, Whitesides GH (1987) Patterns of tropical frugivore diversity. A Rev Ecol Syst 18:91–109

Fujita MS, Tuttle MD (1991) Flying foxes (Chiroptera: Pteropodidae): threatened animals of key ecological and economic importance. Consev Biol 5(4):455–463

Hill JE, Smith JD (1984). Bats. In: A natural history British Museum (Natural History), 243 p

Korad VS (2014) Studies on diversity, distribution and conservation of the bat fauna in Maharashtra state. India Taprobanica 06(1):32–45. ISSN 1800–427X

Korad VS, Yardi KD, Raut RR (2007) Diversity and distribution of bats in Western Ghats. Zoos Print 22(7):2752–2758

Kunz TH (ed) (1988) Ecology and behavioral methods for the study of bats. Smithsonian Institute Press, Washington, DC, p 533

Marshall AG (1983) Bats, flowers and fruits: evolutionary relationships in the Old World. Biol J Linn Soc 20:115–135

Marshall AG (1985) Old World phytophagus bats (Megachiroptera) and their food plants: a survey. Zool J Linnean Soc 83:351–369

Mickleburgh SP, Hutson AM, Racey PA (2002) A review of the global conservation status of bats. Oryx 6(1):18–34

Molur S, Marimuthu G, Srinivasulu C, Mistry S, Hutson AM, Bates PJJ, Walker S, Padma Priya K and AR Binu Priya (2002). Status of South Asian Chiroptera: conservation assessment and management plan (C.A.M.P) workshop report. Zoo Outreach Organisation, CBSG South Asia & WILD, Coimbatore, Viii + 141 pp + CD- ROM

Myers NR, Mittermeier A, Mittermeier CG, da Fonesca GAB, Kent J (2000) Biodiversity hotspots for conservation priorities. Nature 403:853–858

Nameer PO (2008) A note on a checklist of Indian mammals, revised and updated, 2008. Zoos' Print XXIII(8):1–12

Nameer PO, Molur S, Walker S (2001) Mammals of Western Ghats: a simplistic overview. Zoo's Print J 16:629–639

Pierson ED, Rainey WE (1992). The biology of flying foxes of the genus Pteropus; a review. In: Wilson DE, Grahm GN (eds) Pacific Island flying foxes: proceedings of an international conservation conference, USFWS, Biol. Rept. No. 90(23). Washington, DC, pp 1–17

Pradhan MS (2008). Mammalia. In: Fauna of Goa, State Fauna Series 16: 431–496. (Publ.: Director, Zool. Surv. India, Kolkata)

Pradhan MS, Talmale SS (2012). Mammalia. In: Fauna of Maharashtra, State Fauna Series, Zoological Survey of India, Kolkata, 20:21–145

Pradhan MS, Talmale SS (2013). Mammalia. In: Fauna of Karnataka, State Fauna Series, 21:495–595. (Publ.: Director, Zool. Surv. India,Kolkata)

Ronald M and M Nowak (1994). Walker's bats of the world. Pub. The Johns Hopkins University Press. Baltimore, 287pp

Swift SM, Racey PA, Avery MI (1985) Feeding ecology of *Pipistrellus Pipistrellus* (Chiroptera: Vespertilionidae) during pregnancy and lactation. II. Diet J Anim Ecol 54:217–225

Talmale SS, Pradhan MS (2009). A checklist of valid Indian bat species (Chiroptera: Mammalia). Zoological Survey of India, 17 p

Whitekar JO Jr (1993) Bats, beetles and bugs. Bats 11:23

Whitekar JO (1995) Food availability and opportunistic versus selective feeding in insectivorous bats. Bats Res News 35:75–77

Whitekar JO, Master C, Kaller LE (1997) Food habits of bats of Western Oregon North West. Science 51:46–55

Whitekar JO Jr, Sudhakar Issac S, Marimuthu G, Kunz TH (1999) Seasonal variation in the diet of the Indian pygmy bat *Pipistrellus mimus* in Southern India. J Mammal 80(1):60–70

Wilson DE, Reeder DM (eds) (2005) Mammal species of the world, vol 2, 3rd edn. Johns Hopkins University Press, Baltimore, 2141 pp

Status and Distribution of Grizzled Giant Squirrel in Chinnar Wildlife Sanctuary, Kerala, India

11

A. Veeramani, M. Balasubramanian, Sanjayankumar, and John Mathew

Abstract

Grizzled giant squirrel is placed in Schedule I of Indian Wild Life (Protection) Act, 1972, and categorized as "Near Threatened" in 2008 IUCN Red List of Threatened Species. In India, the grizzled giant squirrels are distributed in the Western and Eastern Ghats. The status of their Indian population is vulnerable due to drastic habitat loss, clear-felling, logging, construction of dams, hunting for local consumption, and expansion of agro-industry construction. A survey was conducted in Chinnar Wildlife Sanctuary (CWS) to estimate the population of grizzled giant squirrels, and 34 individuals were recorded within – 106.8 km. The length of the transects varied from 1.3 to 3.7 km. The present study analysis shows an overall density of 7.75 individuals per square kilometer, with standard error of 2.49. The total number of population in the study area was calculated by multiplying density to the total area (34.46 km^2) and obtained 267 individuals/km^2. The analysis based on the low AIC value 47.747 and chi-square P-value 0.51246. The density of the grizzled giant squirrel's nests shows 68.99/km^2 with the standard error of 19.55. A total of 12 tree species were used for nesting by grizzled squirrel, with a height variation of 2.5–35 m. Suitable conservation management suggestions were recommended.

Keywords

Chinnar · Density · Grizzled giant squirrel · *Ratufa* · Status

A. Veeramani (✉)
PG and Research Department of Zoology, Government Arts College (Autonomous), Kumbakonam 612002, Tamil Nadu, India
e-mail: wildveera@gmail.com

M. Balasubramanian · Sanjayankumar · J. Mathew
Periyar Tiger Reserve, Thekkady, Idukki 685536, Kerala, India

C. Sivaperuman, K. Venkataraman (eds.), *Indian Hotspots*,
https://doi.org/10.1007/978-981-10-6605-4_11

11.1 Introduction

Giant squirrels belong to the genus *Ratufa* and are confined only to the Asiatic region. This genus is characterized by three species *Ratufa bicolor* (black or Malayan giant squirrel), *Ratufa macroura* (grizzled giant squirrel), and *R. indica* (Indian or Malabar giant squirrel) (Ellerman 1961; Srinivasulu et al. 2004). There are only two species of giant squirrels found in India which are Malabar giant squirrel (*Ratufa indica*) and grizzled giant squirrel (*Ratufa macroura*). The distributional ranges of giant squirrels vary from evergreen forest to riverine forests. However, its distribution was confined only to forests with tall trees (Ramachandran 1989; Kumara and Singh 2006). There are three subspecies of grizzled giant squirrels, which are found in Sri Lanka, namely, *Ratufa macroura macroura*, *Ratufa macroura dandolena*, and *Ratufa macroura melanochra*. The grizzled giant squirrel (*R. macroura*) is the smallest giant tree squirrel, generally endemic in South Asia, and it is restricted to the forests of Srivilliputhur, Tamil Nadu; Cauvery Valley Karnataka; Chinnar Wildlife Sanctuary, Kerala; and Sri Lanka (Ramachandran 1993; Nowak 1991; Senthilkumar et al. 2007; Vijayakumaran Nair et al. (1997)). This animal has been listed in Schedule I of Indian Wild Life (Protection) Act, 1972, and categorized as "Near Threatened" in 2008 IUCN Red List of Threatened Species.

This species mostly inhabits on high trees in dry deciduous and moist evergreen forests and is rarely coming to the ground. It is diurnal in habit. Its diet consists of fruits, nuts, and insects (Tikader 1983). Studies have shown that habitat loss and hunting lead to decrease its numbers drastically (Joshua and Johnsingh 1992, 1994; Molur et al. 2005). Few studies on the estimation of population of this species in Periyar and Agasthyamalai which included Srivilliputhur Grizzled Squirrel Sanctuary also reported the reduction in number (Joshua 1992; Paulraj et al. 1992; Paulraj and Kasinathan 1993).

Population status and their distribution range in Anamalai and Chinnar Wildlife Sanctuary have been reported by Ramachandran (1993), Senthilkumar et al. (2007), and Joshua et al. (2008). Few individuals have also been reported from Palani Hills of the Western Ghats (Davidar 1989; Sharma 1992). In Eastern Ghats, a small population is reported (Karthikeyan et al. 1992; Kumara and Singh 2006; Baskaran et al. 2011). Few sightings from Cauvery Wildlife Sanctuary and forests in Karnataka and Hosur, Krishnagiri, Tamil Nadu were also reported. There were only few studies about the tree and grizzled giant squirrel interactions and conservational information for maintenance of the squirrel in Srivilliputhur Grizzled Squirrel Sanctuary, Tamil Nadu (Vanitharani et al. 2011). The status of their Indian population is vulnerable because of drastic habitat loss, clear-felling, logging, construction of dam, hunting for local consumption, and expansion of agro-industry construction (Molur et al. 2005). Ramachandran (1989, 1993) carried out an extensive study in Chinnar Wildlife Sanctuary to assess the status, distribution, and population estimation. Jayson and Ramachandran (1996) had studied the habitat utilization of lager mammals in the same area and reported 119 sightings of grizzled giant squirrel. This research study was conducted with the following objectives: (1) estimate the density of grizzled giant squirrel, (2) density of nest and occupancy, (3) preference of tree species for nesting, and (4) relation between nest and height of trees.

11.2 Study Area

CWS is located in the eastern part of the high ranges of southern Western Ghats of Kerala. The sanctuary, which is situated between 10° 15′ to 10° 21′ N latitude and 77° 05′ to 77° 16′ E longitude, has a total area of 90.44 km^2 (Fig. 11.1). The area falls in Marayoor and Kanthalloor Panchayath of Devikulam Taluk in Idukki District and is regarded as one of the important protected areas in Western Ghats. The habitat types range from shola-grassland to dry thorny scrub, across a diverse cultural landscape as well, making the PA unique in comparison with others (Fig. 11.2).

11.2.1 Boundaries

The erstwhile Chinnar Reserve was notified as a sanctuary in 1984. The original notification of the Chinnar Reserved Forest dates back to 1942, and the boundaries follow a jumble of cairn numbers and survey numbers. The boundaries are fully demarcated except in certain areas like Njavala-Ollavayal; thus the status is vague and may not correspond to the situation on the field. The northern and eastern

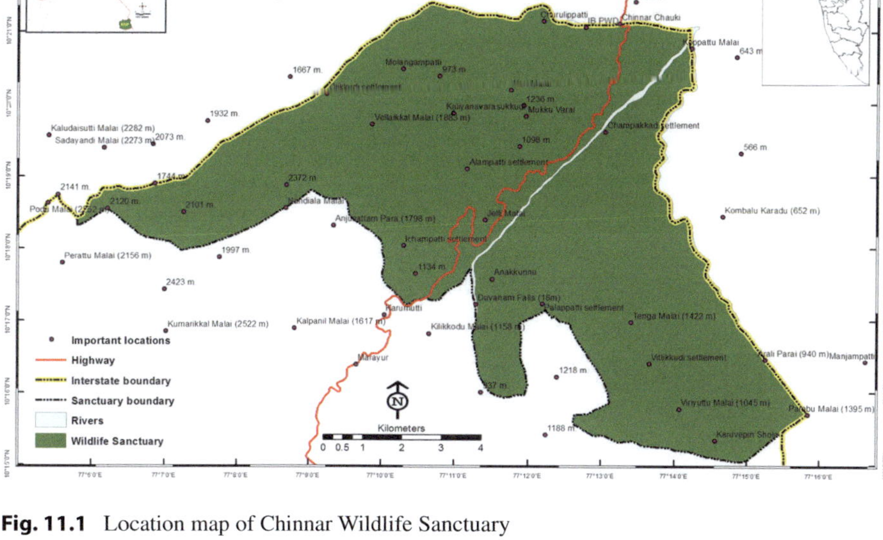

Fig. 11.1 Location map of Chinnar Wildlife Sanctuary

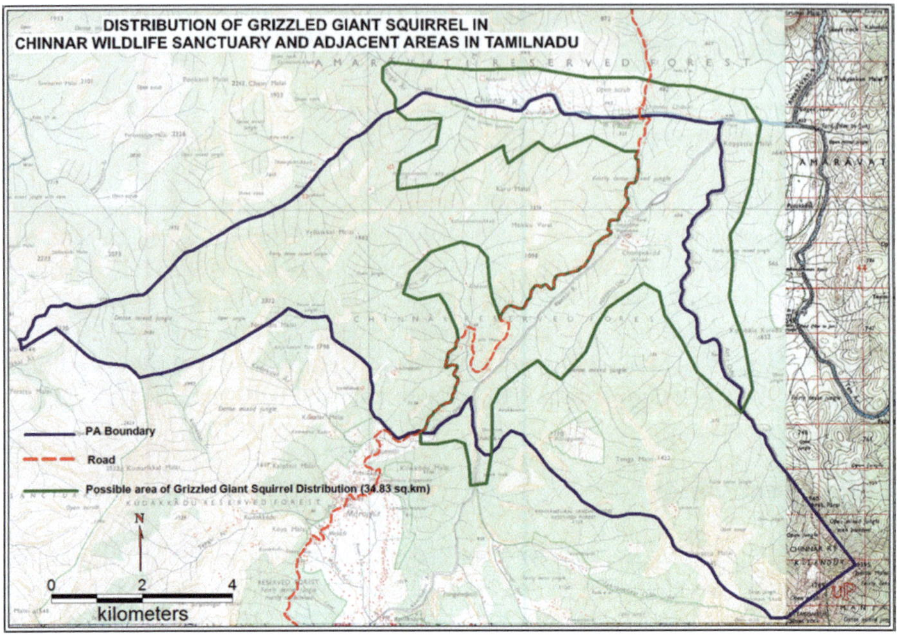

Fig. 11.2 Map showing the possible areas of distribution of grizzled giant squirrel in CWS

boundaries of Chinnar Reserve share 30 km with the Anamalai Tiger Reserve of Tamil Nadu. Toward the west, it is bordered by Eravikulam National Park, and on the southern side, it is bordered by the reserve forests of Marayoor Sandal Division. The park provides an ecological connectivity between the Anamalai Tiger Reserve and Eravikulam National Park.

11.2.2 Water Sources

Chinnar and Pambar rivers are the major perennial water resources in the sanctuary. Chinnar originates from near the Kumarikal Malai and flows through the interstate boundary toward the northwest edge of the sanctuary for 18 km and then to Tamil Nadu as Amaravati River. The Pambar River originates in the Anamudi Hills and is joined by seasonal rivulets and a few perennial streams originating from sholas in the upper reaches. It traverses the Turner's Valley in Eravikulam National Park and flows down into the sanctuary through the Taliar Valley between Kanthalloor and Marayoor villages and eastwards through the sanctuary. These two rivers merge at Koottar and drain into the Amaravati Reservoir in Tamil Nadu. Most of the rivulets and streams inside the sanctuary come alive immediately after the northeast monsoons and dry up soon. The water in the check dams remains for a longer period, but they also dry up during summer months. But a few streams originating from the upper reaches are perennial. The spectacular Thoovanam waterfalls lie deep within the sanctuary on the Pambar River. This breathtaking cascade is a major tourist attraction.

11.2.3 Vegetation

The vegetation shows an entire spectrum ranging from sub-temperate shola to dry scrub of the arid plains. In many areas, vegetation of the sanctuary is highly disturbed mainly due to a combination of factors like earlier fellings and planting, anthropogenic pressures of the settlements inside and on the fringes, and cattle grazing. Therefore in many cases, secondary forest types replace primary types, and an obvious classification of forest types is impracticable. Notwithstanding these, the vegetation of the sanctuary can be broadly classified into the following types according to Champion and Seth (1968) and Chandrasekaran (1962):

1. Southern tropical thorn forest (scrub jungle)
2. Southern dry mixed deciduous forest (dry deciduous forest)
3. Southern moist mixed deciduous forest (moist deciduous forest)
4. Tropical riparian fringing forest (riparian forest)
5. Southern montane wet temperate forest (Hill hoal forest)
6. Southern montane wet grassland (grasslands)

The dominant vegetation is dry deciduous forest followed by scrub forest. Together they constitute about 50% of the total forest area and are located in the low-altitude areas. The riparian fringing forests are linearly distributed along the hill folds and occupy a small but considerable area. Shola forests occupy a tiny fraction of the total area.

11.3 Methods

The population of grizzled giant squirrel and its nest in the study area has been estimated using line transect method (Burnham et al. 1980; Buckland et al. 1993). This method has been effectively used to determine animal densities (Karanth and Sunquist 1992, 1995; Varman and Sukumar 1995; Khan et al. 1996; Biswas and Sankar 2002; Jathanna et al. 2003). A total of ten transects were selected to carry out for the fieldwork. All transects were chosen along the way of riverine stretch with the length between 1.8 and 3.5 km. Three observers walked along each of the transect early morning and late evening. All transects were replicated to the subsequent day again. While walking along the transect, the following parameters were recorded: (1) sighting angle (with a compass), (2) sighting distance (visually estimated), (3) group size, (4) nesting tree species, (5) number of nests and its status, (6) tree height (ocular estimated). The density was estimated using the Distance 6.0 statistical software.

11.4 Results

A total of 40 encounters, which comprise of 34 individuals, were recorded with the effort of 106.8 km. The results from the transect data intended the overall density of grizzled giant squirrel. The name and length of transects used for the survey are shown in Table 11.1, and the length of transect varies from 1.3 to 3.7 km. The output from the line transect survey provided the overall density of 7.75 individuals per square kilometer with the standard error of 2.49 (Table 11.2). The total number of population in the study area was calculated by multiplying density to the total area (34.46 km^2) and obtained 267 individuals. The analysis was based on the low AIC value 47.747 and chi- square P value 0.51246. Figure 11.3 shows the best fit model of half-normal cosine and the component percentage such as cluster size found to be 1.2% with the encounter rate of 87.5% and detection probability of variation of 11.3% (D). The group density and sex ratio were not found because of fewer sightings. The number of nests was also recorded along with the tree species and its height.

The density of the grizzled giant squirrel's nests shows 68.99 km^2 with standard error of 19.55. The number of old nest was lesser than the new nests (Table 11.3). Figure 11.4 shows the best fit model half-normal simple polynomial with the component percentage such as cluster size 0.9% with encounter rate 89.2% and the detection probability of variation 9.9%. The percentage of coefficient variation was 28.34% and 95%. CV is between 39 and 123. A total of 12 tree species were selected

Table 11.1 Name and length of the transects in CWS, Kerala

Sl. No.	Name of the transect	Length
1	Koottar-Athioda	2.4
2	Koottar-Chinnar	2.8
3	Palapatti-Koshuvoda	3.7
4	Thoovanam-Chambakkad	3.5
5	Madhani-Alempetty	1.3
6	Thayannankudi-Churulipatty (Kuttyamma oda)	2.3
7	Madhani-Mangayoda	1.5
8	Surulipatty-Chinnar Bridge	1.8
9	Koottar to Champakkadu	2.8
10	Vashyappara trek path	3.1

Table 11.2 Population densities of grizzled giant squirrel in CWS

Parameter	Point estimate	Standard error	Percent co ef. of variation	95% percent confidence interval	
DS	6.9002	2.1966	31.83	3.4972	13.614
E(S)	1.1232	0.50612	4.51	1.0243	1.2315
D	7.7500	2.4917	32.15	3.9146	15.343
N	8.0000	2.5721	32.15	4.0000	15.000

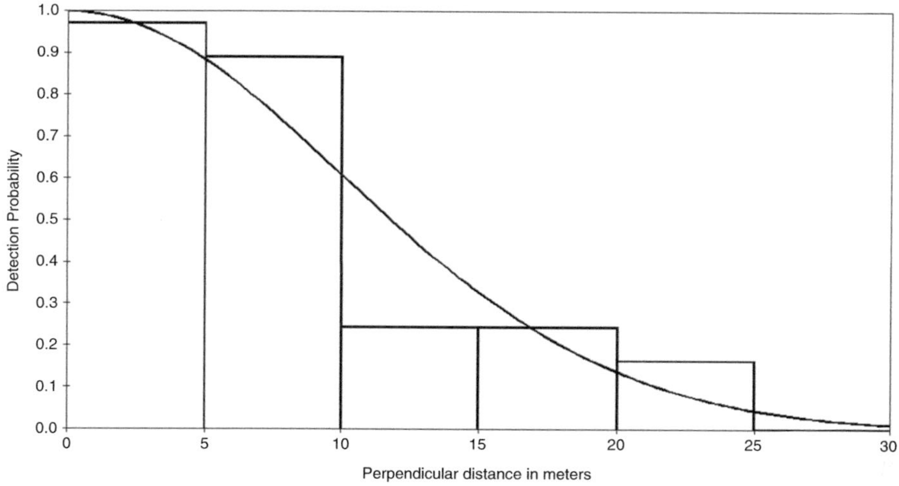

Fig. 11.3 Results of best model fitted in distance to estimate the detection probability and effective strip width for moist deciduous and riparian forests of CWS

Table 11.3 Density of grizzled giant squirrel's nest in CWS, Kerala

Parameter	Point estimate	Standard error	Percent coef. of variation	95% percent confidence interval	
DS	61.571	17.369	28.21	34.554	109.71
E(S)	1.1206	0.30126	2.69	1.0622	1.1823
D	68.998	19.552	28.34	38.651	123.17
N	69.000	19.553	28.34	39.000	123.00

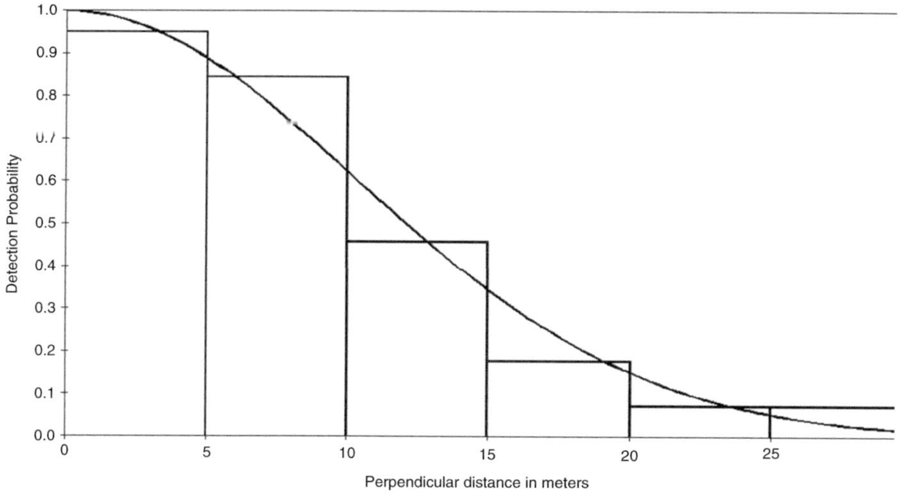

Fig. 11.4 The best model fitted in distance estimating and the detection probability of grizzled giant squirrels' nest in CWS

Table 11.4 Details of tree species, average height of the nest, and the range of the height of grizzled giant squirrel in CWS

Sl. No.	Name of species	Average height (m)	Range (m)
1	*Terminalia cuneata*	10.90	4–20
2	*Mangifera indica*	16.56	10–35
3	*Pongamia glabra*	18.15	8–30
4	*Tamarindus indica*	11.09	6–15
5	*Diospyros oocarpa*	18.08	10–15
6	*Syzygium cumini*	11.25	8–15
7	*Ficus benghalensis*	8.50	2.5–15
8	*Albizia chinensis*	24.50	18–30
9	*Cassine paniculata*	24.50	18–30
10	*Gyrocarpus asiaticus*	26.25	22–30
11	*Azadirachta indica*	22.67	15–35
12	*Cinnamomum verum*	19.25	15–30

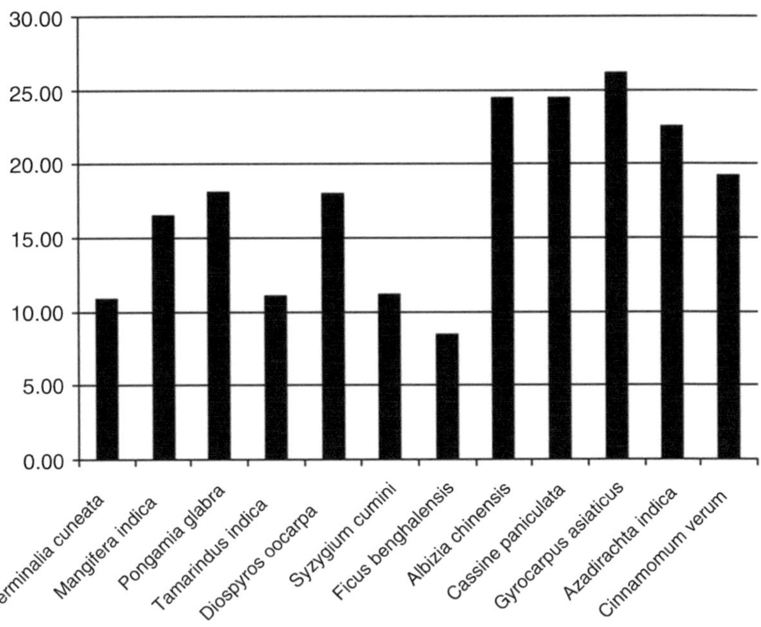

Fig. 11.5 The average height of nest of the grizzled giant squirrel in accordance with the tree species in CWS

for nesting by the grizzled squirrel in different heights ranging between 2.5 and 35 m (Table 11.4, Fig. 11.5). The details of tree species preferred by the grizzled squirrels for nesting are shown in Table 11.5. A total of 32 tree species were recorded for nesting of grizzled squirrel, of which 13 species of trees had only one nest. According to the nest occurrence, the tree species were divided into three categories such as highly preferred, moderately preferred, and less preferred. Compared to the

Table 11.5 Details of the number of nests and tree species used for nest by grizzled giant squirrels in CWS

Sl. No.	Name of species	No. of nest
1	*Terminalia cuneata*	59
2	*Mangifera indica*	37
3	*Pongamia glabra*	14
4	*Tamarindus indica*	13
5	*Diospyros oocarpa*	12
6	*Syzygium cumini*	7
7	*Ficus benghalensis*	6
8	*Albizia chinensis*	5
9	*Cassine paniculata*	5
10	*Gyrocarpus asiaticus*	5
11	*Azadirachta indica*	4
12	*Cinnamomum verum*	4
13	*Ficus racemosa*	4
14	*Alstonia scholaris*	3
15	*Dalbergia sissoides*	3
16	*Acacia planifrons*	2
17	*Drypetes sepiaria*	2
18	*Elaeocarpus serratus*	2
19	*Terminalia paniculata*	2
20	*Bischofia javanica*	1
21	*Butea monosperma*	1
22	*Carallia brachiata*	1
23	*Cassia fistula*	1
24	*Ficus mollis*	1
25	*Ficus tinctoria*	1
26	*Garcinia gummi-gutta*	1
27	*Lannea coromandelica*	1
28	*Lepisanthes tetraphylla*	1
29	*Melia dubia*	1
30	*Phyllanthus emblica*	1
31	*Schleichera oleosa*	1
32	*Sterculia foetida*	1

overall tree species, five species were highly preferred, 14 species were moderately preferred, and the rest was less preferred. Table 11.5 shows the average height of the tree where the nests were built.

11.5 Discussion

This study reveals that the total population of grizzled giant squirrel in CWS is about 260 individuals within the 35 km^2 of riparian vegetation, which indicates the increase in population. The previous studies (Ramachandran 1993, 1995; Senthilkumar et al. 2007) stated that the population was below 150 individuals in the entire potential habitat. The study by Ramachandran (1993) proved that the density of grizzled giant squirrel was 18–23 individuals/ km^2, but the present study states the density is 7.7 individuals/ km^2. The studies of Ramachandran (1989) and Joshua (1992) reported that a total of 50–75 individuals only were estimated. A recent study (Senthilkumar et al. 2007) reported 107 individuals in Chinnar. Moreover the entire study area has been extended a bit more. CWS has been considered as the home for the second viable population of grizzled giant squirrel in South India after the Srivilliputhur Grizzled Giant Squirrel Sanctuary (Ramachandran 1993; Senthilkumar et al. 2007).

During the present survey, Malabar giant squirrel was also sighted which shows that the habitat is being shared by these two giant squirrel species sympatrically. The ecological dynamic state of grizzled giant squirrel with other prey species and predators clearly affirms that the population is abundant. The habitat of grizzled giant squirrels is generally narrow (Ramachandran 1993), and the distribution of this species also occurred along with the stream and riverside. The canopy cover is also continuous along the streamside except some parts. The discontinuity of the tree canopy of the potential areas limits the movement of grizzled giant squirrels.

The feeding habits of the grizzled giant squirrel and the tree density of CWS have been studied by Senthilkumar et al. (2007) in detail. The result of the nest density of the study area shows that the number of old nest was less than the fresh nests. The density of the nest was about 68/km^2. The nests were categorized into two types like fresh and old. The height of the nest in the tree species also shows the range which indicates the grizzled giant squirrels prefer a certain height for each tree species. Sex ratio of grizzled giant squirrel was also not attained because of insufficient data. A broad-range study is required to estimate the statistics of sex ratio of grizzled giant squirrel.

11.5.1 Management Suggestion

Forest fire is one of the major threats to the sanctuary, and every year the fire disturbs and damages not only the forest but also the wild animals. The grizzled squirrel's habitats in the deciduous forests are also getting damaged due to forest fire. The authorities should take necessary action to control the regular forest fires not only for protecting the grizzled squirrel habitats but also the entire forest as such. The grizzled squirrels are seen apart from the riverine habitat and intruded in the deciduous patches. Necessary steps are to be taken to protect the species, and annual enumeration is to be conducted to monitor the population and its distribution. Many ecotourism activities are being carried out by the PA management. Some of the ecotourism activities are carried out along the riverine forests, which negatively affects the squirrel habitat. The preventive measures are to be taken to minimize the disturbance due to visitors and thus reduce the mitigation between visitors and wild animals.

The sanctuary is delimited with the Anamalai Tiger Reserve in Tamil Nadu; therefore, an interstate coordination is relevant to manage the wildlife population. It is being suggested that periodic meetings be conducted at various levels of officers and joint patrolling may be initiated to control illegal activities. The weekly ritual offering at Kodanthur temple situated in Tamil Nadu attracts a number of pilgrims, and they pollute the Chinnar River in various levels, and this affects the survival of wild animals and grizzled squirrels. The authorities should initiate necessary action to control this pollution and rusticate the flow of pilgrims to these areas. A periodic monitoring of the grizzled squirrel in the PA by the forest department staffs of the sanctuary will help the seasonality, status, and movement of wild animals. This can be implemented by giving training to the staffs and forest watchers for monitoring these species in the sanctuary. Long-term studies should be carried out to conduct the detailed investigation of the species in the area.

Acknowledgments We thank the Kerala Forest Department and officials for their enough kindness to provide fund and man power during our survey. We also thank to the resource persons and team members who have actively participated in the survey.

References

Baskaran N, Senthilkumar KS, Saravanan M (2011) A new site record of the grizzled Giant squirrel *Ratufa macroura* (pennant, 1769) in the Hosur Forest division, eastern Ghats, India and its conservation significance. J Threatened Taxa 3(6):1837–1841

Biswas S, Sankar K (2002) Prey abundance and food habit of tigers (Panthera tigris tigris) in Pench National Park, Madhya Pradesh, India. J Zool 411–422

Buckland ST, Anderson DR, Burnham KP, Laake JL (1993) Distance sampling: estimating abundance of biological populations. Chapman and Hall, London

Burnham KP, Anderson DJ, Laake JL (1980) Estimation of density from Line Transect sampling of biological populations, Wildlife monographs, 72. The Wildlife Society, Washington, DC

Champion HG, Seth SK (1968) A revised survey of Forest types of India. Government of India, Delhi

Chandrasekaran C (1962) Ecological study of the forests of Kerala state. Indian For 88(7):473–480

Davidar P (1989) Grizzled Giant squirrel *Ratufa macroura*- distribution in Kudirayar. J Bombay Nat Hist Soc 86(3):437

Ellerman JR (1961) The fauna of India including Pakistan, Burma and Ceylon. Mammalia – 2nd Edition, Rodentia. The Zoological Survey of India, Calcutta, 3: 884

Jathanna D, Johnsingh AJT, Karanth UK (2003) Estimation of large herbivore densities in the tropical forests of southern India using distance sampling. J Zool 285–290

Jayson EA, Ramachandran KK (1996) Habitat Utilization by larger mammals in Chinnar Wildlife Sanctuary. KFRI Research Report

Joshua J (1992) Ecology of the endangered Grizzled Giant Squirrel (*Ratufa macroura*) in Tamil Nadu, South India. Ph.D thesis. Bharathidasan University, Tiruchirapalli, Tamil Nadu

Joshua J, Johnsingh AJT (1992) Status of endangered grizzled giant squirrel and its habitats. In: Singh K, Singh JS (eds) Tropical ecosystems: ecology and management. Willey Eastern Ltd, New Delhi, pp 151–159

Joshua J, Johnsingh AJT (1994) Impact of biotic disturbances on the habitat and population of the endangered grizzled giant squirrel (*Ratufa macroura*) in South India. Biol Conserv 68:29–34

Joshua J, WILDPTS de A Goonatilake, Molur S (2008) *Ratufa macroura*. In: IUCN 2010. IUCN red list of threatened species. Version 2010.4. www.iucnredlist.org. Downloaded on 06 June 2011

Karanth KU, Sunquist M (1992) Population structure, density and biomass of large herbivores in the tropical forests of Nagarahole, India. J Trop Ecol 8:21–35

Karanth KU, Sunquist ME (1995) Prey selection by tiger, leopard and dhole in tropical forests. J Anim Ecol 64:439–450

Karthikeyan S, Prasad JN, Arun B (1992) Grizzled Giant squirrel *Ratufa macroura* Thomas and Wroughton at Cauvery valley in Karnataka. J Bombay Nat Hist Soc 89(3):360–361

Khan JA, Chellam R, Rodgers WA, Johnsingh AJT (1996) Ungulate density and biomass in the tropical dry deciduous forests of Gir, Gujurat, India. J Trop Ecol 12:149–162

Kumara HN, Singh M (2006) Distribution and relative abundance of giant squirrel and flying squirrel in Karnataka, India. Mammalia 70:40–47

Molur S, C Srinivasulu, Srinivasulu B, Walker S, Nameer PO, Ravikumar L (2005) Status of non-volant small mammals: Conservation Assessment and Management Plan (C.A.M.P) Workshop Report. Zoo Outreach Organization/CBSG-South Asia, Coimbatore, India, 618 p

Nowak RM (1991) Walker's mammals of the world, 5th edn. Johns Hopkins University Press, Baltimore

Paulraj S, Kasinathan N (1993) Scanty known grizzled giant squirrel (*Ratufa macroura*) of India: status and conservation. Indian For 119:828–833

Paulraj S, Kasinathan N, Rajendran K (1992) Studies on the biology of Grizzled Giant Squirrel Part I. Population, feeding, home range and activity pattern. Research report, Tamil Nadu State Forest Department

Ramachandran KK (1989) Endangered grizzled giant squirrel habitat. J Bombay Nat Hist Soc 86:94–95

Ramachandran KK (1993) Status survey and distribution of endangered grizzled giant squirrel in Chinnar wildlife sanctuary, Kerala, India. Indian J For 16(3):226–231

Senthilkumar K, Agoramoorthy G, Hsu MJ (2007) Population size, density and conservation status of grizzled giant squirrel in Chinnar wildlife sanctuary, India. Mammalia 71(1):89–94

Sharma N (1992) Status of and ecology of Grizzled Giant Squirrel (Ratufa macroura) in the Palani Hills. M.S. Dissertation, Pondicherry University

Srinivasulu C, Srinivasulu B, Rajesh A, Rao CAN, Nagulu V (2004) Non-volant small mammals of Kasu Brahmananda Reddy National Park, Andhra Pradesh. Zoos' Print J 19(6):1495–1497

Tikader BK (1983) Threatened animals of India. Zoological Survey of India, Calcutta. 307 p

Vanitharani J, Kavitha BB, Margaret VI (2011) Probe in to the conservation status of *Ratufa macroura* through the analysis of plant-animal interaction in the Srivilliputhur grizzled squirrel wildlife sanctuary. J Adv Biotech 10(10):67–69

Varman KS, Sukumar R (1995) The line transect method for estimating densities of large mammals in a tropical deciduous forest: an evaluation of modes and field experiments. J Biosci 20(2):273–287

Vijayakumaran Nair P, Ramachandran KK, Jayson EA (1997) Distribution of mammals and birds in Chinnar Wildlife Sanctuary. KFRI Research Report

Habitat Use of Nilgiri Tahr *Nilgiritragus hylocrius* in Western Ghats, India

12

J. Pandiyan, M. Sandeep, and A. Desai

Abstract

Nilgiri tahr is one of the endangered mammals in the world and also an endemic species of India. The utilization of habitats by Nilgiri tahr was evaluated by pellet analysis method. Four different microhabitats were identified in the Mukurthi National Park. The utilization of the habitat by the species was determined by transect method. Totally 112 transects comprising of 1028 quadrates were evaluated and counted. The pellet characteristics such as pellet groups including fresh pellets and old pellets were assessed. There were no significant variations between pellet groups and microhabitats ($P > 0.05$). But there was a close association between pellet groups, i.e. fresh and old pellets, among the microhabitats ($\chi^2 = 64.12$, df = 3, $P < 0.001$). The present study reveals that the species preferred the habitats which have their preferred/major food items among the availability of the habitats without minding the presence of predator or negative pressures.

Keywords

Habitat utilization · Nilgiri tahr · Pellet analysis · Habitat interactions

J. Pandiyan (✉) · M. Sandeep
Research Department of Zoology & Wildlife Biology, AVC College (Autonomous), Mannampandal, Mayiladuthurai, Tamil Nadu, India
e-mail: dunlinpandiyan@gmail.com

A. Desai
IUCN, New Delhi, India

C. Sivaperuman, K. Venkataraman (eds.), *Indian Hotspots*,
https://doi.org/10.1007/978-981-10-6605-4_12

12.1 Introduction

The Nilgiri tahr (*Nilgiritragus hylocrius*) is the only species of Caprine ungulate that is found south of the Himalayas in India. The animal was formerly named as *Hemitragus hylocrius*. The generic name was changed to *Nilgiri tragus* to be in tune with the latest phylogenic research by Ropiquet and Hassanin (2005).The Nilgiri tahr is an endangered mountain ungulate listed in schedule I of the Indian Wildlife (Protection) Act 1972 and considered as endangered by the IUCN. This species is included in the Red Data Book (1994) as endangered under the criteria EN B1+2acd, C2a on 30.06.2000 (Assessors: CAMP Workshop, India). This species has been assessed as endangered in the Red List of 2008 (Assessors: Alembath, M. and Rice, C.G, Evaluators Harris, R. and Festa-Bianchet, M. (Caprinae Red List Authority). The biogeographical studies according to Daniels (2006) reveal that the geographical range, population size and distribution of the Nilgiri tahr to the mid and high elevation hills of southwestern Karnataka, western Tamil Nadu and Kerala. It has been estimated that there were 50,000–100,000 Nilgiri tahrs at the time that the British invaded the region.

Though there are many investigators, the ecology of ungulates and their critical issues associated with the species and their habitat utilization and their foraging status remain unexplored due to difficulty in carrying out the study in the steep terrains. Nevertheless available information states that the Nilgiri tahr population decreases every year due to intensive predatory pressures, influence of environmental stress and human disturbances. Hence, there is every possibility for the species to get designated and also to get into the category of extinction as suggested by Daniels et al. (2008). The patterns of pellet-group size have been used to assess the habitat use by large herbivores since 1940s (Neff 1968). Pellet-group counts have also been widely used as an indirect method for assessing the movement pattern of ungulates (Bennet et al. 1940; Batcheler 1975; Dzieciolowski 1976; Bailey and Putman 1981; Staines and Ratcliffe 1987; Aulak and Babin' ska-Werka 1990; Plumptre and Harris 1995; Latham et al. 1996; Marques et al. 2001). The present article deals with the movement pattern of Nilgiri tahr in different microhabitats in the Mukurthi National Park and application of faecal pellet count methods in the evaluation and understanding movement pattern.

12.2 Methods

12.2.1 Study Area

The study was carried out in the plateau west of Ootacamund hill station in Nilgiris District, Tamil Nadu state in the Western Ghats mountain range of South India. It is a part of Nilgiri Biosphere Reserve and the entire Mukurthi National Park is under consideration by the UNESCO World Heritage Committee for selection as a World Heritage Site. The Mukurthi National Park faces the west between 11°10′ to 11°22′ N and 76°26′ to 76°34′ E the central location being 11° 16′ N and 76° 32′ E

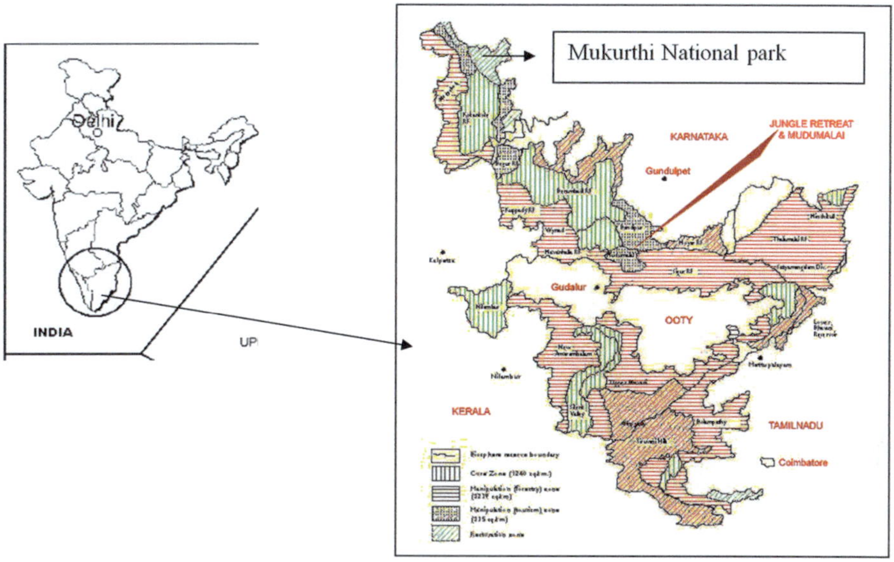

Fig. 12.1 Map showing the microhabitats of Nilgiri tahr at Mukurthi National Park

(Fig. 12.1). It is a grassland dominated by tall or short grasses. The dominant species of grassland include *Themeda tremula*, *Andropogon lividus* and *Eulalia trispicata* and other herbs. Bottom of the hills areas is covered by very long and dense grasses almost completely dominated by species of *Andropogon polyptychus*. Flat areas near streams are not waterlogged and dominated by two species of grass, viz. *Chrysopogon zeylanicus* and *Helictotrichon virescens*. The animals present in the study area include Nilgiri langur, tiger, leopard, wild dog, elephants and Nilgiri tahr.

12.2.1.1 Evaluation of Pellets

The pellet evaluation was assessed by line transect method. Before laying transects, the intensive foraging sites of Nilgiri tahr in the study areas was identified by using preliminary survey method with the help of GPS. On the basis of the survey and assessment, four microhabitats were identified as effective foraging sites and selected for the study in the Mukurthi National Park, Tamil Nadu, Southern India. The length of each transect was 100 m long. In each 100 m transect, there were four sub-quadrates laid, and the size was 1.0 m length of the each transect. Totally 112 transect were laid to cover all the four microhabitats. The presence and the nature of the pellets were recorded as fresh or old and when the pellets were sighted. The newly deposited pellets on the grasslands, which dried within 2 days, were considered as fresh pellets (e.g. <2 days old), and more than 2-day-old pellets were considered as old pellets. The age (number of days) of pellet was determined on the basis of the percentage of moisture content. Repeated surveys were made to assess the usage of the different habitat of Tahr in the Mukurthi National Park. The presence of fresh pellets in the same habitats was used as the indicator to measure the habitat, foraging behaviour and other activities. The length and width of the pellets were measured by using divider, vernier caliper and ruler. The width (breadth) was taken as the whole diameter of the entire pellet measured and the length measured from the anterior point to the posterior end of the Nilgiri tahr pellets.

12.2.2 Data Analysis

The analysis of variance (ANOVA) was performed to know the impact of microhabitats and pellet characteristics such as group, fresh, old, pellet length and widths. However, to understand the relationships between the pellet length and width Pearson's correlations were applied. The proximity and close affinity between the microhabitats and the pellet characteristics such as group, fresh, old, pellet length and widths were determined on the basis of chi-square analysis. All the statistical analyses were done by using SPSS 16.0 and Minitab 19.0.

12.3 Results

12.3.1 Pellet Groups

The present study was carried out in four different microhabitats of Mukurthi National Park. They include, viz. (1) western catchment III, (2) watchtower, (3) western catchment dam II and (4) western catchment rest house. These habitats were identified as the most preferred foraging areas for Nilgiri tahr. Totally 112 transects comprising of 1028 quadrates were monitored, and the pellets were counted as fresh and old pellets. The pellets were cylindrical in shape and they were classified into two regions, viz. anterior end was blunt and the posterior end was pointed. Colour of the pellet was black and greenish black.

Among the four microhabitats, the maximum number of pellets was recorded in the western catchment III (1.4 ± 1.12/m^2) and minimum in the western catchment watchtower (1.3 ± 0.24/m^2). The maximum number of fresh pellet was recorded in the western catchment dam II (0.7 ± 1.02/m^2) and minimum in the western catchment III (0.3 ± 0.75/m^2) (Table 12.1). Besides the range of the pellet size in each group was recorded 1–5 in the four different microhabitats (Fig. 12.2).

Table 12.1 Number of pellets was collected in different microhabitats of Mukurthi National Park during January to March 2010

		Pellet/m^2		
Sl. No.	Microhabitat	Groups (N = 1028)	Fresh (N = 1028)	Old (N = 1028)
1.	Western catchment III	1.4 ± 1.12	0.3 ± 0.75	1.0 ± 1.06
2.	Watchtower	1.3 ± 0.24	0.4 ± 0.89	0.9 ± 1.01
3.	Western catchment dam II	1.1 ± 1.25	0.7 ± 1.02	0.4 ± 0.85
4.	Western catchment rest house	1.2 ± 1.19	0.6 ± 0.85	0.5 ± 0.87

Values are Mean ± 1SD

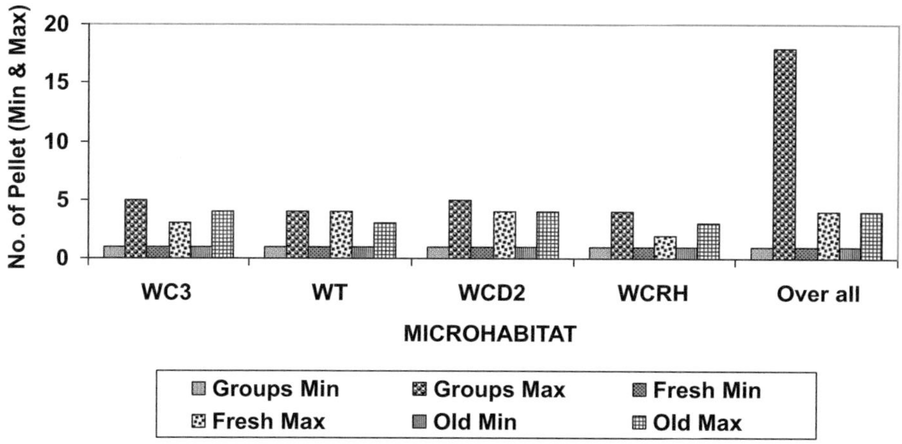

Fig. 12.2 The minimum and maximum pellet group, fresh and old pellets were recorded in the four different microhabitats of Mukurthi National Park. *WC3* western catchment III, *WT* watchtower, *WCD2* western catchment dam II, *WCRH* western catchment rest house

12.3.2 Pellet Size (Length and Width)

The pellet length and width of the Nilgiri tahr's from the four different microhabitats of the Mukurthi National Park were determined. The maximum pellet length was recorded in the western catchment III (9.2 ± 5.09 mm), and the minimum pellet length was in the western catchment II (7.3 ± 6.25 mm). But the width of pellet was maximum in western catchment watchtower (6.5 ± 3.66 mm), and minimum (5.2 ± 4.46 mm) was in the western catchment rest house (Figs. 12.3 and 12.4).

The analysis of the general linear model (GLM) showed that the length and width of Nilgiri tahr pellets showed significant variations among the microhabitats ($P < 0.05$). The chi-square analysis also showed that there were close associations between the pellet length and width among the microhabitats, i.e. length ($\chi^2 = 63.18$, df = 3, $P < 0.001$) and width ($\chi^2 = 57.31$, df = 3, $P < 0.001$) (Fig. 12.5).

12.4 Discussion

12.4.1 Pellet Groups

The presence of pellets (fresh or old) can be used as an indicator to monitor the visits, foraging and behaviour of the animals in any particular habitat. In the present investigation the population, activity and foraging behaviour were assessed on the basis of the study of the pellets of Nilgiri tahr. The pellet groups, fresh and old pellets, of Nilgiri tahr were collected and counted during January to March 2010. There were significant differences between the pellets groups and among the four

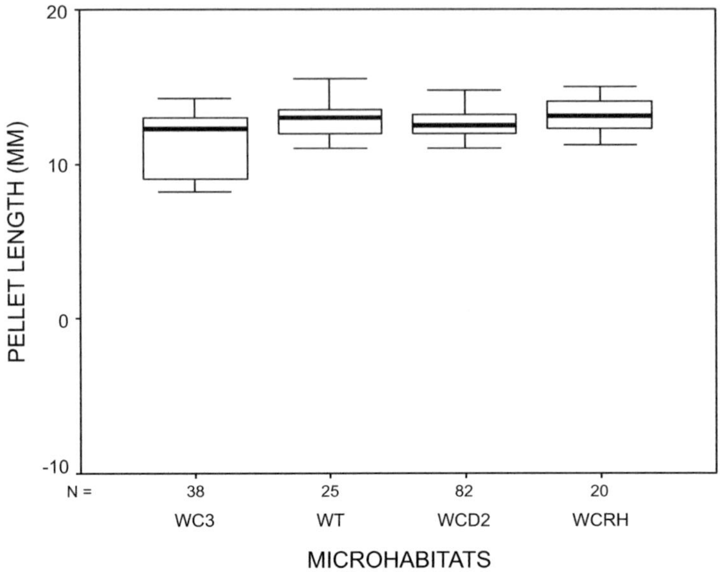

Fig. 12.3 Mean size of pellet length of Nilgiri tahr at Mukurthi National Park. *WC3* western catchment III, *WT* western catchment watchtower, *WCD2* western catchment dam II and *WCRH* western catchment rest house

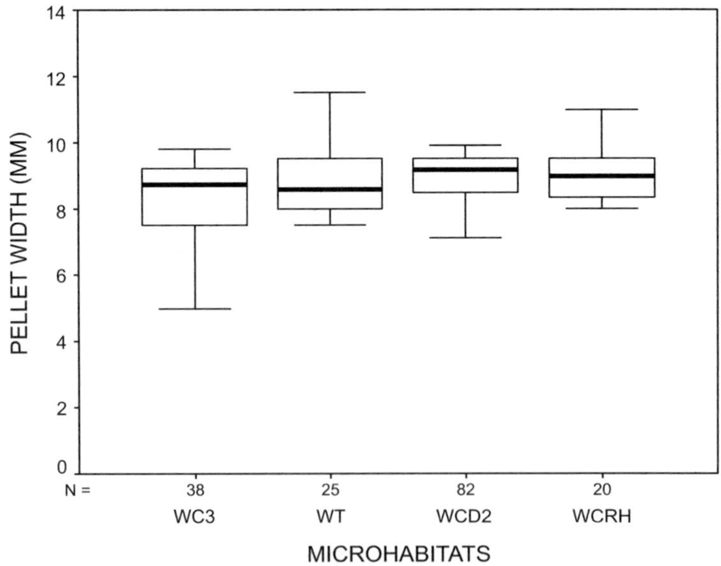

Fig. 12.4 Mean size of pellet width of Nilgiri tahr at Mukurthi National Park. *WC3* western catchment III, *WT* western catchment watchtower, *WCD2* western catchment dam II and *WCRH* western catchment rest house

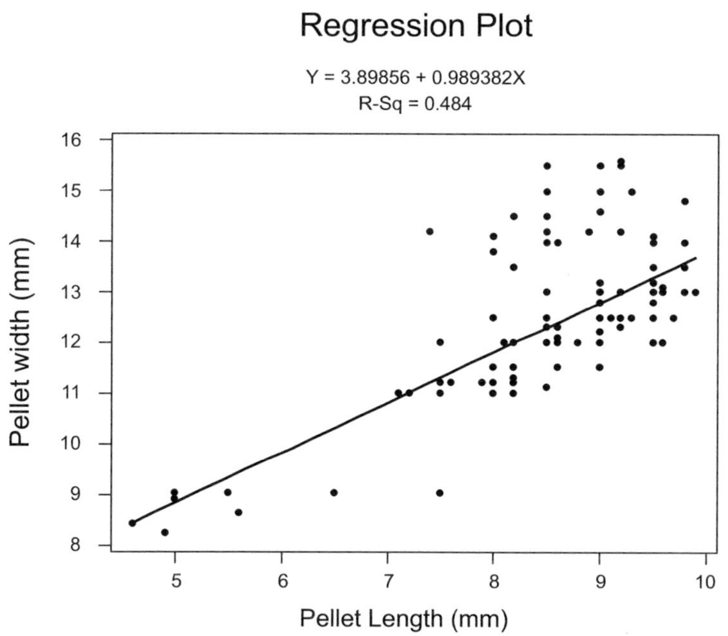

Fig. 12.5 Interpretation of Nilgiri tahr's pellet length and width recorded in the four microhabitats at Mukurthi National Park

microhabitats. The western catchment III was the most preferred foraging and feeding grounds when compared to other microhabitats for the Nilgiri tahr (Table 12.1, Fig. 12.2 and Appendix I).

It has been frequently reported that the pellet group varied, depending mainly on the sex and age of the animals, habitat type, season and the plants foraged (Sawyer et al. 1990). In general the pellet size purely depends on the availability of food, quality of food, nature of the animal and other edaphic factors. The animals will produce maximum number of pellets when they forage more quality food items. The determination of getting quality of food items depends on the availability of quality habitats and the availability of foraging plants. However, the eating of more food items by herbivores depends on not only availability of food items but also the presence of predatory pressures. In the present study, both disturbed and minimum disturbed areas among the four microhabitats have been determined merely basis of observation. The minimum disturbed area was the western catchment III, and the other three microhabitats showed maximum predatory pressures. Besides, 39 species of plants from the 4 microhabitats were recorded in the present study (Appendix I). But the western catchment III showed only 9 species out of 39 when compared to the other 3 microhabitats. But the western catchment watchtower showed 29 species of plants. All these plant species recorded in the 4 different microhabitats have been justifiably determined as the most preferred species for Nilgiri tahr.

Appendix I Most preferred food plant species of Nilgiri Tahr, Mukurthi National Park

S.No	Species name	Family	Habitat
1	*Anaphalis wightiana*	Asteraceae	Herb
2	*Bamboo spp*	Papilionaceae	Herb
3	*Cytisus scoparis*	Asteraceae	Shrub
4	*Erigeron mucronatum*	Asteraceae	Herb
5	*Eupatorium glandulosum*	Ericaceae	Herb
6	*Gaulthria fragrantissima*	Asteraceae	Herb
7	*Helichrysum buddleoides*	Asteraceae	Herb
8	*Hypochaeris glabra*	Acanthaceae	Herb
9	*Strobilanthus gossypinus*	Acanthaceae	Shrub
10	*Strobilanthus kunthianas*	Acanthaceae	Shrub
11	*Syzigium spp*	Gramineae	Shrub
12	*Andropogon lividus*	Gramineae	Grass
13	*Chrysopogon zeylanicus*	Gramineae	Grass
14	*Eualila trispicata*	Gramineae	Grass
15	*Isachna kunthiana*	Gramineae	Grass
16	*Themeda tremula*	Gramineae	Grass
17	*Tripogon ananthaswamianus*	Gramineae	Grass

Table 12.2 Pellet size was recorded in the different microhabitats of Mukurthi National Park during January to March 2010

		Morphology	
Sl. no.	Microhabitats	Length (mm)	Width (mm)
1	Western catchment III	9.2 ± 5.09	6.5 ± 3.66
2	Watchtower	8.8 ± 6.08	6.2 ± 4.28
3	Western catchment dam II	7.3 ± 6.25	5.2 ± 4.46
4	Western catchment rest house	8.2 ± 6.57	5.5 ± 4.42

Values are Mean ± 1SD

Generally animals prefer to forage in quality habitats which are without predatory pressures. The present results also reveal that more numbers of pellet groups were found in the undisturbed area, i.e. western catchment III, while only fresh pellets were found in the disturbed habitats (Table 12.2). It seems that the animals were using undisturbed habitats for both foraging and resting, whereas the disturbed areas were cautiously used by the animal for the purpose of foraging. Obviously the animals used the disturbed habitats because of the presence of their preferred food plants in spite of the predatory pressures such as the presence of dangerous predator like tiger, panther, wolves, etc., (author's unpublished data). The present study strongly suggests that the Nilgiri tahr visits the habitats for their preferred food plant species without minding the predatory pressures. It is because of the fact that the survival instinct compels them to forage even when there is predatory pressures.

12.4.2 Pellet Size (Length and Width)

The size of the pellets is generally directly based on the quality of the food plant species which in turn indirectly dependent on the quality of the habitat such as edaphic factors, including soil quality, moisture and fertility to sustain the plant species and other biological factors such as age and sex of the animal. There was a significant relationship between the length and width of the pellets of Nilgiri tahr ($P < 0.001$). Definitely the length and width is influenced by the availability of food items, nature of the animal and other environmental factors. The study areas are having abundant food items to meet out their day to day survival. The length and width of the pellets are directly proportional to the availability and quality of the food, the nature and body condition of the animal. The length and width differed significantly among the habitats, i.e. length ($\chi^2 = 63.18$, df = 3, $p < 0.001$) and width ($\chi^2 = 57.31$, df = 3, $p < 0.001$). The overall results showed that the pellet length and width were maximum in the western catchment III than the other micro-habitats studied (Figs. 12.3 and 12.4). It indicates that the Nilgiri Tahr preferred their habitat for foraging based on the availability of more number of preferred food plant species, less predatory pressures and favourable other environmental conditions.

Nevertheless the fragmentation of habitat leading to habitat loss and population segregation have been addressed in different studies. The evaluation of quality of habitats for their long-term suitability and sustainability has been recommended. Scientific means of improving the available forage (as against the total grass/shrub biomass) has to be seriously explored and adopted. The Tamil Nadu Forest Department is removing the exotic monocultures along the periphery of the Mukurthi National Park (Daniels 2006).

12.4.3 Management Implications

The results of the present study revealed that the endangered species of Nilgiri tahr use a selected habitat from among other available habitats. To ensure species population and sustain the same, we have to assess the carrying capacity of the wildlife habitats and increase the carrying capacity by way of increasing the population of foraging plant species so as to prevent the predatory pressures and to ensure the sustainability of the animal. Nilgiri tahr is one of such animal which is under Red List category to be given attention so as to prevent the animal to enter into the category of 'Extinct'.

Acknowledgements We are very thankful to WWF-India for having financial assistance to carry out the work and our special thanks to Mr. Sundararaju, IFS., PCCF and Chief Wildlife Warden, Tamil Nadu State Government, India, for giving permission to carry out this study in Mukurthi National Park, Tamil Nadu, Southern India. Our sincere thanks to Dr. R. Saravanamuthu, Former Head of the Department of Botany, AVC College (Autonomous), Mannampandal for his critical comments on the manuscript.

References

Aulak W, Babin ska-Werka J (1990) Estimation of roe deer density based on the abundance and rate of disappearance of their faeces from the forest. Acta Theriol 35:111–120

Bailey RE, Putman RJ (1981) Estimation of Fallow deer (*Dama dama*) populations from faecal accumulation. J Appl Ecol 18:697–702

Batcheler CL (1975) Development of a distance method for deer census from pellet-groups. J Wildl Manag 39:641–652

Bennet LG, English PF, McCain R (1940) A study of deer populations by use of peller group counts. J Wildl Manag 4:399–403

Daniels RJR (2006) The Nilgiri Tahr: an endemic South Indian Mountain goat. Macmillan India Ltd, New Delhi

Daniels RJR, Easa PS, Ramkumar K, Mammen PC Alembath M (2008) How may Nilgiri Tahr (*Himitragus hylocrius*) are there in the wild? 4th world congress on mountain Ungulates held at Munnar (India) between September 12 and 15 2006 (Abstract)

Dzieciolowski R (1976) Roe deer census by pellet-group counts. Acta Theriol 21:351–358

Latham J, Staines BW, Gorman ML (1996) The relative densities of red (*Cervus elaphus*) and Roe (*Capreolus capreolus*) deer and their relationship in Scottish plantation forests. J Zool (Lond) 240:285–299

Marques FFC, Buckland ST, Goffin D, Dixon CE, Borchers DL, Mayle BA, Peace AJ (2001) Estimating deer abundance from line transect surveys of dung: Sika deer in southern Scotland. J Appl Ecol 38:349–363

Neff DJ (1968) The pellet-group count technique for big game trend, census and distribution: a review. J Wildl Manag 32:597–614

Plumptre AJ, Harris S (1995) Estimating the biomass of large mammalian berbivores in a tropical mountane forest: a method of faecal counting that avoids assuming a 'steady state' system. J Appl Ecol 32:111–120

Ropiquet A, Hassanin A (2005) Molecular evidence for the polyphyly of the genus Hemitragus (Mammalia, Bovidae). Mol Phylogenet Evol 36(1):154–168

Sawyer TG, Marchinton RL, Lentz WM (1990) Defecation rates of female white-tailed deer in Georgia. Wildl Soc Bull 18:16–18

Staines BW, Ratcliffe PR (1987) Estiating the abundance of red deer (*Cervus elaphus* L.) and roe deer (*Capreolus capreolus* L.) and their current status in Great Britain. Symp Zool Soc Lond 58:131–152

Biology, Ecology, and Conservation of Golden Langur, *Trachypithecus geei*

13

Debahutee Roy and Rajarathinavelu Nagarajan

Abstract

The Golden Langur *Trachypithecus geei* is a rare colobine monkey with a very restricted range being confined to Western Assam in India and Bhutan. In Assam, at least 19 fragmented areas now hold the species, which was originally a single habitat. The langur distributed from the subtropical forests of Western Assam to the broadleaf forests of Bhutan. The individuals of Golden Langur in Manas, Ripu, and to some extent Chirang have link among themselves and also with Bhutanese populations, and the remaining populations found in fragmented areas have no link and became isolated. The estimated total population in India is 5600 individuals, and most of the information on this species was available in the form of short-term studies; most of them are about the status and distribution. None of the population was monitored for longer period of time, and the population inhabited both in India and Bhutan are declining. Golden Langur is folivorous in nature and feeds on a variety of food plants. Activity pattern showed a bimodal diurnal activity pattern in feeding with peaks in the morning and evening hours. Golden Langur is threatened by habitat fragmentation. In fragmented areas, they cause considerable damages to food crops. To maintain its populations for the future, long-term and large-scale planning is necessary.

D. Roy
PG and Research Department of Zoology and Wildlife Biology,
A.V.C. College (Autonomous), Mayiladuthurai, India

R. Nagarajan (✉)
PG and Research Department of Zoology and Wildlife Biology,
A.V.C. College (Autonomous), Mayiladuthurai, India

Washington Singer Laboratories, Centre for Research in Animal Behaviour (CRAB),
School of Psychology, University of Exeter, Exeter, UK
e-mail: r.nagarajan@exeter.ac.uk

C. Sivaperuman, K. Venkataraman (eds.), *Indian Hotspots*,
https://doi.org/10.1007/978-981-10-6605-4_13

Keywords
Conservation · Distribution · Ecology · Golden Langur · *Trachypithecus geei*

13.1 Introduction

The colobine monkeys have provided primatologists with a rich array of ethological and ecological data upon which many models have been generated on the evolution of primate behavior. The colobines are medium-sized primates with long tails and diverse colorations. They are characterized by large molar teeth with high pointed cusps to breakdown leaves and a special sacculated stomach adapted to harness bacterial colonies that enable them to digest cellulose and tannins by process of fermentation. In this group of monkey most of the research information is available on two species viz., (a) Hanuman Langur and (b) on Red Colobus monkey. The genus *Trachypithecus* is the most diverse langur taxon, having a broad distribution including India, Sri Lanka, Bangladesh, Southwestern China, and Southeast Asia. It is phylogenetically embedded within the family Cercopithecidae and closely related to Semnopithecus.

13.2 Golden Langur

Golden Langur *Trachypithecus geei* (Khajuria 1956) is one of the least studied primate species of Northeast India. The Golden Langur is a rare colobine monkey with a very restricted range being confined to Western Assam in India and Bhutan only. At the time of discovery in the early 1950s, it was believed that it is confined to the forests along Indo-Bhutan Border only but in fact it was found all over the Western Assam as is apparent from its present extent of occurrence. The Golden Langur is in real danger all over its range in India. Never before had it faced such dangers – from rapid loss of habitat, occasional poaching, and unreported accidental deaths. The local tribes, particularly Bodos, usually do not kill langurs, however some individuals of younger generation develop taste for its meat. Some individuals of langurs that particularly live in fragmented habitats have even started raiding crops due to loss of natural habitat.

At least 19 fragmented areas now hold the species, which was originally a single habitat. Except for Manas, Ripu, and to some extent Chirang, the remaining populations have no link with the larger and more secure Bhutanese populations. Many of these fragmented populations have little possibility of long-term survival. Thus, primates that specialize in one primary habitat are more likely to go extinct. Wangchuk et al. (2003) described a new subspecies, *T. g. bhutanensis*, from Bhutan in the north of the range, with *T. g. geei* occurring in the south, the two taxa being separated by the Main Frontal Thrust (MFT) of the Indian plate hitting into the Himalayas. However, *T. g. bhutanensis* is not recognized here as a valid subspecies, as it has not been described according to International Commission on Zoological Nomenclature (ICZN) rules.

The reduction of habitat into fragments is increasingly one of the most important issues threatening the existence of primates in Tropical Asia. Most of the primate populations which once occupied large contiguous habitats are now reduced to small isolated patches of forest which again are under various degree of anthropogenic pressure. Once the population is isolated due to fragmentation of the habitat, a host of derogatory conditions are created which are interruptions of the gene flow, inbreeding depression, and truncation of the habitat matrix and intrusion of edges. This leads to a rapid or gradual deterioration of the local ecological conditions amounting to a loss of continued supply of resources, e.g., food, cover, and mates (Saunders et al. 1991). Continued isolation may also result in saturation of the forest remnants by some species, altering interspecific interaction, overexploitation of resources, increased competition and predation, and changes in fecundity and all leading to significant reductions in population sizes of many species and to a potential collapse of the remaining population.

The time of fragmentation of the intact forests could be deduced for some of the sites from the establishment history of British-owned tea estates in the vicinity, starting from the 1840s. The histories of human settlements near the forests, however, are not well documented. In 1996, the Supreme Court of India imposed a ban on tree-felling without approved working plans in all forests in Northeast India. The transport of timber outside the region was also prohibited. Poor implementation of existing forest laws, political interference, and a lack of ecological responsibility of the wood industries are some extent largely responsible for the extensive degradation of forests. An armed insurgent group called the United Liberation Front of Assam (ULFA) has been active in the state since 1980s. The law and order problems created by this insurgency have further facilitated the deforestation of the reserves through felling and encroachments.

13.3 Different Names of Golden Langur

Golden Langur is commonly known as Sonali Bandar ('Sonali-Golden, Bandar-Langur') in Assamese due to its golden coat color. For many local Hindu people, it is sacred, a living incarnation of the God hanuman, who played a key role in the triumph of good over evil. The various local names of Golden Langur are in Table 13.1 (Chetry and Chetry 2009).

Table 13.1 List of vernacular names of Golden Langur in different languages (adopted from Chetry and Chetry 2009)

Sl. no.	Language/community	Vernacular names
1.	Assamese	Sonali Bandar/Baga Bandar
2.	Bengali	Dhala Bandar/Sada Bandar
3.	Bodo	Mukhra Gufur
4.	Rabha	Kawai Bulung
5.	Bhutanese	Sugriv
6.	Hindi	Sunhera Bandar
7.	Rajbanshi	Check Bandar
8.	Nepali	Dheru/Sugriv
9.	Garo	Rangol
10.	Adivasi	Haanu/Hanuman

Fig. 13.1 Showing sexual dimorphism in Golden Langur with genital difference marked with an arrow. (**a**) An adult female with a juvenile, (**b**) an adult male

13.4 Morphological Features

Golden Langur is a beautiful and a charismatic species and is characterized by its coat color. They are sexually dimorphic, and the genital organs are distinct (Fig. 13.1). The coat of adult Golden Langur ranges from creamy white to golden color, gaining a more reddish tinge in winter season. On their flanks and chest, the hairs are darker and often rust colored with a black face. Neonates vary from creamy white to pale in color which gradually turns into golden color at the age of 9–15 months. Subadults can be distinguished from the adults by their relatively smaller body sizes and females lack visible teats. The overall shape of this primate species is slim, with long limbs and tail. The tail has a bobble on the end and is notably larger in males than in females. Adult males are slightly larger than adult females. Females are 490 mm and males are 640–720 mm in length. Tail length is 713 mm in females and 780–940 mm in males. The body weight also varies among the two sexes. Female weighs 9.5 kg against the 10.9 kg of male.

13.5 Distribution and Population

Our planet supports a variety of flora and fauna, yet all species are not uniformly available in every part of the globe. There are species which have wide distribution and some species with limited distribution. Golden Langur (*Trachypithecus geei*) is one of the most endangered primate species of South Asia, endemic to adjacent areas of India and Bhutan. Given the limited distribution range of Golden Langur

Table 13.2 Population status of Golden Langur in India

Place/Area	Area (km^2)	Population	Survey period	Source
Abhaya Rubber Garden	1	146	2013–2014	Current study
		112	2008	Ghosh (2009)
Kakoijana Reserve Forest	17	502	2013–2014	Current study
		507	2012	Horwich et al. (2013)
		144	2008	Ghosh (2009)
Bamungaon Reserve Forest	10.07	42	2013–2014	Current study
		30	2007	Horwich et al. (2013)
Ripu Reserve Forest	605.25	2294	2008	Ghosh (2009)
Chirang Reserve Forest	592.54	11	2008	Ghosh (2009)
		1473	2008	Ghosh (2009)
Chakrashila Wildlife Sanctuary	45.58	501	2008	Ghosh (2009)
Naddengiri Reserve Forest	10.2	66	2008	Ghosh (2009)
Nakkati Reserve Forest	28.46	8	2008	Ghosh (2009)
		37	2008	Ghosh (2009)
Bhumeshwar Reserve Forest	14.16	14	2007	Horwich et al. (2013)
		53	2012	Horwich et al. (2013)
Khoragaon Reserve Forest	6.3	8	2008	Ghosh (2009)
		36	2012	Horwich et al. (2013)
Bhairab Reserve Forest	36.04	26	2008	Ghosh (2009)
		49	2007	Horwich et al. (2013)
Manas Reserve Forest (a part)	625.25	476	2008	Ghosh (2009)
	213	263	2009	Ghosh (2009)

both in Bhutan and India and the current trend of habitat destruction in India, even small local populations are valuable and should be protected wherever possible. The Golden Langur is distributed in India and Bhutan. The rivers and mountains in the distribution range have limited its distribution by posing as natural barriers. In India, the total estimated population is 5600 individuals (Table 13.2).

13.6 Global Distribution of the Species

India and Bhutan are the only two countries in the world where Golden Langur occurs (Fig. 13.2). Again within these two countries also the distribution is restricted. In Bhutan, it is found in Central Bhutan ranging between the Sankosh River and Chamkar-Mangde, Manas River complex. The Black Mountains is the limit of its distribution in the north. The furthest north is from below the Trongsa Dzong at 2353 m (Wangchuk 2005). In the west, they have been recorded from Chendebji at 2600 m (Wangchuk 2005). In Bhutan the distribution range of Golden Langur covers an area of 4,782.27 km^2 (Wangchuk 2005).

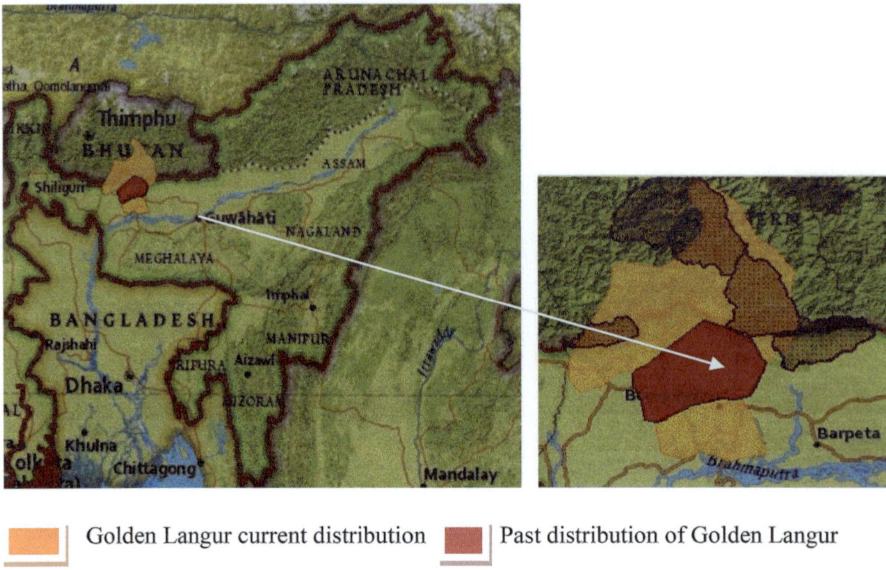

Fig. 13.2 The map showing the distribution of Golden Langur (*Trachypithecus geei*) in India and Bhutan

13.7 Distribution in India

This species occurs only in Northeast India (Assam). It is confined to a forest belt in Western Assam between the Manas River in the east, Sankosh in the west, and Brahmaputra in the south along the Indo-Bhutan Border (Medhi et al. 2004). Its distribution in Bhutan is limited to the foothills of the Black Mountains (Srivastava et al. 2001). The total known range of Golden Langur in both India and Bhutan is less than 30,000 km^2, and much of it is not a suitable habitat (Srivastava et al. 2001). The population in India is highly fragmented, with the southern population completely separated from the northern population due to the effects of human activities (Fig. 13.3).

13.8 Species Status

The Golden Langur is classified as endangered (EN) [A2c; C2a (i)] on the IUCN Red List 2008 and is listed on Appendix I of CITES. Listed as endangered because of a serious population decline, estimated to be more than 50% over the last three generations (30 years), inferred from observed reduction in the extent of its habitat

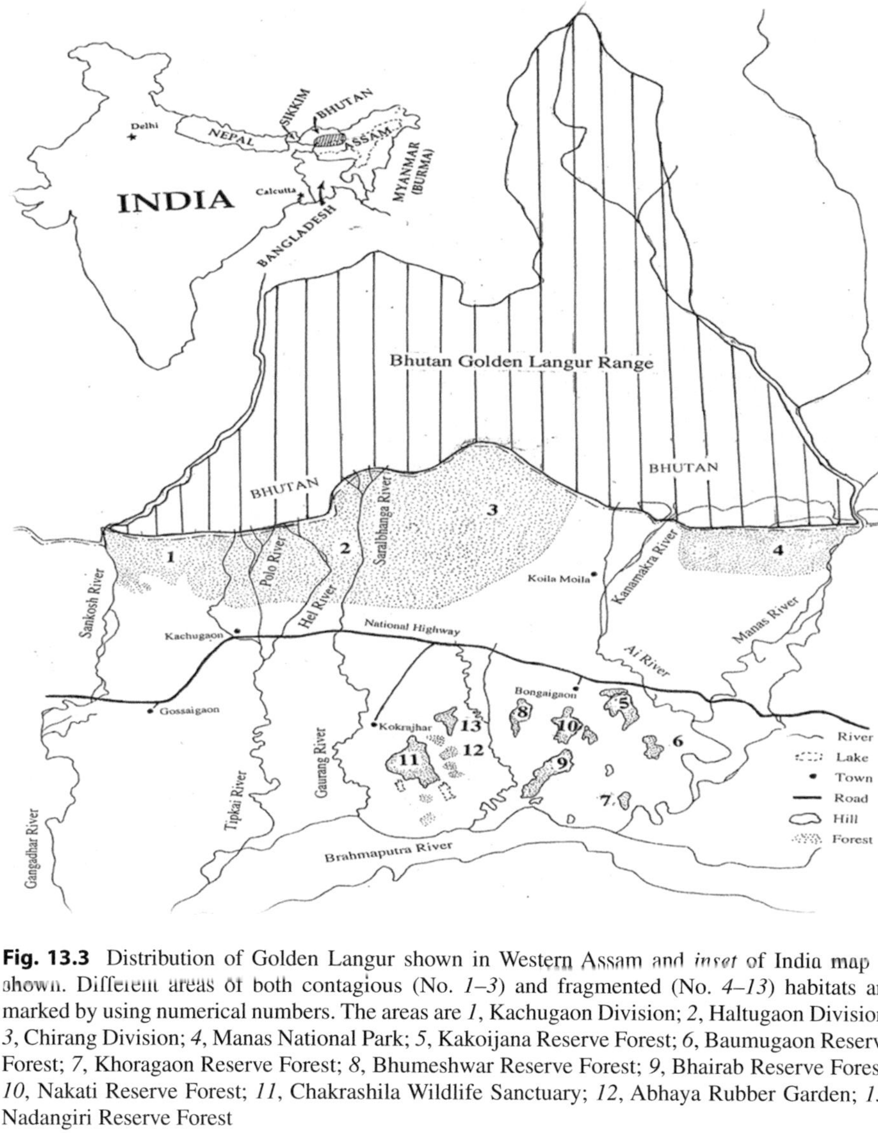

Fig. 13.3 Distribution of Golden Langur shown in Western Assam and *inset* of India map is shown. Different areas of both contagious (No. *1–3*) and fragmented (No. *4–13*) habitats are marked by using numerical numbers. The areas are *1*, Kachugaon Division; *2*, Haltugaon Division; *3*, Chirang Division; *4*, Manas National Park; *5*, Kakoijana Reserve Forest; *6*, Baumugaon Reserve Forest; *7*, Khoragaon Reserve Forest; *8*, Bhumeshwar Reserve Forest; *9*, Bhairab Reserve Forest; *10*, Nakati Reserve Forest; *11*, Chakrashila Wildlife Sanctuary; *12*, Abhaya Rubber Garden; *13*, Nadangiri Reserve Forest

and its population size is estimated to number fewer than 2500 mature individuals, and there is an observed continuing decline in the number of mature individuals. Manas Biosphere Reserve, covering the Ripu Reserve Forest, Chirang Reserve Forest, Manas Reserve Forest, and western range of Manas National Park, yielded the presence of 4231 individuals in 449 troops during 2009.

Fig. 13.4 A troop of Golden Langur sitting on *Terminalia chebula* tree found in "Peacock Island" or "Umananda," and the photograph was taken during January 2015

13.9 Golden Langur of Peacock Island

Apart from the wild populations, a small introduced population which is isolated and semi-habituated can also been seen in Umananda Island (Figs. 13.4 and 13.5). Situated in the midst of the Brahmaputra River in Kamrup district at Guwahati, Assam, a river island of Brahmaputra popularly known as "Peacock Island" or "Umananda" with an area of 4.961 ha. In the year 1984, a pair of male and female rescued Golden Langur was introduced by a Hindu devotee in the Umananda temple. Since then, the pair has bred six times, and the size of the troop has increased to eight individuals (Chetry and Chetry 2009). A survey made during January 2015 revealed that there are nine individuals in the island (Debahutee Roy personal observation).

13.10 Habitat

The habitat of Golden Langur ranges from tropical semievergreen to tropical moist deciduous forests. The sal (*Shorea robusta*) dominated forests of Western Assam, and secondary forests also support Golden Langur (Fig. 13.6). It occurs from flat plains to low hills inside Assam, while inside Bhutan, it inhabits warm broadleaf forests between 1000 and 2400 m and subtropical forests between 200 and 1000 m (Chetry and Chetry 2009).

Fig. 13.5 An adult male Golden Langur feeding on the leaves of fern (*Dryopteris* sp.) during the nonbreeding season, and the photograph was taken in March 2015

Fig. 13.6 Landscape showing the habitat of Golden Langur in Kakoijana Reserve Forest of Assam, a fragmented habitat with adjacent paddy fields. The langur population found in this habitat is isolated

Golden Langur, being a habitat specialist, is found to thrive in an altered habitat of Nayakgaon Rubber Plantation (Fig. 13.7) where they are found to feed on young leaves, fruits, and nuts inside the rubber seeds by breaking the hard cover of the dry seed. Golden Langur of tropical evergreen forests is found to inhabit the top middle strata of the forests. Observation made till date clearly indicates that Golden Langur can adjust in changing environment. The recent habitat fragmentation and changing environment compel them to use the lower strata of the forest or even to come down to ground.

Fig. 13.7 Landscape showing the habitat of Golden Langur in Nayakgaon Rubber Garden of Assam, a fragmented habitat. The langur population found in this habitat is also isolated

13.11 Social Organization

Golden Langurs are social animal, like other colobines. Golden Langur's social systems or troops are male centered. Depending on the number of males, Golden Langur societies has been divided into five types, viz., uni-male with multi-female troop, bi-male with multi-female troop, multi-male with multi-female troop, all male bands, and lone or solitary male. The average group size ranged from 5 to 19 individuals (Debahutee Roy, personal record).

13.12 Behavior

The activity pattern of Golden Langur shows a bimodal diurnal activity pattern in feeding peaks in the morning and evening, whereas resting peaked in middle of the day. Our observations revealed that the predominant activity of Golden Langur during daytime was resting, followed by feeding, moving, and other activities. These results are compatible with the findings of Zhou et al. (2007), who suggest that leaf-eating primates minimize energy expenditure in order to cope with the low protein content found in leaves. Resting time in the winter is shorter because they have 10 h on average to devote to feeding activities or moving between feeding sources. By contrast, resting time in summer may be longer owing to the increased ingestion of leaves in their diets. In primates that rely heavily on relatively low-quality (e.g., high-fiber) food, however, the capacity of the forestomach and the speed with which food residues are eliminated from it can influence the maximum amount of food that can be consumed.

Fig. 13.8 (**a**) Allogrooming behavior of Golden Langur, (**b**) playing behavior of juvenile Golden Langur, and (**c**) vigilant behavior of an adult male of Golden Langur, and the photographs were taken during November 2013

These patterns in feeding and resting are also consistent with findings in other studies of *Trachypithecus* species (Huang et al. 2002; Zhou et al. 2007). Higher feeding activities in the morning and evening were recorded (Chetry and Chetry 2009), and it suggests bimodal feeding peak. Feeding at these times allows the leaf-eating primates to compensate for the energy lost through the long hours of non-feeding during the night. In addition, the primates can make use of the cooler temperatures at these times, spending the middle of the day, when the temperatures are highest, resting and conserving energy in the shade (Clutton-Brock 1977; Huang et al. 2002).

Social behavior includes mating, grooming (Fig. 13.8a), and playing (Fig. 13.8b). A vigilant Golden Langur sits straight in a typical posture with its head raised, legs spread wide, and penis erect and directs its head toward the source of danger

(Fig. 13.8c) (Debahutee Roy personal observation). Golden Langurs are mainly diurnal and arboreal folivorous primate. But recent habitat fragmentations and human interference have made them to raid crops which leading to human-primate conflict.

13.13 Sleeping Behavior of Golden Langur

Primates spend a momentous part of their lives at sleeping sites. The selection of a secure and stable sleeping tree can be crucial for individual survival and fitness. The selection of secure and stable sleeping places and the abundance of such sites in the forest can impact each individual's chances of survival and reproductive success (Cheyne et al. 2013). Sleeping site selection in primates has commonly been explained through five nonexclusive hypotheses: predation avoidance, food access, parasite avoidance, comfort/thermoregulation, and range/resource defense. Among these, the assumption of optimized food access often features prominently despite little quantitative data. A troop of Golden Langur (*Trachypithecus geei*) at Kakoijana Reserve Forest, a fragmented and isolated habitat in Assam, India, was found to use 20 different plant species of 14 different families and 9 orders as sleeping sites (Table 13.3) (Roy and Nagarajan 2013, 2016). The Golden Langur preferred to sleep trees

Table 13.3 Tree species used for sleeping by Golden Langur in Kakoijana Reserve Forest, Assam, India

Sl. no.	Order[a]	Family[a]	Species[a]
1	Fabales	Caesalpiniaceae	*Tamarindus indica*
2	Fabales	Mimosaceae	*Albizia lebbeck*
3	Fabales	Papilionaceae	*Erythrina* sp.
4	Genitales	Apocynaceae	*Alstonia scholaris*
5	Lamiales	Verbenaceae	*Gmelina arborea*
6	Lamiales	Verbenaceae	*Tectona grandis*
7	Laurales	Lauraceae	*Litsaea sebifera*
8	Malpighiales	Euphorbiaceae	*Mallotus philippensis*
9	Malpighiales	Moraceae	*Hevea brasiliensis*
10	Malvales	Bombacaceae	*Bombyx* sp.
11	Malvales	Dipterocarpaceae	*Shorea robusta*
12	Poales	Poaceae	*Bambusa tulda*
13	Rosales	Moraceae	*Ficus glomerata*
14	Rosales	Moraceae	*Ficus hispida*
15	Rosales	Moraceae	*Ficus religiosa*
16	Rosales	Moraceae	*Ficus rumphii*
17	Sapindales	Anacardiaceae	*Mangifera indica*
18	Sapindales	Anacardiaceae	*Odina wodier*
19	Sapindales	Anacardiaceae	*Spondias pinnata*
20	Sapindales	Meliaceae	*Cedrela toona*

The list is prepared based on alphabetical order of plant orders
[a]Classification is based on Bentham and Hooker (1962–1983), Gamble (1935), Saldanha (1995), and Saldanha and Nicolson (1976)

belonged to families of Moraceae and Anacardiaceae. Interestingly, trees were used as a sleeping tree when they bear the food items such as fruits and young leaves, and on the other hand, they have avoided such trees when they do not have food items.

13.14 Home Range and Ranging

Home range is defined as the area the animals were known to use during a specified period. In other words, the total area over which the study group was seen moving and foraging during a specified period is the home range. Home range is considered as the animal's feeding, resting, and sleeping site. The area within the home range which is most frequently used and where the animals stay for longer period of time has been defined as core area or preferred area (Chetry and Chetry 2009). Studies of ranging behavior in forest primates showed that complex interaction and decision-making as "where to go" and "what to eat" occur in forest monkeys. In a number of primate species, individuals travel together during the day and also lodge together during the night to avoid predators as they often fall prey to carnivorous mammals and snakes like python (Chetry and Chetry 2009). Hence, they have to select the roosting tree that is safe enough from the predators.

The size of the home range is dependent on several factors. Ecological factors affect the size of the home range in different species of primates (Stanford 1991). Clutton-Brock (1977) illustrated that rather small differences in feeding ecology might affect markedly in ranging pattern of primates. The ranging pattern is also influenced by the location of sleeping tree, group movement in the previous day, tree phenology, and distribution and abundance of food trees (Stanford 1991).

Study of ranging behavior is useful for understanding the behavior and ecology of primates in a particular area. In primates, ranging behavior is closely related to a number of factors including availability and abundance of food, group size and composition, population density, and predator pressure besides other environmental pressures. Our observations also indicated that the size of home range of the Golden Langur was 10 ha in one isolated fragmented reserve forest, Kakoijana, and 7.7 ha in one altered rubber plantation in Nayakgaon. In Kakoijana, the groups spent most of the time outside the park, and hence the home range is larger. In a degraded habitat, home range is larger as the langurs have to travel long distances to get food.

13.15 Food Habits

Golden Langurs are generally folivorous, and young leaves are the major component in its diet followed by mature leaves, fruits, shoot, seeds, and flowers. Golden Langurs are selective feeders and feed on variety of food plants throughout their lifetime. Thus, the varied food species selection enables the species to colonize a wide range of vegetation types from tropical evergreen forests to sal-dominated forests. Golden Langur is observed to feed on 130 spp. of plants belonging to 46 families and 27 orders in fragmented forests of Assam (Table 13.4). Roy and

Table 13.4 List of food plants of Golden Langurs identified in three different fragmented forests, viz., Kakoijana Reserve Forest, Bamungaon Reserve Forest, and Nayakgaon Rubber Garden of Assam, India

Sl. No.	Category	Species[a]	Order[a]	Family[a]
1	Tree	*Moringa oleifera*	Brassicales	Moraceae
2		*Alangium chinense*	Cornales	Cornaceae
3		*Saurauia roxburghii*	Ericles	Actinidiaceae
4		*Schima wallichii*	Ericles	Theaceae
5		*Callicarpa arborea*	Ericles	Verbenaceae
6		*Bauhinia vahlii*	Fabales	Caesalpiniaceae
7		*Bauhinia purpurea*	Fabales	Caesalpiniaceae
8		*Bauhinia racemosa*	Fabales	Caesalpiniaceae
9		*Bauhinia variegata*	Fabales	Caesalpiniaceae
10		*Bauhinia variegata*	Fabales	Caesalpiniaceae
11		*Caesalpinia pulcherrima*	Fabales	Caesalpiniaceae
12		*Cassia fistula*	Fabales	Caesalpiniaceae
13		*Delonix regia*	Fabales	Caesalpiniaceae
14		*Tamarindus indica*	Fabales	Caesalpiniaceae
15		*Pterocarpus dalbergioides*	Fabales	Fabaceae
16		*Derris elliptica*	Fabales	Fabaceae
17		*Archidendron* sp.	Fabales	Fabaceae
18		*Acacia auriculiformis*	Fabales	Mimosaceae
19		*Acacia auriculiformis*	Fabales	Mimosaceae
20		*Albizia procera*	Fabales	Mimosaceae
21		*Albizia lebbeck*	Fabales	Mimosaceae
22		*Albizia stipulate*	Fabales	Mimosaceae
23		*Erythrina* sp.	Fabales	Papilionaceae
24		*Butea parviflora*	Fabales	Papilionaceae
25		*Dalbergia sissoo*	Fabales	Papilionaceae
26		*Alstonia scholaris*	Genitales	Apocynaceae
27		*Holarrhena antidysenterica*	Genitales	Apocynaceae
28		*Anthocephalus cadamba*	Genitales	Rubiaceae
29		*Hymenodictyon excelsum*	Genitales	Rubiaceae
30		*Oroxylum indicum*	Lamiales	Bignoniaceae
31		*Sterospermum chelonoides*	Lamiales	Bignoniaceae
32		*Nyctanthes arbor-tristis*	Lamiales	Oleaceae
33		*Gmelina arborea*	Lamiales	Verbenaceae
34		*Tectona grandis*	Lamiales	Verbenaceae
35		*Cinnamomum* sp.	Laurales	Lauraceae
36		*Litsea glutinosa*	Laurales	Lauraceae
37		*Litsea monopetala*	Laurales	Lauraceae
38		*Litsea sebifera*	Laurales	Lauraceae
39		*Phoebe goalparensis*	Laurales	Lauraceae
40		*Michelia champaca*	Magnoliales	Magnoliaceae
41		*Michelia oblonga*	Magnoliales	Magnoliaceae
42		*Gynocardia odorata*	Malpighiales	Achariaceae

(continued)

Table 13.4 (continued)

Sl. No.	Category	Species[a]	Order[a]	Family[a]
43		*Bischofia javanica*	Malpighiales	Euphorbiaceae
44		*Endospermum diadenum*	Malpighiales	Euphorbiaceae
45		*Macaranga denticulata*	Malpighiales	Euphorbiaceae
46		*Mallotus philippensis*	Malpighiales	Euphorbiaceae
47		*Phyllanthus emblica*	Malpighiales	Euphorbiaceae
48		*Hevea brasiliensis*	Malpighiales	Moraceae
49		*Salmalia malabarica*	Malvales	Bombacaceae
50		*Bombyx* sp.	Malvales	Bombacaceae
51		*Shorea robusta*	Malvales	Dipterocarpaceae
52		*Kydia calycina*	Malvales	Malvaceae
53		*Pterospermum acerifolium*	Malvales	Sterculiaceae
54		*Grewia multiflora*	Malvales	Tiliaceae
55		*Terminalia chebula*	Myrtales	Combretaceae
56		*Terminalia bellerica*	Myrtales	Combretaceae
57		*Lagerstroemia parviflora*	Myrtales	Lythraceae
58		*Lagerstroemia speciosa*	Myrtales	Lythraceae
59		*Syzygium fruticosum*	Myrtales	Myrtaceae
60		*Syzygium jambos*	Myrtales	Myrtaceae
61		*Psidium guajava*	Myrtales	Myrtaceae
62		*Syzygium cumini*	Myrtales	Myrtaceae
63		*Daubanga grandiflora*	Myrtales	Sonneratiaceae
64		*Averrhoa carambola*	Oxalidales	Averrhoaceae
65		*Artocarpus chaplasha*	Rosales	Moraceae
66		*Artocarpus lakoocha*	Rosales	Moraceae
67		*Ficus drupacea*	Rosales	Moraceae
68		*Ficus glomerata*	Rosales	Moraceae
69		*Ficus hispida*	Rosales	Moraceae
70		*Ficus racemosa*	Rosales	Moraceae
71		*Ficus religiosa*	Rosales	Moraceae
72		*Ficus retusa*	Rosales	Moraceae
73		*Ficus rumphii*	Rosales	Moraceae
74		*Ficus* sp.	Rosales	Moraceae
75		*Ziziphus rugosa*	Rosales	Rhamnaceae
76		*Ziziphus jujuba*	Rosales	Rhamnaceae
77		*Trema orientalis*	Rosales	Thymelaeaceae
78		*Mangifera indica*	Sapindales	Anacardiaceae
79		*Odina wodier*	Sapindales	Anacardiaceae
80		*Spondias pinnata*	Sapindales	Anacardiaceae
81		*Melia sempervirens*	Sapindales	Meliaceae
82		*Melia azadirachta*	Sapindales	Meliaceae
83		*Cedrela toona*	Sapindales	Miliaceae
84		*Chukrasia tabularis*	Sapindales	Miliaceae
85		*Dysoxylum binectariferum*	Sapindales	Miliaceae
86		*Dysoxylum procerum*	Sapindales	Miliaceae

(continued)

Table 13.4 (continued)

Sl. No.	Category	Species[a]	Order[a]	Family[a]
87		*Garcinia cowa*	Theales	Clusiaceae
88		*Dillenia pentagyna*	Unplaced	Dilleniaceae
89		*Dillenia indica*	Unplaced	Dilleniaceae
90	Shrub	*Eupatorium odoratum*	Asterales	Asteraceae
91		*Mimosa rubicaulis*	Fabales	Mimosaceae
92		*Rauvolfia serpentina*	Genitales	Apocynaceae
93		*Wrightia tomentosa*	Genitales	Apocynaceae
94		*Jasminum* sp.	Lamiales	Oleaceae
95		*Holmskioldia sanguinea*	Lamiales	Verbenaceae
96		*Antidesma acidum*	Malpighiales	Euphorbiaceae
97		*Homonoia riparia*	Malpighiales	Euphorbiaceae
98		*Flueggea leucopyrus*	Malpighiales	Phyllanthaceae
99		*Hibiscus rosa-sinensis*	Malvales	Malvaceae
100		*Microcos paniculata*	Malvales	Theaceae
101		*Bursera serrata*	Sapindales	Burseraceae
102	Orchid	*Bulbophyllum kaitens*	Asparagales	Orchidaceae
103		*Dendrobium lituiflorum*	Asparagales	Orchidaceae
104	Herb	*Evolvulus nummularius*	Solanales	Convolvulaceae
105	Grass	*Bambusa pallida*	Poales	Poaceae
106		*Bambusa tulda*	Poales	Poaceae
107	Fern	*Dryopteris* sp.	Dryopteridales	Dryopteridaceae
108		*Asplenium nidus*	Polypodiales	Aspleniaceae
109	Climber	*Mikenia micrantha*	Asterales	Asteraceae
110		*Trichosanthes cucumeroides*	Cucurbitales	Cucurbitaceae
111		*Momordica cochinchinensis*	Cucurbitales	Cucurbitaceae
112		*Tetrameles nudiflora*	Cucurbitales	Datiscaceae
113		*Dioscorea alata*	Dioscoreales	Dioscoreaceae
114		*Dioscorea bulbifera*	Dioscoreales	Dioscoreaceae
115		*Dolichos lablab*	Fabales	Papilionaceae
116		*Vigna catjang*	Fabales	Papilionaceae
117		*Hoya parasitica*	Genitales	Rubiaceae
118		*Thunbergia grandiflora*	Lamiales	Acanthaceae
119		*Smilex* sp.	Liliales	Rubiaceae
120		*Bridelia stipularis*	Malpighiales	Euphorbiaceae
121		*Aristolochia* sp.	Piperales	Aristolochiaceae
122		*Cissampelos pareira*	Ranunculases	Menispermaceae
123		*Paederia cruddasiana*	Rubiales	Rubiaceae
124		*Argyreia nervosa*	Solanales	Convolvulaceae
125		*Ipomea lacunosa*	Solanales	Convolvulaceae
126		*Merremia* sp.	Solanales	Convolvulaceae
127		*Merremia vitifolia*	Solanales	Vitaceae
128		*Tetrastigma thomsonianum*	Vitalis	Vitaceae
129		*Cissus repanda*	Vitalis	Vitaceae
130		*Vitis latifolia*	Vitalis	Vitaceae

[a]Classification is based on Bentham and Hooker (1962–1983), Gamble (1935), Saldanha (1995), and Saldanha and Nicolson (1976)

Fig. 13.9 Sleeping posture of an adult Golden Langur on Rubber (*Hevea brasiliensis*) tree in Kakoijana Reserve Forest, Assam, India, and the photograph was taken during May 2013

Nagarajan (2013) assessed and enlisted 130 plant species belonging to 46 families and 28 orders for food consumption and 20 different plant species of 14 different families and 9 orders as sleeping sites by Golden Langur in the fragmented forests of Assam. The Golden Langur preferred to sleep on trees belonged to families of Moraceae and Anacardiaceae. Interestingly, trees were used as a sleeping tree (Fig. 13.9) when they bear the food items such as fruits and young leaves, and on the other hand, they have avoided such trees when they do not have food items (Roy and Nagarajan 2013). Das et al. (2013) reported 91 plant species as food species for Golden Langur comprising both trees and climbers in Chirang Reserve Forest. In the Kakoijana Reserve Forest of western Assam, the golden langur's food comprises 67% of trees, followed by 18% of climber, 10% of shrub, 3% of grass, and 2% of orchid. Thus trees, climber, and shrub constitute over 95% of its food, in which the major proportion of food tree species belong to the family Caesalpiniaceae, followed by Moraceae and Mimosaceae (Roy and Nagarajan submitted). Mukherjee and Saha (1974) reported *Terminalia bellirica* at Jamduar, *Lagerstroemia parviflora*, flowers of *Salmalia malabarica*, and fruits of *Bridelia retusa* in the Manas sanctuary as the most preferred plants during winter. Golden langurs also feed on crops including corn (*Dolichos lablab*), shoots of bamboo (*Bambusa tulda*), etc., causing great damage in the villages adjacent to the reserve forest areas of western Assam, especially in the Kakoijana Reserve Forest area. Though they acquire water from the plant leaves, being a requisite drinker, langur drinks water from available streams during dry seasons. Moreover, Golden Langurs use rubber seeds as "fall-back" food during the scarcity of nutritious wild food (Roy and Nagarajan 2015). Rubber seed is a poor source of both calcium and iron but is potential protein source. It contains a toxic factor, cyanogenic glucoside (18.6 mg/100 g), which might be a hindrance to use as a food source. The iodine value of rubber seeds was 28.07. A study in Sri Lanka found that the inclusion rates of Rubber Seed Meal (RSM) in feeds of mature chicken did not affect the egg production, but egg size, shell thickness, hatchability, and chick weight were reduced (Eka et al. 2010).

13.16 Food Handling Techniques

Handling of the food items by Golden Langurs has been observed in the field, while they are feeding on different plants. They forage mainly on the top or middle strata, depending on the weather conditions (rainy/sunny) and habitat characteristics (canopy continuity). They sometimes come to the ground and forage (e.g., Nayakgaon Rubber Plantation). It has been observed that with variation of food items, the techniques of eating them also vary. For a leafy food item, Golden Langurs pluck the leaf and eat by hand, and sometimes they pluck a handful of leaves and put directly to their mouth. In cases of small fruits like *Syzygium cumini*, they collect the fruits one by one with their fingers and put them into their mouth. Figs of *Ficus hispida* are eaten directly from trees. Larger fruits are held in hands and then eaten. The different ways in which the Golden Langurs deal with the food items are as follows:

1. Golden Langurs generally pluck the food items like leaf, which may be a single leaf (e.g., leaf of *Gmelina arborea*) or clump of leaves (e.g., leaves of *Albizia procera*), and put the items directly to their mouth after plucking (Fig. 13.10g, h). In case of fruit (e.g., *Mangifera indica*), shoots of *Bambusa tulda*, or flower (e.g., *Bombax ceiba*), Golden Langurs hold the items with their right hand/left hand or both and then eat with average bouts of 8–12 per minute (Fig. 13.10e, f). We call this as "pluck holding."
2. During a rainy or hot mid-day, Golden Langurs have been observed to tow the branch of food plants (e.g., *Ficus hispida*, *Hevea brasiliensis*, etc.) close to them and either pick with their lips or pluck with their fingers and eat (Fig. 13.10c, d). We term this as "towing."
3. Occasionally, Golden Langurs bend themselves toward the branch of food plants and either collects the food item from trees and eats, or they eat directly from the branches (Fig. 13.10b). We consider this to be as "bending."

Uncommon ground feeding behavior on eating fallen dry fruits of rubber tree *Hevea brasiliensis* by Golden Langurs (Fig. 13.10a) was recorded in Kokrajhar District of Assam, and this is one of the main food items during scarcity of preferred food items. The langurs were observed to come down to the ground using the tree trunk and collect the fruit from the ground. After collecting, they sit on the

Fig. 13.10 Various food handling techniques of Golden Langur. (**a**) Golden Langur feeding on dry rubber (*Hevea brasiliensis*) fruit on the ground. (**b**) Golden Langur feeding on *Ficus* bending itself toward the branch. (**c**) Golden Langur feeding on *Ficus* without plucking it, directly taking from the branch. (**d**) Golden Langur feeding on young leaves of rubber (*Hevea brasiliensis*), towing the branch toward itself. (**e**) Golden Langur feeding on bamboo shoot (*Bambusa tulda*) holding it with left hand. (**f**) Golden Langur feeding on seeds of *Bauhinia* holding with both hands and opening the fruit cover. (**g**, **h**) Golden Langur collecting leaves of *Hevea brasiliensis* and *Gmelina arborea* with hand and feeding

(a)
Debahutee Roy
(b)
Debahutee Roy
Debahutee Roy
(d)
Debahutee Roy
(e)
(f)
Debahutee Roy
(g)
(h)
Debahutee Roy

ground, crack the hard covering, and feed on the nuts. The adult male initiated the ground feeding followed by the subadults and adult females. Most of the feeding bout lasted from 40 to 50 s/seed, and some of them extended up to 1 min. While eating bamboo shoots, they hold it with either hand or in between the grip of their feet, break the soft shoots with teeth, and then eat (Debahutee Roy personal observation).

13.17 Reproduction

Reproduction is an essential feature of all living organisms. Most of the primate species breeds throughout the year. As far as reproduction is concerned, there are two groups of colobines – the first group includes the Hanuman Langur species (*Semnopithecus entellus*) which breeds throughout the year, while the second group includes the species like Capped Langur (*Trachypithecus pileatus*), Nilgiri Langur (*Trachypithecus johnii*), and Golden Langur (*Trachypithecus geei*) which have definite breeding seasons (Chetry and Chetry 2009).

Observations were made on Golden Langur from 2012 to 2014 that the males remain sexually active throughout the year and on the other hand females become sexually active only in certain period of the year. It has been observed that the adult females become receptive during the post monsoon (June to September) and retreating monsoon (October to January) seasons. During the breeding seasons, maximum mating takes place in the months of September and October which declines gradually from December (Debahutee Roy personal observation). Gestation period is estimated to be 168–180 days. Pregnant females have a black line on the belly (Fig. 13.11). They give birth to a single infant at a time. January to June is the birth season for Golden Langur (Chetry and Chetry 2009).

Fig. 13.11 Adult pregnant female feeding shows the *black line* in its belly shown with a *red arrow mark*

13.18 Sexual Life in Golden Langur

Apart from copulation behavior of mating individuals, there exist some important dimensions among non-mating individuals which interfere or harass a mating pair. This type of response to mating is called "harassment of mating" (Chetry and Chetry 2009). It has been seen that the harassers are mostly the infants and sometimes the females both adult and subadult.

13.19 Female Competition

Competition among the receptive females to entrée the reproductive male of the troop has been noticed. However, females in a troop ovulate in synchrony during the mating season and try to engage sexually with the resident male, but ultimately it is the choice of the male to accept or reject a female.

13.20 Female-Infant Bonding

In nonhuman primates, care for the offspring is more intense. Neonates depend on its mother for milk till 6 months of age. Mother remains very protective toward the infant in the first few months.

13.21 Allomothering Behavior

Allomothering or aunt behavior is a unique and interesting behavior in colobines. It has been observed that females of a troop shows intense love and care for the neonates in the trop. Adult females have been seen to engage in infant transferring from the lap of one another. This behavior has been seen to be more when the adult mother is busy on feeding. She transfers the infant to another female in the troop and engages herself on feeding. According to some primatologists, allomothering behavior enhances the social bonding in a troop. Kinship may influence the likelihood of allomothering (Chetry and Chetry 2009).

13.22 Mortality

Information on mortality rates is not available. Carnivores like leopard (*Panthera pardus*) and wild dog (*Cuon alpinus*) are prominent predators in nature (Chetry and Chetry 2009). Domestic dogs (*Canis lupus familiaris*) are the major enemies of langurs at Umananda (Chetry and Chetry 2009).

13.23 Threats

13.23.1 Forest Loss in Northeast India

Northeast India falls in the Eastern Himalaya ecoregion, a priority biodiversity hotspot as well as an Endemic Bird Area (WWF Global 200 Ecoregions, Conservation International Hotspots, and Birdlife). It has over 7500 angiosperm species (Ramakantha et al. 2003). In the 2 years between 2001 and 2003, a total of 2788 km^2 (17.6%) of dense forest in Assam had been converted into open forest (FSI 2004). This large-scale degradation has left most forest species restricted to small, remnant forests where "island" or fragment dynamics and external factors such as the nature of the matrix and hunting pressures will determine whether they will survive at all. In India, forests, despite being extensively fragmented, harbor the more charismatic ape species the Hoolock Gibbon (*Bunopithecus* (*Hylobates*) *hoolock*) which is the symbol of the rain forest.

13.24 Logging of Forest

Researchers have conducted many studies to examine responses of primate species to logging in tropical forests (Johns and Skorupa 1987; Cowlishaw and Dunbar 2000). However, many of the studies give contradicting findings. For example, of the 38 tropical primate species Johns and Skorupa (1987) considered, 71% declined after logging, while 22% increased, and 6.7% were apparently unaffected. Though some of the differences may reflect differences in feeding guilds of the primates concerned, e.g., folivores vs. frugivores, in general, the importance of such comparisons is limited by factors such as differences in field methods used, primate-specific compositions and densities, logging intensity and incidental damage to forest trees, vegetation types adjacent to logged areas, forest age, tree-specific composition and density before logging, and natural variation in habitat types within the forest and in densities of large terrestrial herbivores. The factors can influence the responses of primates to logging accounted differently in several studies and even when the same primate species responded differently in different areas.

Ideally, understanding how primate populations respond to logging would require a prelogging study to provide baseline data against which one can measure any change in primate populations after logging (e.g. Cowlishaw et al. 2009). Practically, this has been impossible because logging companies are not obliged to inform primatologists or conservation groups about their activities. Second, getting a good understanding of the primate population in an area to be logged requires several years of observations, which may be incompatible with logging schedules. Consequently, the most practical way to study the responses of primate populations to habitat changes caused by logging has been via comparisons of primate populations in logged areas with the ones in adjacent unlogged parts of the forest. This may be fair enough, assuming that before the logging, the logged and unlogged forests were similar in primate populations and habitat conditions.

Fig. 13.12 Selective logging of *Lagerstroemia parviflora* in Ripu-Chirang Reserve Forest, Assam, India

However, it may not always be the case, which may account for contradictory results on the responses of primate populations to logging.

Frequent logging (Figs. 13.12 and 13.13), pitfall traps (Fig. 13.14), illegal fishing (Fig. 13.15), and extensive firewood collection by local villagers (Fig. 13.16) are the major direct threats to Golden Langur in India. Apart from these, natural calamities like storm (Fig. 13.17) cause threats to arboreal dwellers. It is regular and common to come across any temporary logging sites by illegal woodcutters inside the forests of Ripu-Chirang Reserve Forest of Assam, India (Fig. 13.18).

The risk of disease transmission between humans and nonhuman primates is even higher for the primates because they are phylogenetically closer to humans. For example, in Gombe National Park (which is essentially a forest fragment), Tanzania, 14 chimpanzees from the Kasekera community died of suspected pneumonia between 1968 and 1987, another 11 chimpanzees in the neighboring Mitumba community died of a respiratory disease in 1996 (Reynolds 2006). In the Virunga volcanoes, ≥81% of the gorillas in seven groups used for research and tourism suffered from an influenza-like disease, and six adult females succumbed to the disease during 1988 (Reynolds 2006). The observations highlight the dangers to which nonhuman primates can be exposed as forests are fragmented further, which is not studied in details in Assam. However, with primate populations and their interactions with changing habitats, 41 disease transmission occurred (identified so far) in both directions, and in which the nonhuman primates are at greater risk than humans. Diseases in humans can be detected faster and controlled or eliminated, whereas in wild nonhuman primates it can be an insurmountable task. Moreover, many primate species live in small populations, in which case a single outbreak can eradicate the

Fig. 13.13 Image of logging of a mature tree of *Shorea robusta* in Ripu-Chirang Reserve Forest, Assam, India

Fig. 13.14 Pit fall trap in Ripu-Chirang Reserve Forest, Assam, India

Fig. 13.15 Illegal fishing using a traditional fishing gear inside a cannel in Ripu-Chirang Reserve Forest, Assam, India

Fig. 13.16 Firewood collection by local villagers for self consumption in Kaoijana Reserve Forest, Assam, India

Fig. 13.17 Image of felling of *Alstonia scholaris*, a food plant of Golden Langur due to heavy storm in Kakoijana Reserve Forest, Assam, India

Fig. 13.18 Temporary camps for logging by illegal woodcutters in Ripu-Chirang Reserve Forest, Assam, India

Fig. 13.19 Showing bamboo shoots (*Bambusa tulda*) consumed by Golden Langur in human habitation in Bamungaon Reserve Forest, Assam, India, as a sign of damage to cultivated crops which leads to human-primate conflict

entire population. Infections of laboratory and zoo workers and bushmeat hunters by Simian Immunodeficiency Viruses (SIV) and other simian retroviruses have occurred (Wolfe et al. 2004). Accordingly, by increasing human contact with non-human primates via forest fragmentation, we also lope the risk of divulging ourselves to new diseases.

Straying into human habitations and crop damage especially to bamboo shoots (Fig. 13.19), corns, and seasonal vegetables is serious problems caused by Golden Langurs especially near smaller and fragmented habitats, leading to human-primate conflict. Village people drives away the Golden Langurs which frequently enter into the human habitation for crop raiding with stones by using Slingshot or Catapult (Fig. 13.20) and sometimes by bursting crackers. Organochlorine insecticide poisoning in Golden Langurs *Trachypithecus geei* has been reported from Chakrashila Wildlife Sanctuary (CWS) where a strip of agricultural land of about 100 m wide separates the CWS from the rubber plantation. Golden Langurs frequently visit the plantation area in search of food. This report is a record of organochlorine insecticide poisoning in three Golden Langurs in CWS (Pathak 2011). Another incident occurred in November, 2012, at Kakoijana Reserve Forest where an adult male was found dead (Fig. 13.21), and the cause was found to be poisoning (Debahutee Roy personal observation).

Electrocution is another threat to these arboreal dwellers. Incidence from Kakoijana Reserve Forest of Assam, India, that an adult male of Golden Langur came into the village areas to feed on bamboo shoots during March, 2014, while moving along the vicinity of the village from canopy to canopy, it got electric shock from a high-voltage power line and fell down on the ground. Immediately it was reported to Nature's Foster (a local NGO) by some local villagers, and the Golden Langur was taken to the nearest veterinary hospital and was treated. After its recovery from the shock, he was released back to the wild from where he was. Though in this case the langur is saved, but in the future, power lines may fetch a positive threat to this species.

Fig. 13.20 Villagers driving the Golden Langurs using Slingshot or Catapult which frequently enter into the human habitation for crop raiding in Bamungaon Reserve Forest, Assam, India. Slingshot or Catapult is a ballistic device used to launch a projectile a great distance using stones or pebbles

13.25 Conservation and Management Perspective

As most primate species live in tropical forests (Chapman et al. 2006), protection of forest habitats should be high on the agenda for primate conservation. However, conserving the world's tropical forests is not an easy task for several reasons. First, the forest habitats are mostly fragmented and scattered in many different countries. Hence, international organizations that are key players in conservation, need to collaborate with many national governments, each with its own priorities and problems. The countries with tropical forests that are rich in primate diversity are economically poor nations. Some of which are politically unstable. It is difficult to expect such countries to make forest protection as a high priority. Even if foreign assistance was available, it is difficult to deliver during war. Population growth rate in developing nations, particularly in Africa and Asia, is high, and most people depend directly on natural resources such as land, forest, etc. for survival. Therefore, the need to clear forests for agriculture is high. This does not promise well for forest protection. Given the problems, it is clear that saving forest habitats will need full commitment from governments and people of poor nations by international

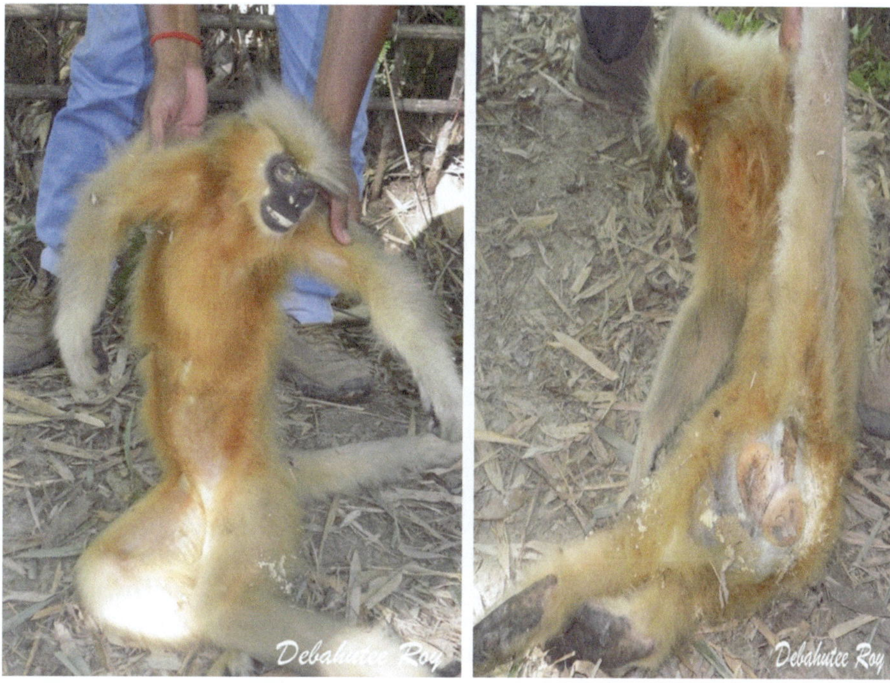

Fig. 13.21 Golden Langur died due to organochlorine insecticide in Kakoijana Reserve Forest, Assam, India

organizations, governments and people of rich nations. Assam, one of the critical geographical areas covering two biodiversity hotspots, namely, the Himalayas and Indo-Burma hotspots, occupying a special place in Northeastern India, is located between 24° 44′ N to 27° 45′ N latitude and 89° 41′ E to 96° 02′ longitude, covering 2.4% of the geographical area of the country, i.e., 78,438 km^2 area in the eastern Himalayas.

According to Forest Survey of India (2004), the recorded forest of Assam is 26,832 km^2, which is 34.21% of the total geographical of the state. Reserve forests constitute 66.58% and unclassed forest, 33.42%. The six major forest types occurring in the state are tropical wet evergreen, tropical semievergreen, tropical moist deciduous, subtropical broad-leaved hill, subtropical pine, and littoral and swamp forest (FSI 2004).

Assam is one of India's richest states in primate diversity. As per estimation, nine species of primate have been recorded. Although nonhuman primates do survive in the forest of Assam, their habitats are under severe pressure. Most of the reserved forest which had once been a rich primate habitat has been degraded, and primate populations are declining and barely able to subsist and in rapid decline. Therefore, most of the species in Assam are threatened, and their legal status is inadequately addressed by various conservation agencies. According to IUCN over one-third of primate species of the world are listed as critically endangered, endangered, or vulnerable. The primates of Assam dependent on the forests that are now threatened by legal and illegal logging, habitat fragmentation, and shifting for cultivation. Much of the forest is highly disturbed, with encroachment. Another major threat is hunting. Local people hunt primate for food. Until recently, hunting was regulated by rituals and taboos. In some parts of Assam, primates are considered sacred in Hindu religion.

Conservation efforts in recent years have not significantly reduced the impact of hunting and habitat disturbance of primates in the region. Many important studies have been conducted in recent years to evaluate species conservation and management strategies. It has become essential to document the adverse effect of destructive anthropogenic activities on primates (Narasimmarajan et al. 2011). Among the primates, langurs are unusually vulnerable to extirpation and extinction via hunting because of their inopportune combination of large body size, sluggishness, and often low level of visual alertness.

Although studies on other langur species found in the state are ongoing on aspects to behavior, ecology, and conservation, there is no research attention is given to Golden Langur despite considerable threat faced by this endangered species. Therefore, it has become a difficult task for conservation planning and management of this species. The major threats to Golden Langurs throughout their range are habitat loss, degradation, and fragmentation.

Without adequate protection of forests, Golden Langur and all other forest-dependent species will continue to decline. It is an established fact that protected areas form the critical foundation for any effective conservation effort. At the same time, these protected areas face a wide array of problems for effective conservation of Golden Langur and other primates in the region. It takes much more than the legal acquisition and demarcation of land to effectively protect any species of primate.

For the formulation of conservation action plan of an arboreal primate species like Golden Langur, concrete scientific database on various aspects of the species and its habitat are essential. Without a clear understanding of the status, distribution, and behavioral ecology of the species, it is difficult to visualize further prospects and conservation action plan of the species. The present study offers the relevant insight on Golden Langur which will have to be addressed in formulating a viable conservation action plan centering Golden Langur. At present in India, ten community-based organizations (CBOs), with the Forest Department (Fig. 13.22), patrol the habitats of Golden Langur (Horwich et al. 2013).

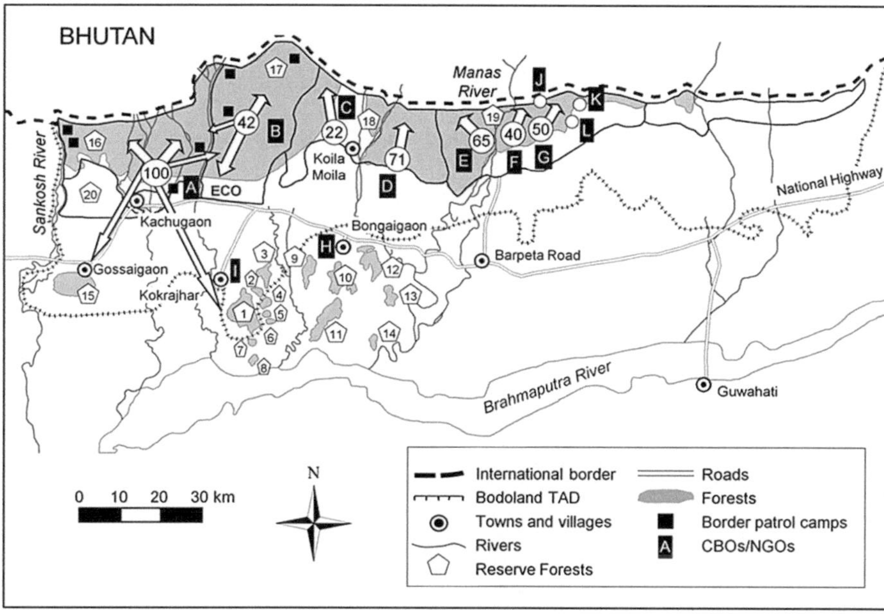

Fig. 13.22 The range of the Golden Langur *Trachypithecus geei* in Western Assam, India. The Manas Biosphere Reserve (outlined in *black*) extends along the Bhutan Border within the Bodoland Territorial Administrative District (TAD). Ten community-based organizations (CBOs), represented by letters, have forest protection forces (Represented by circles with the force number inside) that patrol the areas where the *arrows point*: (*a*) Green Forest Conservation; (*b*) Biodiversity Conservation Society; (*c*) New Horizon; (*d*) Raigajli Ecotourism and Social Welfare Society; (*e*) Panbari Manas National Park Protection and Ecotourism Society; (*f*) Swarnkwr Mithinga Onsai Afut (Bansbari range); (*g*) Manas Maozigendri Ecotourism Society; the two NGOs are Nature's Foster (*h*) and Green Heart Nature Club (*i*) based in Bongaigaon and Kokrajhar, respectively; (*j*) Manas Bhuiyanpara Conservation and Ecotourism Society; (*k*) Manas Souchi Khongkar Ecotourism Society; (*l*) Manas Agrang Society. The reserve forests (RF) and other areas are represented by numbers in pentagons: *1* Chakrashila Wildlife Sanctuary, *2* Nayekgaon, *3* Nadangiri RF, *4* rubber plantation, *5* Bheskamari RF, *6* Bangaldoba RF, *7* Singram RF, *8* Sampamon RF, *9* Dhumeshwar RF, *10* Nakkati RF, *11* Bhairab RF, *12* Kakoijana RF, *13* Bamungaon RF, *14* Kharagaon RF, *15* Guma RF, *16* Ripu RF, *17* Chirang RF, *18* Manas RF, *19* Manas National Park, *20* Kachugaon RF. ECO indicates an area that is being reforested (Source: Horwich et al. 2013)

Acknowledgments We thank the Assam Forest Department for the permission to conduct research within reserve forests of Assam, India. We thank Bombay Natural History Society (BNHS), India, for their help in identifying plant species and A.V.C. College (Autonomous), Department of Zoology and Wildlife Biology, Tamil Nadu, India, for their support. We thank Nature's Foster (NGO, based at Bongaigaon Assam, India) for their moral support. We thank our field assistants Mr. Kartik and Mr. Bhuban for their help in the field.

References

Bentham G, Hooker JD (1962–1983) Genera plantarum ad exemplaria imprimis in herbariis kewensibus servata defnita, 3 vols. Reeve & Co., London

Bentham G, Hooker JD (1983) Genera plantarum. Lovell, Reeve and Co, London, p 475

Chapman CA, Lawes MJ, Eeley HAC (2006) What hope for African primate diversity? Afr J Ecol 44(2):116–133

Chetry R, Chetry D (2009) The golden langur. Gibbon Conservation Centre. Gibbon Wildlife Sanctuary, Mariani, p 206

Cheyne SM, Rowland D, Hoing A, Husson SJ (2013) How orangutans choose where to sleep: comparison of Nest-site variables. Asian Primates J 3(1):13–17

Clutton-Brock TH (1977) Some aspects of intraspecific variation in feeding and ranging behavior in primates. In: Clutton-Brock TH (ed) Primate ecology: studies of feeding and ranging behavior in lemurs, monkeys and apes. Academic Press, London, pp 539–556

Cowlishaw G, Dunbar RI (2000) Primate conservation biology. University of Chicago Press, Chicago

Cowlishaw G, Pettifor RA, Isaac NJB (2009) High variability in patterns of population decline: the importance of local processes in species extinctions. Proc R Soc B Biol Sci 276(1654):63–69

Das R, Sinha H, Sahu HK, Choudhury K (2013) Golden Langur *Trachypithecus geei* (Khajuria, 1956) feeding on Cryptocoryne retrospiralis (Roxb.) Kunth (Family: Araceae): a rare feeding observation in Chirang reserve Forest, Assam, India. J Threatened Taxa 5(15):5013–5015

Eka HD, Aris T, Nadiah W (2010) Potential use of Malaysian rubber (*Hevea brasiliensis*) seed as food, feed and biofuel. Int Food Res J 17:527–534

Forest Survey of India (FSI) (2004) Forest cover change matrix. Assam

Gamble JS (1935) The flora of the Presidency of Madras. Adlard and Son's Ltd, London, p 2017

Ghosh S (2009) Report on the distribution and population of golden langur (*Trachypithecus geei*) in Bodoland territorial council. Assam, India, pp 1–44

Horwich RH, Das R, Bose A (2013) Conservation and the current status of the golden langur in Assam, India, with reference to Bhutan. Primate Conserv 1(27):77–83

Huang C, Wei F, Li M, Li Y, Sun R (2002) Sleeping cave selection, activity pattern and time budget of white-headed langurs. Int J Primatol 24(4):813–824

Johns AD, Skorupa JP (1987) Responses of rain-forest primates to habitat disturbance: a review. Int J Primatol 8(2):157–191

Khajuria H (1956) A new langur (Primates: Colobidae) from Goalpara District, Assam. Ann Mag Nat Hist 12(9):86–88

Medhi R, Chetry D, Bhattacharjee PC, Patiri BN (2004) Status of Golden Langur, *Trachypithecus geei* in rubber plantation in Western Assam, India. Int J Primatol 25(6):1331–1337

Mukherjee RP, Saha SS (1974) The Golden Langurs (*Presbytis geei* Khajuria, 1956) of Assam. Primates 15(4):327–340

Narasimmarajan K, Nagarajan R, Kumaraguru A (2011) Some observations on demography and edible plants of Lion-tailed Macaques (*Macaca silenus*) in the rain forest fragmented habitats of Anamalai Hills, Western Ghats. J Res Biol 1(5):352–362

Pathak DC (2011) Organochlorine insecticide poisoning in golden langurs *Trachypithecus geei*. J Threatened Taxa 3(7):1959–1960

Ramakantha V, Gupta AK, Kumar A (2003) Biodiversity of Northeast India – an overview. In: Gupta AK, Kumar A, Ramakantha V (eds) ENVIS bulletin: wildlife and protected areas, conservation of rainforests in India, pp 25–40

Reynolds V (2006) Threats and protection of the chimpanzees of the Budongo forest reserve. In: Newton-Fisher NE, Notman H, Paterson JD, Reynolds V (eds) Primates of Western Uganda. Springer, New York, pp 391–403

Roy D, Nagarajan R (2013) Food plants and sleeping trees used by Golden Langur (*Trachypithecus geei,* Khajuria, 1956) in fragmented forests of Assam, India. Mayur 9:72–77

Roy D, Nagarajan R (2015) Dry seeds of rubber (*Hevea brasiliensis*)-a fall-back food for golden langur with respect to habitat fragmentation in Assam, India. Students Conference on Conservation Science on 8–12 Sept 2015 at IISC, Bangalore, India
Roy D, Nagarajan R (2016) Use of sleeping trees by golden langurs (*Trachypithecus geei*). Taprobanica 8(1):33–36
Roy D, Nagarajan R (submitted) Plant diversity in the diet of Golden Langur *Trachypithecus geei* in Kakoijana Reserve Forest, Western Assam, India. J Threatened Taxa
Saldanha CJ (1995) Flora of Karnataka. Oxford and IBH Pub Co, New Delhi, p 1984
Saldanha CJ, Nicolson DH (1976) Flora of Hassan District, Karnataka, India. Amerind Pub Co, New Delhi, p 915
Saunders DA, Hobbs RJ, Margules CR (1991) Biological consequences of ecosystem fragmentation: a review. Conserv Biol 1(5):18–32
Sristava A, Biswas J, Das J, Bujarbarua P (2001) Status and distribution of Golden Langurs (*Trachypithecus geei*) in Assam, India. Am J Primatol 55(1):15–23
Stanford CB (1991) The diet of the capped langur (*Presbytis pileata*) in a moist deciduous forest in Bangladesh. Int J Primatol 3(12):199–216
Wangchuk T (2005) The evolution and phylogeography, and conservation of the Golden Langur (*Trachypithecus geei*) in Bhutan. D.Phil. thesis, University of Maryland
Wangchuk T, Inouye DW, Hare MP (2003) A new subspecies of Golden Langur (*Trachypithecus geei*) from Bhutan. Folia Primatol 72(2):104–108
Wolfe DN, Switzer WM, Carr JK, Bhullar VB, Shanmugan V, Tamoufe U (2004) Naturally acquired simian retrovirus infections in central African hunters. Lancet 363(9413):932–937
Zhou Q, Wei F, Huang C, Li M, Ren B, Luo B (2007) Seasonal variation in the activity patterns and time budgets of *Trachypithecus francoisiin*. The Nonggang Nature Reserve. China. Int J Primatol 28(3):657–671

14 Why Are Lion-Tailed Macaques Rare?

R. Krishnamani and Ajith Kumar

Abstract

The intermittent distribution of various animal and plant species in the world, has long been at the core of conservation biology and community ecology. It is therefore vital to know the processes and factors that impact rarities and endemism in the plant and animal world. The mechanism that regulates this rarity and limits a species' distribution and abundance is still not fully understood. We are now beginning to understand that certain plants play an important role during periods of food scarcity. Generally, the plants that sustain animals during periods of food shortage are themselves not abundant (relative to other food trees), and during periods of food abundance, these plants may not be eaten at all. But *Cullenia* and *Ficus* species are the most sought after by the lion-tailed macaques and are important food resources throughout the year. Here we show how these much-favored food trees govern the distribution and very existence of this primate.

Keywords

Ficus spp. · Lion-tailed macaque · *Macaca silenus* · *Cullenia exarillata* · Rainforests

R. Krishnamani (✉)
The Rainforest Initiative, 51/2, 50 Feet Road, Krishnaswamy Nagar,
Ramanathapuram PO, Coimbatore, 641045 Tamil Nadu, India.
e-mail: rkmani@rainforest-initiative.org

A. Kumar
Center for Wildlife Studies, National Center for Biological Sciences,
Tata Institute of Fundamental Research, Bellary Road,
Bangalore, 560065 Karnataka, India.

C. Sivaperuman, K. Venkataraman (eds.), *Indian Hotspots*,
https://doi.org/10.1007/978-981-10-6605-4_14

14.1 Introduction

This world is replete with rare and endemic animal and plant species that are patchily distributed. Understanding the processes that cause this rarity has always been a moot point in community ecology and conservation biology (Soulé 1986). While various hypotheses and theories have been put forward to explain this paradigm (Hubbell and Foster 1986; Preston 1962a, b; Rabinowitz et al. 1986), the mechanisms that regulate and limit a species' distribution and abundance are still not fully understood. The "top-down" school holds that predators limit herbivores and thereby prevent them from overexploiting vegetation. "Bottom-up" proponents stress the role of plant chemical defenses in limiting plant depredation by herbivores. Although "top-down" and "bottoms-up" proponents battle on the synecological front (Polis and Strong 1996; Terborgh et al. 2001), ecological and evolutionary forces that govern animals and/or plants are not fully understood. Of particular interest is why certain animals are always rare?

Generally, animals at the top of an ecological pyramid are scarce when compared to the secondary consumers at the bottom of the pyramid. This is true of predators that are at the apex of a food chain because they are controlled by the abundance of their herbivorous prey. Then again smaller species tend to be more abundant than their larger counterparts. Moreover, the distribution and abundance of an animal are dictated by the "quality" of the habitat it occupies. In among these generalizations, if any animal is a large predator, then it is all more rare. Similarly, altitudinal changes could also make an animal uncommon. Equally, folivorous species tend to be more abundant than frugivores and nectarivores. Frugivores generally have large home ranges because of the ephemeral and patchy nature of the availability of the fruits (Fleming 1992). Animals with low reproductive rates and higher inter-birth intervals are also rare. Taking into consideration all these factors, our attention was drawn toward the lion-tailed macaque (*Macaca silenus*).

Although the modern primates and angiosperms appear to have a very tight coevolutionary relationship (Sussman 1991), Howe (1986) opined that it was more of a co-occurrence than otherwise. Angiosperms made their appearance in the early Cretaceous (65 million years ago (MYA)), and the modern rainforests along with modern primates appeared around the same time in mid-Eocene (50 MYA: Sussman 1991). The adaptive radiation and eventual dominance of angiosperms during the Cretaceous opened up a variety of dietary opportunities for the primates (Regal 1977). If present-day primates are any indication, early primates appear to have taken strong advantage of arboreal plant foods since almost all potential food comes from dicotyledonous species using the C_3 carbon pathway and not from the monocotyledonous species (like grass) using the C_4 pathway (Milton 1987).

The genus *Macaca* made its appearance around the mid-Miocene (10 Ma), and the lion-tailed macaque may have evolved around 3 Ma. It is generally considered the ancestor of all extant Asian macaques (Fooden 1975) and hence is the earliest and longest resident in the tropical rainforests of the Western Ghats of southern India.

The lion-tailed macaque is one of the most endangered primates of India. It inhabits the dense tropical rainforests of the Western Ghats between 100 and 1500 m (msl). It is an obligate frugivore and has adapted to a highly arboreal life. Nearly 60% of its diet consists of fruits, and the balance is made up of seeds, flowers, invertebrates, and vertebrates. Females stay with the natal group throughout their lives while males migrate between groups. The home range varies from 1 km^2 to around 7 km^2 depending upon the "quality" of the habitat. Lion-tailed macaques live in groups of 8–40 animals, with an average of about 18 animals. Typically, groups have one adult male, one subadult male, five to seven adult females, and the remaining being juveniles and infants. The average adult sex ratio is about five females to one male. Births occur throughout the year with a peak between late November and February (Kumar 1995; Lindburg 1987). Females become primiparous at an average age of 6.6 years, and the inter-birth interval is around 2.5 years, which is considerably higher compared to other macaques. The mean mortality rate irrespective of age/sex classes is around 0.045 per year, which is considerably lower than other macaques. The low population growth of these macaques is due to a delayed primiparity and very low birth rates, but it is compensated by its relatively high survival rate (Kumar 1995). Hence the lion-tailed macaques have always been rare when compared to the other primates of India. Recent population surveys suggest that there are about 4000 individuals in the wild. Here we show how the distribution and existence of a rare and endemic primate, the lion-tailed macaque, are governed by the abundant and much favored food tree, *Cullenia exarillata*, and trees of the genus *Ficus*. The lion-tailed macaque is endemic to the Western Ghats of southern India, whereas the tree, *Cullenia*, is present in the Western Ghats (Kadambi 1954) and Sri Lanka.

14.2 Methods

We surveyed the vegetation of the lion-tailed macaque habitats in the rainforests of the Western Ghats, during 1999–2000 with particular reference to *Cullenia* and *Ficus*. The floristic composition was studied using 48 belt transects by dividing their habitat into two regions based on the vegetation patterns. The area of each belt transect was 0.25 ha (250 × 50 m), and the distance between the first and the 48th belt transect was around 720 km. The lion-tailed macaque habitat was split into two regions and the Palakkad Gap, at around 11° N (Fig. 14.1). The number of individuals and basal areas of each woody plant species were calculated – representing different altitudinal and latitudinal gradients. The woody plant species are represented by trees (≥30 cm gbh) and lianas (≥10 cm gbh). The altitudinal range was between 100 and 1500 m, and the latitudinal range was between 8°3′ and 14°17′ N. The *Ficus* densities were calculated using the software Distance 3.5 (Thomas et al. 1998).

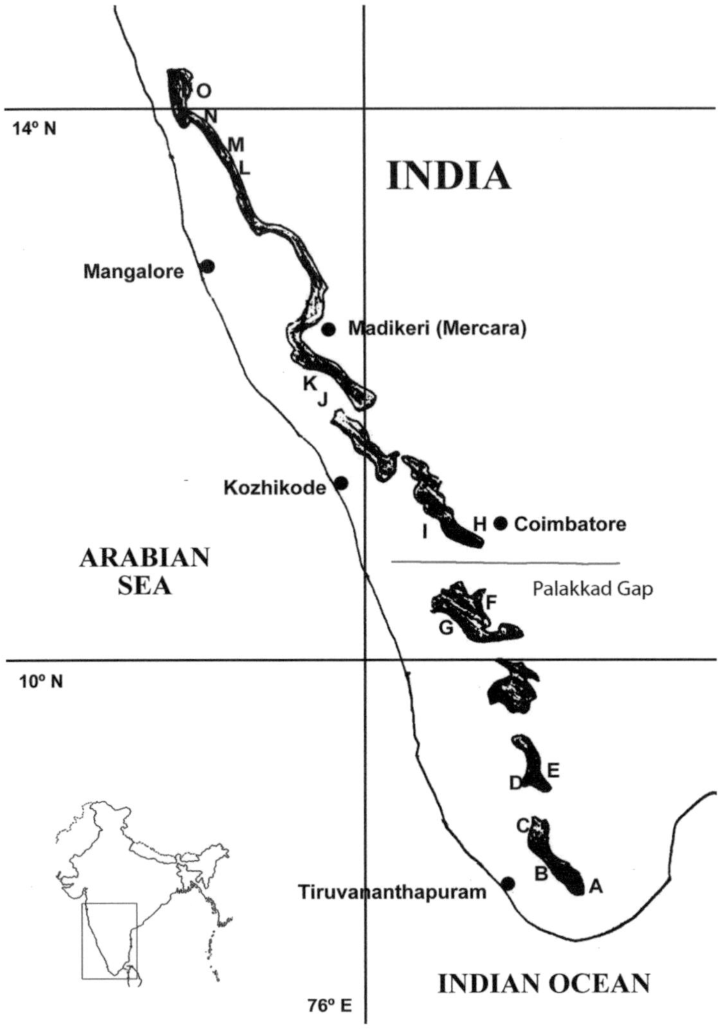

Fig. 14.1 The distributional range of the lion-tailed macaques and the location of the study areas

14.3 Results and Discussion

We sampled 391 woody plant species consisting of 341 trees and 50 lianas (hereinafter referred to as "sampled species") in the lion-tailed macaque habitats, of which 114 woody plant species were lion-tailed macaque food species consisting of 10 lianas and 104 trees (hereinafter referred to as "food species"). This formed 29.16% of the known food trees and lianas. The 15 most important species that contributed to a large portion of the lion-tailed macaque's diet are only 11.23% of the total number of individuals of the sampled trees, and their basal area contributed to only 20.11% of the total.

Studies in different parts of the Western Ghats show that a total of 218 plants are used as food resource by the lion-tailed macaques from 61 plant families (Krishnamani and Kumar 2000). For the lion-tailed macaque, leaves constitute less than 1% of its diet, and nearly 60% of its food items are fruits. The number of individuals in the sampled species and food species did not vary among zones, and there was no north-south gradation (one-way ANOVA: $F = 0.854$, 1.179, df = 3, $P > 0.0$), whereas the basal areas of the sampled species showed variation among zones, and the basal areas were higher in the sampled species ($F = 6.92$, df = 3, $P < 0.01$), but the basal areas of the food species showed none ($F = 2.484$, df = 3, $P > 0.05$). The number of individuals of all the species of the sampled species decreased with altitude ($r = -0.371$, $n = 48$, $P < 0.01$), and it increased with latitude ($r = 0.317$, $n = 48$, $P < 0.05$), both being weakly significant. But the number of individuals of food species did not show any altitudinal or latitudinal pattern ($r = 0.099$, 0.124, $n = 48$, $P > 0.05$), whereas the basal areas of the sampled species and food species showed marked patterns, with both altitude ($r = 0.431$, 0.547, $n = 48$, $P < 0.01$) and latitude ($r = -0.558$, −0.301, $n = 48$, $P < 0.01$, <0.05).

Cullenia is an important food tree of the lion-tailed macaque. It is the predominant tree in the southern region, and its dominance progressively decreases toward the north and is scarcely present in the northern region (Table 14.1). In fact the northern most limit of *Cullenia* is at 11°75′ N (Pascal 1988; Pascal et al. 1982), and the northern limit of the lion-tailed macaque habitat is at 14°5′ N. Despite this tree's rarity and/or absence in the northern region, the lion-tailed macaques are present there. The lion-tailed macaques feed on *Cullenia* flowers and seeds for most of the year. Flowering in *Cullenia* occurs during periods of fruit scarcity (February–April) and also when most of the other plant species do not flower (Ganesh and Davidar 1997; Green and Minkowski 1977). During this time the lion-tailed macaque feeds heavily on the flowers of *Cullenia*. Although the macaques do depend on nectar, because of its low volume, the fleshy and sweet sepals are readily consumed. The dominance of *Cullenia* results in an overabundance of flowers during this period and is the only species producing such high-quality flowers in the southern region. It is also a remarkably predictable food resource, flowering with the same intensity year after year.

Primates generally give birth when the food resources are abundant (Lindburg 1987), whereas the lion-tailed macaques give birth during December to February, and their favored food trees are not in fruit during February to April. During this time the macaques depend on the flowers of *Cullenia* and the aseasonal and nonsynchronous fruits of *Ficus*. Seasonal reproduction allows animals to be exposed to a varying environment to anticipate and prepare for changes that would adversely affect their fitness. The high survival rates of the neonates in these macaques prove that the macaques may not face problems associated with nonavailability of food resources during the time when they are born. However it is difficult for the lion-tailed macaques to proliferate since the lean season seems to be a bottleneck. The

Table 14.1 The top-ranked food species (*n* = 5–10) contribute to a major portion of a primate's diet. Taxonomy and nomenclature follow Rowe 1996

Primate species	Top-ranked food species (n)	% age of its diet	Source
Blue monkey,	10	69.23	Rudran (1978)
Cercopithecus mitis stuhlmanni	5	34.70	Struhsaker (1978)
Tana river red colobus, *Procolobus rufomitratus* (= *Colobus badius rufomitratus*)	10	>50	Medley (1993)
Red colobus, *Procolobus badius tephrosceles* (= *Colobus badius tephrosceles*)	5	34.3	Struhsaker (1978)
Black and white colobus, *Colobus guereza occidentalis*	5	59.8	Struhsaker (1978)
Gray-cheeked mangabey, *Lophocebus albigena johnstonii* (= *Cercocebus albigena johnstoni*)	5	50.9	Struhsaker (1978)
Red-tailed monkey, *Cercopithecus ascanius schmidti*	5	36.5	Struhsaker (1978)
Bonnet macaque, *Macaca radiata*	5	54.9	Krishnamani (1994)
Barbary macaque, *Macaca sylvanus*	8	≈65	Ménard and Vallet (1996)
Sulawesi crested black macaque, *Macaca nigra*	5	≈50	O'Brien and Kinnaird (1997)

severity of this impediment might vary temporally and spatially depending the "quality" of the habitat. Coping up with this selection pressure may be very crucial to the very survival of these macaques. Considering the fact that these macaques had always been rare (Kumar 1985), we believe these ecological pressures seem to have operated in the same way in the past.

The lion-tailed macaques living in the northern region are at a double disadvantage. The phenological patterns exhibited for the northern region are unimodal, whereas for the southern region, it is bimodal. Hence the availability of fruits in the northern region is limited to a shorter period due to an increased dry period, whereas in the southern region, fruits are available for a considerable longer period. Of the 84 plant species used as food plants, 23 species (27.38%) accounted for a whopping 70.9% (mean for 2 years, SE = 1.7) of the lion-tailed macaque's annual diet (Kumar 1987). Other researchers have also observed the fact that a small percentage of the food species contribute to a major portion of a primate's diet (Table 14.1). This is of paramount importance since only a few food trees sustain the survival of the lion-tailed macaques and the species diversity of the sampled species is lesser in the northern region compared to the south. The sampled species richness for zones 1 and 2 in the northern region was 164 and 183, respectively, and for zones 3 and 4, in the southern region, were 187 and 210, respectively. This means that the southern

Table 14.2 Densities and basal areas of sampled species, food species, *Cullenia*, and *Ficus* spp. in the two regions of the lion-tailed macaque habitats

		Southern region	Northern region	Overall
Sampled area (ha)		6	6	12
Sampled species	Density/ha	430.00	477.17	453.58
	Basal area/ha (m^2)	50.59	41.59	45.99
Food species	Density/ha	245.17	255.50	250.33
	Basal area/ha (m^2)	33.28	24.99	29.14
Cullenia exarillata	Density/ha	28.00	4.83	16.42
	Basal area/ha (m^2)	6.59	0.85	3.72
Ficus spp.	Density/ha	3.01	4.4	3.49
	No. of species	9	9	11

region is more species-rich when compared to the northern region. Also the nonavailability of *Cullenia* in most of the areas in the northern region leaves the macaques with only one choice: *Ficus*. Fortunately, the *Ficus* density in the lion-tailed macaque habitat is higher for areas north of Palakkad Gap compared to the southern region (Table 14.2). The *Ficus* density in the Western Ghats (3.49 trees/ha) is comparable to other areas (Krishnamani and Kumar 2000), whereas the *Ficus* density at the southern extreme of its distribution is very low (1 tree/ha) (Ganesh and Davidar 1997).

Two main attributes determine *Ficus* species as an important food resource for primates during periods of fruit scarcity. Firstly, fruiting patterns of *Ficus* species exhibit spatiotemporal patchiness (Gautier-Hion and Michaloud 1989; Janzen 1979; Milton et al. 1982) ensuring that some individual *Ficus* species are in fruit throughout the year within the same habitat. Secondly, *Ficus* species are usually present at low densities, when compared to other trees in forest ecosystems, and individual trees are usually clumped (Heithaus and Fleming 1978; Gautier-Hion and Michaloud 1989). Owing to these characteristics, it appears that *Ficus* species can be exploited as a major fruit resource only by an animal with a relatively large home range (Borges 1993).

Ficus species form a major portion of lion-tailed macaque's diet. *Ficus* species contain high amounts of amino acids, such as leucine, lysine, valine, and arginine and minerals, such as potassium, calcium, magnesium, sodium, and phosphorous (Wendeln et al. 2000). Although no single species of *Ficus* may be sufficient to sustain frugivores; a mix of *Ficus* species can provide a complete set of nutrients (Wendeln et al. 2000). There are more than 750 species of *Ficus* in this world (Berg 1989), and southern India is the home to 30 *Ficus* species (Sasidharan and Augustine 1999). Of this, 16 species are present within the range of the lion-tailed macaques, and they eat at least 15 species. *Ficus* is a keystone genus that supports a large number of frugivores during periods of fruit scarcity (Terborgh 1986a) since their asynchronous fruiting habit assures that crops will ripen at all times of the year (Milton 1980; Morrison 1978). Some non-*Ficus* fruits also act as keystone resources, although quantitatively they are of minor importance (Terborgh 1986b), but here we see that a non-*Ficus* species like *Cullenia* seems to be very important for the frugivore community. Usually a species is considered to be a "keystone" when its effect is large compared to its biomass or because of its large density (Jordán et al. 1999). Here *Cullenia* and *Ficus* fit this description. Although carnivorous predators limit the population densities of primary consumers like primates, it is these keystone food resources that regulate the carrying capacity of the frugivorous community during periods of food (fruit) scarcity (Terborgh 1986a). Since plant genera evolve far more slowly than animal genera, "keystone plant species" may act over an evolutionary time as a decisive factor in the evolution of whole faunal assemblages (Terborgh 1986b).

The density of fruit trees, especially some *Ficus* species, regulates the overall density of the orangutans (van Schaik and Djojosudharmo 1992). Sumatran orangutans (*Pongo abelii*) are rare at higher altitudes where *Ficus* densities are also low along with other soft-pulped fruit trees. Similarly, the rarity of a tamarin species (*Saguinus fuscicollis*) has been attributed to two keystone species. The flowers of *Combretum assimile* and *Quararibea cordata* provide them with nectar during periods of food scarcity (Terborgh and Stern 1987).

In the southern region of the Western Ghats, *Ficus* trees are at low densities, and hence *Cullenia* is the major keystone species (Ganesh and Davidar 1997). In the northern region, *Ficus* densities are relatively higher, and hence the lion-tailed macaques do not suffer during periods of fruit scarcity even in the absence of *Cullenia.* Hence the absence of *Cullenia* cannot be a limiting factor for the lion-tailed macaques, in the northern region. The only difference is that the densities of *Cullenia* far outweigh the densities of *Ficus*, but the asynchronous and year-round fruiting pattern of the *Ficus* may compensate for the predictable flowering in *Cullenia.* It is likely that *Ficus* is the keystone species for the lion-tailed macaque in the northern parts and *Cullenia* in the southern parts. The high survival rates of the infant lion-tailed macaques prove that the macaques may not face problems associated with nonavailability of food resources during the time when they are born; however, it is difficult for them to proliferate since the lean season seems to be a bottleneck. Hence, these tree species could well be the limiting factors on which the densities of these macaques depend.

Acknowledgments We thank the Forest Departments of Tamil Nadu, Kerala, and Karnataka for giving us the necessary permits to do this study. We thank Drs. Ramachandra Swamy, Sasidharan, and Ravikumar for identifying the plant specimens and Appu and Ronnie for their assistance in the field. This study was supported by Chicago Zoological Society, Primate Conservation, Inc., International Primatological Society, Wildlife Conservation Society, and National Geographic Society.

References

Berg CC (1989) Classification and distribution of *Ficus*. Exp Dermatol 45:605–611

Borges RM (1993) Figs, Malabar giant squirrels, and fruit shortages within two tropical Indian forests. Biotropica 25:183–190

Fleming TH (1992) How do fruit- and nectar-feeding birds and mammals track their food resources? In: Hunter MD, Ohgushi T, Price PW (eds) Effects of resource distribution on animal-plant interactions. Academic, San Diego, pp 355–391

Fooden J (1975) Taxonomy and evolution of lion tail and pigtail macaques (Primates: Cercopithecidae). Fieldiana (Zoology) 67:1–169

Ganesh T, Davidar P (1997) Flowering phenology and flower predation of *Cullenia exarillata* (Bombacaceae) by arboreal vertebrates in Western Ghats. India J Trop Ecol 13:459–468

Gautier-Hion A, Michaloud G (1989) Are figs always keystone resources for tropical frugivorous vertebrates? A test in Gabon. Ecology 70:1826–1833

Green SM, Minkowski K (1977) The lion-tailed macaque and its south Indian rainforest habitat. In: Prince Rainier HSH III, Bourne GH (eds) Primate conservation. Academic, New York, pp 290–337

Heithaus ER, Fleming TH (1978) Foraging movements of a frugivorous bat, *Carollia perspicillata* (Phyllostomatidae). Ecol Monogr 48:127–143

Howe HF (1986) Seed dispersal. In: Murray DR (ed) Seed dispersal by fruit-eating birds and mammals. Academic, Sydney, pp 123–189

Hubbell SP, Foster RB (1986) Conservation biology: the science of scarcity and diversity. In: Soulé ME (ed) Commonness and rarity in a neotropical forest: implications for tropical tree conservation. Sinauer Associates, Massachusetts, pp 205–231

Janzen DH (1979) How to be a fig. Annu Rev Ecol Syst 10:13–51

Jordán F, Takács-Sánta A, Molnár I (1999) A reliability theoretical quest for keystones. Oikos 86:453–462

Kadambi K (1954) *Cullenia excelsa*, Wight. (*C. zeylanica*, Gardner, *Durio zeylanicus* Gardner). Indian For 79:443–445

Krishnamani R (1994) Diet composition of the bonnet macaque (*Macaca radiata*) in a tropical dry evergreen forest of southern India. Trop Biodivers 2:285–302

Krishnamani R, Kumar A (2000) Phyto-ecology of the lion-tailed macaque (*Macaca silenus*) habitats in Karnataka, India: floristic composition and density of food trees. Primate Rep 58:27–56

Kumar A (1985) The lion-tailed macaque: status and conservation. In: Heltne PG (ed) Patterns of extinction in India, Sri Lanka, and elsewhere in southeast Asia: implications for lion-tailed macaque wildlife management and the Indian conservation system. Alan R. Liss, New York, pp 65–89

Kumar A (1987) The ecology and population dynamics of the lion-tailed macaque (*Macaca silenus*) in south India. Unpublished Ph.D. thesis. University of Cambridge, Cambridge, 174 p

Kumar A (1995) Birth rate and survival in relation to group size in the lion-tailed macaque, Macaca silenus. Primates 36:1–9

Lindburgh DG (1987) Comparative primate biology, vol 2: behavior, cognition and motivation. In: Mitchell G, Erwin J (eds) Seasonality of reproduction in primates. Alan R. Liss, New York, pp 167–218

Medley KE (1993) Primate conservation along the Tana River, Kenya: an examination of the forest habitat. Conserv Biol 7:109–121

Ménard N, Vallet D (1996) Evolution and ecology of macaque societies. In: Fa JE, Lindburg DG (eds) Demography and ecology of Barbary macaques (*Macaca sylvanus*) in two different habitats. Cambridge University Press, Cambridge, pp 106–131

Milton K (1980) The foraging strategy of howler monkeys: a study in primate economics. Columbia University Press, New York

Milton K (1987) Food and evolution: toward a theory of human food habits. In: Harris M, Ross EB (eds) Primate diets and gut morphology: implications for Hominid evolution. Temple University Press, Philadelphia, pp 93–115

Milton K, Windsor DM, Morrison DW, Estribi MA (1982) Fruiting phenologies of two neotropical *Ficus* species. Ecology 63:752–762

Morrison DW (1978) Foraging ecology and energetics of the frugivorous bat *Artibeus jamaicensis*. Ecology 59:716–723

O'Brien TG, Kinnaird MF (1997) Behavior, diet, and movements of the Sulawesi crested black macaque (*Macaca nigra*). Int J Primatol 18:321–351
Pascal J-P 1988. Wet evergreen forests of the Western Ghats of India: ecology, structure, floristic composition and succession. Pondicherry: French Institute, 346 p
Pascal J–P, Shyamsundar S, Meher-Homji VM (1982) Forest map of south India: Mercara-Mysore. Forest Departments of Karnataka and Kerala and French Institute, Pondicherry
Polis GA, Strong DR (1996) Food web complexity and community dynamics. Am Nat 147:813–846
Preston FW (1962a) The canonical distribution of commonness and rarity. Ecology 43:185–210
Preston FW (1962b) The canonical distribution of commonness and rarity. Ecology 43:410–432
Rabinowitz D, Cairns S, Dillon T (1986) Conservation biology: the science of scarcity and diversity. In: Soulé ME (ed) Seven forms of rarity and their frequency in the Flora of the British Isles. Sinauer Associates, Massachusetts, pp 182–204
Regal PJ (1977) Ecology and evolution of flowering plant dominance. Science 196:622–629
Rudran R (1978) Socioecology of the blue monkeys (*Cercopithecus mitis stuhlmanni*) of the Kibale Forest, Uganda. Smith Contrib Zool 249:1–88
Sasidharan N, Augustine J (1999) *Ficus caulocarpa* Miq. and *F. costata* Ait. (Moraceae) – addition to the flora of India. Rheedea 9:77–80
Soulé ME (1986) Conservation biology: the science of scarcity and diversity. Sinauer Associates, Massachusetts
Struhsaker TT (1978) Recent advances in primatology. Vol. 1: behaviour. In: Chivers DJ, Herbert J (eds) Food habits of five monkey species in the Kibale Forest, Uganda. Academic, London, pp 225–248
Sussman RW (1991) Primate origins and the evolution of angiosperms. Am J Primatol 23:209–223
Terborgh J (1986a) Frugivores and seed dispersal. In: Estrada A, Fleming TH (eds) Community aspects of frugivory in tropical forests. Dr. W. Junk Publishers, Dordrecht, pp 371–384
Terborgh J (1986b) Conservation biology: the science of scarcity and diversity. In: Soulé ME (ed) Keystone plant resources in the tropical forest. Sinauer Associates, Massachusetts, pp 330–344
Terborgh J, Stern M (1987) The surreptitious life of the saddle-backed tamarin. Am Sci 75:260–269
Terborgh J, Lopez L, Nuñez P, Rao M, Shahabuddin G, Orihuela G, Riveros M, Ascanio R, Adler GH, Lambert TD, Balbas L (2001) Ecological meltdown in predator-free forest fragments. Science 294:1923–1926
Thomas L, Laake JL, Derry JF, Buckland ST, Borchers DL, Anderson DR, Burnham KP, Strindberg S, Hedley SL, Burt ML, Marques F, Pollard JH, Fewster RM (1998) Distance 3.5. Research Unit for Wildlife Population Assessment, University of St. Andrews, St. Andrews. Available at http://www.ruwpa.st-and.ac.uk/distance/
van Schaik CP, Djojosudharmo S (1992) Why are orang utans so rare in the highlands? Altitudinal changes in a Sumatran forest. Trop Biodivers 1:11–22
Wendeln MC, Runkle JR, Kalko EKV (2000) Nutritional values of 14 fig species and bat feeding preferences in Panama. Biotropica 32:489–501

Ranging and Spacing Behaviour of Asian Elephant (*Elephas maximus* Linnaeus) in the Tropical Forests of Southern India

15

Nagarajan Baskaran, R. Kanakasabai, and Ajay A. Desai

Abstract

Larger body mass and associated life history traits of large mammals pose high risks of anthropogenic extinction. Given the wide ranging nature and the synergistic impacts of habitat loss and fragmentation, the living elephants are among the most threatened mammals in the world. Therefore, information on ranging and space use pattern are extremely important for conservation planning, especially in the case of long-ranging species. We studied the ranging and spacing behaviours of Asian elephant (*Elephas maximus*) radio-collaring three clans and two bulls between 1991 and 1995 in Nilgiri Biosphere Reserve, southern India, to understand their implications for conservation. Home-range size varied considerably among the clans (range 562–800 km^2), bulls (range 211–375 km^2), and between them. Clans ranged over larger areas (mean 677 ± 69 km^2) compared to bulls (293 ± 82 km^2). Clan ranged in degraded, poor quality habitat with low annual rainfall had larger home range (800 km^2) than those ranged in high rainfall optimal habitats (562 km^2). The smaller home ranges of bulls were possibly due to nonrepresentation of *musth* during the study period. The dry season

N. Baskaran (✉)
Bombay Natural History Society, Hornbill House, S.B. Singh Road,
Mumbai 400023, Maharashtra, India

Department of Zoology and Wildlife Biology, A.V.C. College (Autonomous),
Mannampandal, Mayiladuthurai 609305, Tamil Nadu, India
e-mail: nagarajan.baskaran@gmail.com

R. Kanakasabai
Department of Zoology and Wildlife Biology, A.V.C. College (Autonomous),
Mannampandal, Mayiladuthurai 609305, Tamil Nadu, India

A. A. Desai
Bombay Natural History Society, Hornbill House, S.B. Singh Road,
Mumbai 400023, Maharashtra, India

84, BC Camp, Belgaum 590001, Karnataka, India

C. Sivaperuman, K. Venkataraman (eds.), *Indian Hotspots*,
https://doi.org/10.1007/978-981-10-6605-4_15

movements of the clans and the bulls were restricted around the perennial water sources, while the wet season movements were extended to areas with temporary water sources. Seasonal home ranges of clans were large during wet seasons (401 ± 64 km^2) than in the dry season (308 ± 20 km^2). On the other hand, bulls ranged over larger areas during the dry season (231 ± 47 km^2) than in the wet seasons (141 ± 35 km^2). All the clans (excepting one) and bulls showed strong fidelity to their home and seasonal ranges. One of the clans shifted its range during the second year of the study. This clan had a major part of the range in the reserved and revenue forests areas, which continued to experience severe biotic pressure resulting in devoid of access to perennial water source. Therefore, the home-range shift of this clan could be attributed to habitat loss and degradation of major parts of its original range. Clans, unlike bulls, appeared to space themselves out and mostly avoided meeting each other, despite extensive overlap in space. Observations on interclan interactions further suggest that there is hierarchy among clans in space use. Such a hierarchy and its resultant spacing among the overlapping clans influence the seasonal habitat use pattern, which in turn could result in dominant clans having priority access to good quality habitat and food, thereby show better survival and reproductive success.

Keywords
Asian elephant · Fidelity · Home range · Hierarchy · Spacing

15.1 Introduction

Larger body mass and associated life history traits of large mammals pose high risks of anthropogenic extinction. Given the wide ranging nature and the synergistic impacts of habitat loss and fragmentation, the living elephants are among the most threatened mammals in the world. Therefore, information on ranging and space use pattern are extremely important for conservation planning, especially in the case of long-ranging species. Movements of an individual or a group of animals are generally restricted to an area named as home range or territory. Seton (1909) had originally suggested the concept of home range. Burt (1943) defined it as that area traversed by the individual in its normal activities of food gathering, mating and caring for young. This excludes the area of uncharacteristic and erratic wanderings outside the normal range. Earlier studies on the home range of elephants in Africa (Douglas-Hamilton 1972; Leuthold 1977; Viljoen 1989a) and in Asia (Easa 1988; Sukumar 1989; Desai 1991) were based on re-sighting individually identified elephants based on the natural characteristic features. This method has severe limitations when studying elephants, which range over large areas and dense wooded forests, especially in Asia. Difficulties in re-sighting the same individual on a regular basis result in inadequate sample size to define home range reliably. To overcome this problem, VHF telemetry came as a handy tool, which has been widely used to study the ranging behaviour of elephants in Africa (Douglas-Hamilton 1972; Leuthold and Sale 1973; Leuthold 1977; Dunham 1986; Viljoen 1989a, b; Thouless and Dyer 1992; Thouless 1995, 1996; Tchamba et al. 1995; De Villiers and Kok 1997). In Asia, Olivier (1978) first used telemetry; thereafter, only in the recent past there have been studies that are more detailed on the ranging behaviour of elephants using telemetry (Baskaran et al. 1995; Joshua and Johnsingh 1995; Desai and Baskaran 1996; Baskaran 1998).

One of the major drawbacks of the studies on ranging behaviour has been the inability to ascertain how a home range can be reliably defined. Some studies on home range of elephants in Africa (Leuthold 1977; Viljoen 1989a; Thouless 1996; De Villiers and Kok 1997) and in Asia (Baskaran et al. 1995) used location-area curve to examine whether the home range has been well defined. With increasing sample size, an increase in range reaches an asymptotic value. Most studies on Asian elephants (Sukumar 1989; Easa 1988; Desai 1991) have not used this method, and this coupled with low sample sizes has often resulted in an underestimation of home range. There is thus a lack of detailed information on ranging behaviour for many of the Asian elephant population. Information on ranging behaviour is essential primarily to provide adequate space for long-term conservation of the species and in designing the protected areas so that administrative boundaries can coincide with ecological boundaries. This information is especially crucial for species like elephant that range so widely and in conflict with humans frequently due to continual loss of their ranges. The present study was carried out between 1991 and 1995 using radiotelemetry to study the home-range size and to find out extent of fidelity shown by elephants to their ranges in Nilgiri Biosphere Reserve, southern India.

15.2 Methods

15.2.1 Study Area

Nilgiri Biosphere Reserve (76° 0′E and 77° 15′ E and 12° 15′ N and 10° 45′ N), spread over an area of 5520 km^2, is situated at the junction of three southern states – Tamil Nadu, Karnataka and Kerala. It has an undulating terrain with an average elevation of 1000 m above MSL. Rivers such as Nugu, Moyar and Bhavani and most of their tributaries are perennial and drain the area. The reserve has a diverse climate due to its varied reliefs and topography. The temperature ranges from 7 °C in December to 37 °C in April and receives rainfall both from the southwest (May to August) and northeast (September to December) monsoons. The mean annual rainfall varies from 600 (in the eastern side) to 2000 mm (in the western side). The dry season is from January to April. Corresponding to the gradient in rainfall, the vegetation varies from southern tropical thorn forest in the east to moist deciduous forest in the west with dry deciduous forest in between the two forest types (Champion and Seth 1968). NBR along with its adjoining natural habitats has remarkable faunal diversity and is well known for supporting the largest population of Asian elephants with an estimated population of 5750 individuals (Project Elephant 2007; Baskaran 2013) and relatively undisturbed. Overgrazing by domestic cattle and fire wood collection are serious problems in the eastern fringes of NBR (Baskaran et al. 2012).

15.2.2 Methods

Data on the ranging behaviour was collected from three clans and two bulls monitoring the movements of five elephants (three adult females from three different clans and two adult males), fitted with conventional (VHF) radio transmitter manufactured by the Telonics Inc., USA. The three collared females were named as *Priyanka*, *Wendy* and *Hariny*, and the clans were also named after the collared females as *Priyanka* Clan, *Hariny* Clan and *Wendy* Clan. The adult males were named as *Salim Ali* and *Admiral*. The collared elephants were regularly tracked from February 1991 to April 1994, and thereafter data was collected at a lower intensity. An attempt was made to collect a minimum of eight locations per month for each elephant. During the greater part of the study, this minimum sample size was achieved. The only exception was the clan *Wendy* which was tracked only twice a month as the distance involved in reaching the clan was extremely large (> 400 km) round trip from the base camp across the Nilgiri Mountain. The bull, *Salim Ali,* was tracked until June 1992 when its transmitter stopped functioning. The other bull *Admiral* was tracked until September 1993 when it was shot dead by a local villager.

The locations of the collared animals were plotted on 1:50,000 scale topographic maps (Survey of India), and the data was analysed using the Spatial Ecology Analysis System (SEAS) software developed by John Carey, Wisconsin University, USA. The home range was estimated using minimum convex polygon (MCP) method (Jennrich and Turner 1969) by pooling the data for each individual for the entire study period. Seasonal range was estimated using the pooled data for all the study animals except for *Wendy's* clan, which was tracked only twice a month, and consequently the smaller sample sizes do not permit such analysis. Home-range fidelity was determined by calculating the activity centre (defined here as arithmetic mean of all animal locations) for consecutive years and estimated the distance between activity centre in consecutive years in order to know how far the activity centre shifted between years. A similar method was adopted for determining the seasonal range fidelity. To determine whether different areas were used during different seasons, the activity centre for each season in each study year and the distance between activity centres for sequential seasons over the years (i.e. dry 1991 to first wet 1991 to second wet 1991, dry 1992 to first wet 1992 to second wet 1992, etc.) were calculated.

15.2.3 Definitions Used in the Study

The following abbreviations and definitions have been used in the context of elephant social units and legal status of the forest areas in the present study. Clan: A clan is defined as a group of elephants consist of related females, and their offspring from sub-adults, juveniles to calves of both sexes, which associate regularly and show coordinated activity and movement (Moss 1988).

Bulls/adult males: Males leave their natal clan at puberty and mostly lead solitary life with weak social bonds with clans and other males (Douglas-Hamilton and Douglas-Hamilton 1975; Moss 1988; Desai and Johnsingh 1995).

Protected areas (PAs): Forest areas that have been designated as national park or wildlife sanctuaries.

National park (NP): An area designated for wildlife conservation. This area comes under the management of wildlife wing (a part of forest department).

Wildlife sanctuary (WS): An area designated for wildlife conservation, but it enjoys a lower legal status than the NP. This area is also managed by the wildlife wing of the forest department.

Reserve forest (RF): The territorial wing of the forest department manages these forest areas. Wildlife wing has no control over this area. These forests are open to normal forestry operations and not legally designated as wildlife conservation area.

Revenue land (REVF): These include forested and non-forested (agriculture and settlements) lands. A part of the land is privately owned, and the rest is the public land under the revenue department (mostly forested land).

15.3 Results

15.3.1 Location-Area Curve

To determine whether the home ranges of the study animals are defined, the home-range sizes were plotted in a chronological order against time axis (monthly) for each study animal separately (Fig. 15.1a). The area curves attained asymptotic value for *Hariny*, *Priyanka* and *Wendy* indicating that home ranges were defined in the case of clans. Similarly, the area curve in the case of the bull *Admiral* also shows asymptotic value (Fig. 15.1b). However, as males during *musth* period have much larger home range than non-*musth* (Desai and Johnsingh 1995), the home range has not been defined in this bull, as this bull did not come into *musth*. Area curve in the case of *Salim Ali* is yet to stabilize (Fig. 15.1b), suggesting that the home range is not defined for this bull also.

15.3.2 Home-Range Size

The clan *Hariny* had the smallest home range of 562 km^2 (Table 15.1). In contrast, the clan *Wendy* ranged over the largest area of 800 km^2. The home range of clan *Wendy* included vast inaccessible areas of steep hills, human settlements and cultivated lands. Including these inaccessible areas, the home-range size was estimated to be 1665 km^2. As the entire area was not accessible, the actual area available to the clan was only 800 km^2. An intermediate home-range size of 670 km^2 was recorded for the clan *Priyanka*. The mean home range of the clans was 677 ± 69 km^2. Among the bulls, *Salim Ali* ranged over a larger area of 375 km^2, but the other bull *Admiral*

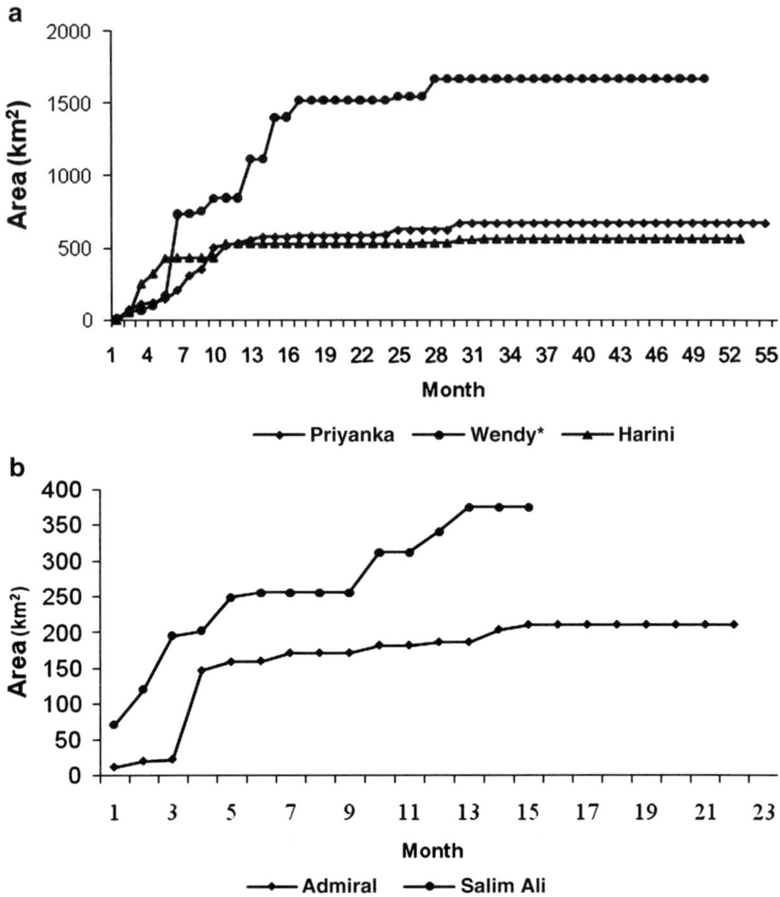

Fig. 15.1 Cumulative monthly increase in home-range size of Asian elephant clan (**a**) and bull (**b**) in Nilgiri Biosphere Reserve, southern India

Table 15.1 Home-range size of elephant clan and bull in Nilgiri Biosphere Reserve, southern India, estimated using minimum convex polygon method

Elephant ID		Duration of tracking (month)	# of locations (*n*)	Home-range size (km^2)
Hariny	(Clan)	47	574	562
Priyanka	(Clan)	52	578	670
Wendy	(Clan)	48	131	800
Admiral	(Bull)	23	257	211
Salim Ali	(Bull)	17	114	375

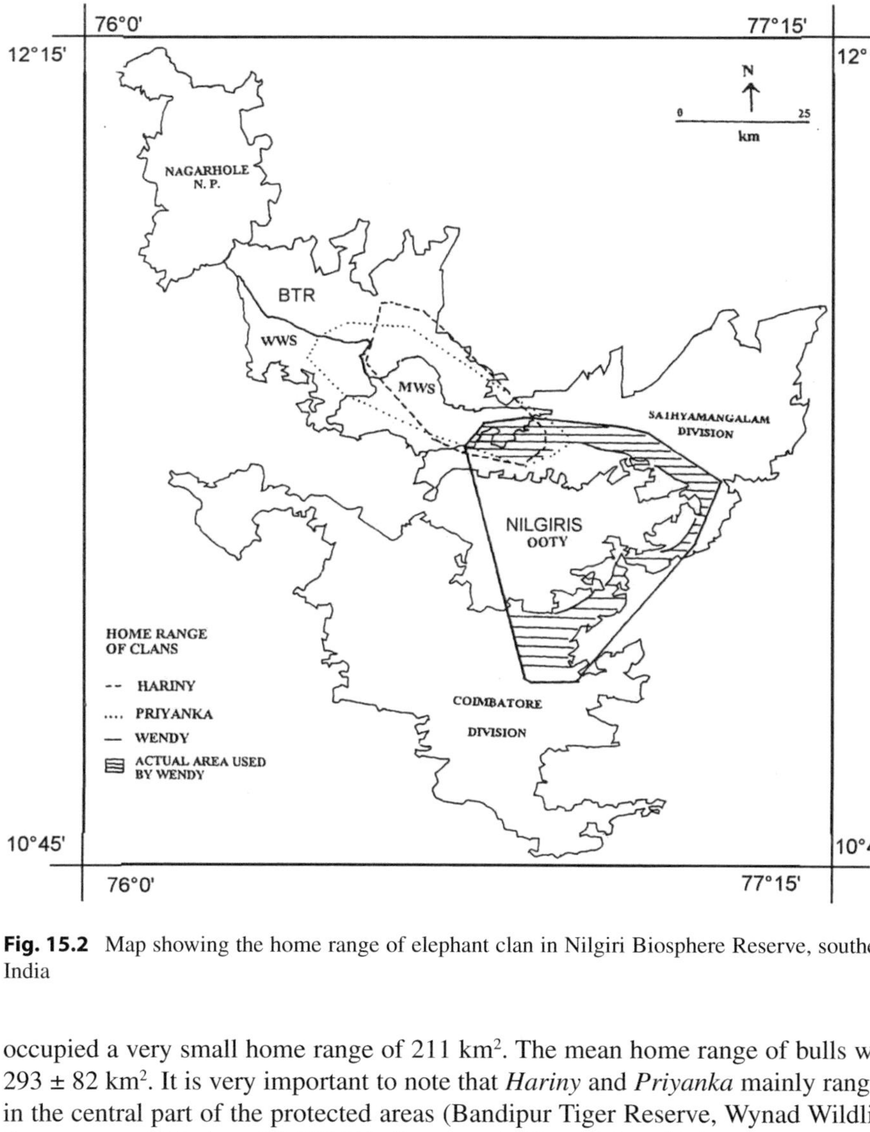

Fig. 15.2 Map showing the home range of elephant clan in Nilgiri Biosphere Reserve, southern India

occupied a very small home range of 211 km^2. The mean home range of bulls was 293 ± 82 km^2. It is very important to note that *Hariny* and *Priyanka* mainly ranged in the central part of the protected areas (Bandipur Tiger Reserve, Wynad Wildlife Sanctuary and Mudumalai Wildlife Sanctuary) (Fig. 15.2), which constituted 87% and 84% of *Hariny's* and *Priyanka's* home ranges, respectively. The rest of the clans' ranges consisted of reserve and revenue forests. On the other hand, *Wendy* clan ranged mainly in the dry thorn forests habitats of reserve and revenue forests that constituted 96% of its home range, while the rest was in the protected areas of Mudumalai Wildlife Sanctuary. Data collected before the radio-collaring of this clan also suggests a similar trend of ranging pattern. However, the *Wendy* clan shifted its original range by the second year. Its original range, 96% of which consisted of dry thorn forests habitat of the reserve and revenue forests, was getting highly degraded, and some parts of its range were lost for agriculture and human

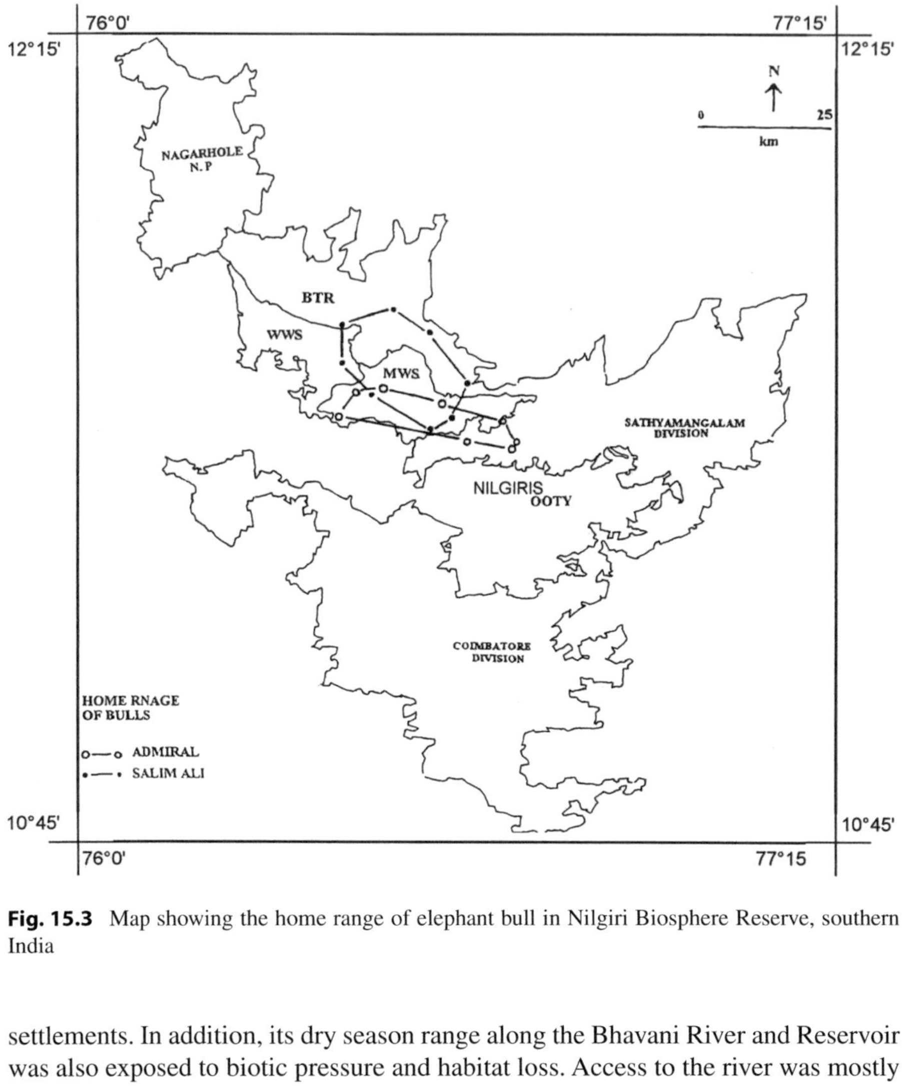

Fig. 15.3 Map showing the home range of elephant bull in Nilgiri Biosphere Reserve, southern India

settlements. In addition, its dry season range along the Bhavani River and Reservoir was also exposed to biotic pressure and habitat loss. Access to the river was mostly cut off by agricultural operations, making a large part of its summer range devoid of water. It is possible that this clan could not sustain itself in the original range, and thus it was forced to shift its range during the second year of this study to a new area towards southwest of the original range, where the elephant density appeared to be much lower than its original range.

A major part (99.7%) of the bull *Salim Ali's* home range was within the protected areas, only a fraction (0.3%) falling within the revenue forests (Fig. 15.3). The other bull *Admiral* had 76% of its home range within the protected areas (Mudumalai Wildlife Sanctuary) and the rest in revenue forests (15%) and reserve forests (9%). The bull *Salim Ali*, a young adult (18–20 year old), did not come into regular *musth* during the study period. The other bull, *Admiral*, a large adult male, also did not come into *musth*, possibly because of loss of condition from gunshot injuries it had received. The estimated home ranges of both the bulls therefore may represent only their non-*musth* ranges. Desai and Johnsingh (1995) reported that males, during the *musth* period, range over greater areas than when they are not in *musth*. Hence, the present home-range size of the bulls without *musth* range has been taken only as their minimum range. The clan *Wendy* and bull *Admiral* that had considerable part of their range in the reserve and revenue forests (degraded habitats) raided crops, and both elephants were shot by the villagers during the study period. These results suggest that elephant clans and bulls that lost part of their range to agriculture and with the remaining range being extensively degraded may end up in conflict with humans.

15.3.3 Seasonal Range Size

Both the clans ranged extensively and used major part of their home range during second wet season (Table 15.2). Unlike clans, bulls ranged extensively and used large part of their home ranges during the dry season compared to the wet seasons. Patterns of range sizes in different seasons were not uniform among the clans and bulls. For example, *Hariny* clan had the smallest range (216 km^2) during the first wet season unlike the *Priyanka* clan that had smallest range during dry season (288 km^2). Seasonal movement of elephants from the dry season ranges to first wet season ranges (towards western side) at the end of dry or the beginning of the first wet season and from there towards eastern side by second wet season coincided with the onset of southwest and northeast monsoons, respectively. In turn, from the second wet season range, they returned to the dry season ranges almost at the beginning of the dry season. The above results show that the seasonal range size varies within individual clans and bulls between seasons and between individual clans and bulls in any given season.

15.3.4 Home-Range Fidelity

Home-range fidelity analysis was done for two clans (*Hariny* and *Priyanka*) and a bull (*Admiral*). In the case of clan *Wendy* and bull *Salim Ali*, due to inadequate data for home-range analysis in different years, the activity centre was not calculated.

Table 15.2 Seasonal home range of elephant clan and bull in Nilgiri Biosphere Reserve, southern India, estimated by minimum convex polygon method

	Seasonal home range (km^2)		
Elephant ID	Dry season	First wet season	Second wet season
Hariny	328	216	511
Priyanka	288	420	457
Admiral	184	72	137
Salim Ali	277	238	115

In the case of *Hariny*, the shift in activity centre was 0.4 km between the first and the second year, 1.9 km between the second and the third year and 2.4 km between the third and the fourth year. The overall mean shift in activity centre was 1.6 ± 0.6 km, and the area enclosed by four activity centres was 0.9 km^2, which is just 0.16% of the home range defined for this clan. In *Priyanka*, the shift in activity centre was 3.8 km from the first to the second year, 6.4 km from the second year to the third year and 5.3 km from the third year to the fourth year, and the overall mean distance was 5.2 ± 0.75 km. The area enclosed by these four activity centres was 13.4 km^2 that is just 2% of the home range defined for this clan. For the bull *Admiral,* the activity centre was calculated for 2 years, and the distance between these 2 years was 6.6 km. As only two points were available, the area enclosed could not be calculated. The small shift between the activity centres of different years and the small enclosed area of activity centres suggest that same area being used over the years, indicative of very strong fidelity shown by the clans and bull to their home ranges.

15.3.5 Seasonal Range Fidelity

To know whether clans and bulls have distinct seasonal ranges, the activity centre year-wise for each season and the distance between activity centres for sequential seasons were calculated over the years. It can be inferred from the results that clan *Hariny* differs from *Priyanka* in having dry and first wet season ranges closest (Table 15.3), while in the latter clan, the two ranges that were closest were dry and second wet. This indicates differences between the clans in their strategy of space use. Similarly, bulls also adopted different strategy of space use (Table 15.3). Such difference in space use pattern among clans and bulls may be a strategy of spacing

Table 15.3 Shift in activity centre (km) between consequent seasons arrived plotting core area of each season and estimating the distance between two consequent seasons to understand whether different area used in different season

Elephant ID	Season	1991	1992	1993	1994	Mean
Hariny	Dry–first wet	–	1.3	0.6	–	0.95
	First wet–second wet	6.4	4.7	4.03	–	5.04
	Second wet–dry	6.2	2.6	1.4	–	3.40
Priyanka	Dry–first wet	7.3	15.6	18.9	–	13.9
	First wet–second wet	14.6	21.6	21.4	–	19.2
	Second wet–dry	–	4.9	1.4	2.6	2.97
Admiral	Dry–first wet	–	11.17	21.4	–	16.43
	First wet–second wet	–	15.73	–	–	15.73
	Second wet–dry	9.47	4.48	–	–	6.9
Salim Ali	Dry–first wet	9.49	–	–	–	9.49
	First wet–second wet	1.10	–	–	–	1.10
	Second wet–dry	11.72	–	–	–	11.72

Table 15.4 Shift in activity centre between/among years in the same season to understand whether same area used in every season (range fidelity)

Elephant ID	Season	Distance between activity centres (km)	Mean distance between/among activity centre (km)
Hariny	Dry 92–dry 93	1.3	1.8
	Dry 93–dry 94	2.3	
	First wet 91–first wet 92	1.4	1.8
	First wet 92–first wet 93	2.2	
	Second wet 91–second wet 92	2.4	2.0
	Second wet 92–second wet 93	1.6	
Priyanka	Dry 91–dry 92	5.4	4.57
	Dry 92–dry 93	5.9	
	Dry 93–dry 94	2.4	
	First wet 91–first wet 92	2.9	2.75
	First wet 92–first wet 93	2.6	
	Second wet 91–second wet 92	4.9	3.8
	Second wet 92–second wet 93	2.7	
Admiral	Dry 92–dry 93	9.09	9.09
	First wet 92–First wet 93	1.29	1.29
	Second wet 91–second wet 92	4.85	4.85
Salim Ali	Dry 91–dry 92	3.0	3.0

among the range overlapping species to reduce conflicts. To know whether individuals show fidelity to their seasonal ranges, activity centre of the same season between years (dry 1991 to dry 1992, dry 1992 to dry 1993, dry 1993 to dry 1994, etc.) and distance between activity centres were calculated. A comparison of mean distance of activity centres of different seasons with that of the same season among years shows that shift was more between seasons (Table 15.3) than in the same season of different years for all the clans and bulls (Table 15.4). These findings indicate that clans and bulls had different seasonal ranges but used specific areas every year in the same season with strong fidelity to seasonal ranges.

15.3.6 Home Range Overlap

An extensive spatial overlap in the home range was observed among clans and between bulls and among clans and bulls (Table 15.5). A total of 466 km^2 was used by both *Priyanka* and *Hariny* clans that formed nearly 83% of the *Hariny* clan's range and 70% of the *Priyanka* clan's range, indicating that the *Priyanka* clan had a large area not overlapped by the *Hariny* clan. It may be noted here that the 466 km^2 area is common to both the clans and thus both had access to space and resources available within. The home range of the clan *Wendy* did not overlap much with the other two clans (though not calculated) as this clan ranged at the periphery of the ranges of the other two clans. Among the bulls, 31% of the home

Table 15.5 Percentage of home range overlap between elephant clan and bull in Nilgiri Biosphere Reserve, Sothern India

Elephant ID	Hariny	Priyanka	Admiral	Salim Ali
Hariny	–	83.1	21.8	52.1
Priyanka	69.6	–	20.3	53.8
Admiral	58.2	64.6	–	30.8
Salim Ali	78.3	96.2	17.3	–

range of the *Admiral* overlapped with that of *Salim Ali*, and only 17.3% was vice versa, with a common area of 65 km^2 available to both the bulls. The percentage range overlap between bulls was, thus, smaller than the overlap recorded between clans. The range of the bull *Salim Ali* overlapped the clan's ranges much more than that of *Admiral*.

15.3.7 Seasonal Range Overlap and Spacing

Overall, the seasonal range overlap between clans and among clans and bulls was highest during the dry and second wet seasons as compared to the first wet season (Table 15.6). There was no overlap between the bulls in any season, as *Admiral* did not intensively use the central part of its range (where its annual range overlapped with *Salim Ali*), and therefore this area was not part of any seasonal range estimated, using harmonic mean distance method with 75% of locations. The area of 466 km^2 available commonly to both the clans was used mainly during the dry and second wet seasons as shown by the highest overlap in these two seasons. These results clearly suggest that these two clans overlapped in space during all the seasons but more significantly during dry and second wet seasons.

15.3.8 Interclan Encounter

Despite the fact that *Priyanka* and *Hariny* clans overlapped extensively in space maximum to the tune of 100% (*Hariny* clan's range by *Priyanka*) during dry season and 55% during second wet season (*Hariny* clan's range by *Priyanka*), the number of times the two clans observed together or in the vicinity of each other was on only one occasion during the 4 years of observations. The interclan encounter was observed in the feeding ground, in which a large female from the *Priyanka* clan, that was operating more than 100 m away from the *Hariny* clan, moved with raised head and extended ears towards the *Hariny* clan, preventing the feeding of an adult female belonging to *Hariny* clan, by jabbing with her tushes and by depriving the food from the same clump of browse. The adult female of *Hariny* clan crouched herself and consequently moved away from that place with her clan members without any resistance. The next day, the *Hariny* clan showed a displacement of more than 5 km from the encounter spot, while the *Priyanka* clan continued operating

Table 15.6 Percentage of seasonal range overlap between elephant clan and bull in Nilgiri Biosphere Reserve, Southern India

	Range overlap (%)			
Elephant ID	Hariny	Priyanka	Admiral	Salim Ali
Dry season				
Hariny	–	100	0	100
Priyanka	13.8	–	37.4	43.1
Admiral	0	75.1	–	0
Salim Ali	14	43.6	0	–
First wet season				
Hariny	–	47.7	0	0
Priyanka	8.8	–	2.9	19.8
Admiral	0	19.6	–	0
Salim Ali	0	37.1	0	–
Second wet season				
Hariny	–	54.6	14	0
Priyanka	40.3	–	13.2	0
Admiral	78.2	100	–	0
Salim Ali	0	0	0	–

close to the same area. The results on range overlap and the observations on interclan encounter suggest that despite extensive spatial overlap, clans mostly avoided each other and there appears to be aggression and hierarchy among them when they rarely encounter with each other.

15.4 Discussion

Within the species, the home range varies widely between areas and within the area and between the sexes and individuals. These variations have been attributed to different factors. Variations in the home-range size of elephants have been related to habitat quality, both in Asia and Africa. Olivier (1978) and Easa (1988) found the home ranges of the Asian elephants to be larger in the primary forest and smaller in secondary forest. In Africa, the smaller home ranges of elephants have been related to higher rainfall (Leuthold 1977; Thouless 1996) and better habitat quality (Douglas-Hamilton 1972). In the present study, the clans had home-range sizes that varied from 562 to 800 km^2 with a mean size of 677 km^2. The home-range size of the *Wendy* clan (800 km^2) was much larger than the one estimated for the *Hariny* and *Priyanka* clans. The major part (96%) of the *Wendy* clan's range was outside the protected area, a rain shadow area mainly dry thorn forests, which was subjected to high biotic pressure. The availability of grass biomass was very low in dry thorn forests when compared to dry deciduous forest (Baskaran 1998; Baskaran et al. 2010). Thus, the larger range of clan *Wendy* could possibly be ascribed to low rainfall, poor quality of the habitat and consequently limited food supply. The home-range size estimated for the three clans in the present study was larger than reported earlier studies in this region (Sukumar 1989; Desai 1991), other parts of India (Easa 1988; Joshua and Johnsingh 1995) and other parts of Asia (Olivier 1978). The small range sizes estimated by the earlier Asian studies could be either due to smaller sample size (re-sightings) resulting in underestimation of home range or due to compression of original range due to developmental activities and biotic pressure as reported in Asia (Joshua and Johnsingh 1995) or physical barriers as reported in Africa (Douglas-Hamilton 1972; Dunham 1986).

Home ranges estimated for bulls in the present study represented only non-*musth* range. Bulls, during *musth* period, are said to move extensively in search of oestrous females (Barnes 1982; Desai and Johnsingh 1995; Joshua and Johnsingh 1995). Therefore, the present home range of bulls without the *musth* range could be treated as the minimum range size for bulls. Though, in the present study, the home ranges of bulls could not be defined completely, it was still larger compared to earlier studies in Asia (Olivier 1978; Sukumar 1989; Desai 1991; Joshua and Johnsingh 1995). Bulls being solitary animals are expected to range shorter than clans as speculated by Olivier (1978) as food requirements of a solitary individual per unit time is lesser compared to clans. However, for adult males in polygynous species especially during the reproductive period, oestrous females are the most important resources that are scarcely available. Males might enhance the probability and frequency of encountering potential mates and hence increase their reproductive success by enlarging their home range (Lindstedt et al. 1986). Thus, the need for locating oestrous females, a resource scarcely available in a given breeding period, imposes on the bulls to cover equal or a much larger ranges than clans. Therefore, bulls may have equal or much larger ranges than clan as shown by other studies in Asia (Daniel et al. 1995; Joshua and Johnsingh 1995) and Africa (Leuthold and Sale 1973; Leuthold 1977; Viljoen 1989a).

Seasonal home ranges estimated for the clans *Hariny* and *Priyanka* were generally larger during the wet season than in the dry season. The trends of seasonal range size recorded in the present findings are consistent with earlier studies in Asia (Easa 1988; Daniel et al. 1995) and Africa (Leuthold 1977; Viljoen 1989a; De Villiers and Kok 1997). The dry season ranges of elephants in Nilgiri Biosphere Reserve were restricted to areas along the perennial water sources, since the temporary water sources dry up during summer. Therefore, the relatively small dry season ranges of the clans could be attributed to restricted availability of water. Desai and Baskaran (1996) found that the clans *Hariny* and *Priyanka* significantly preferred areas close to water. Movements of elephants in Africa also indicated a sedentary nature in dry season, followed by dispersal and scattering during rains (Rodgers and Elder 1977; Jachmann 1988). During dry season, apart from restricted water availability, food quality and quantity also are very low compared to wet season. It would be better choice for elephants to restrict themselves along the perennial water sources and exploit whatever food resources that are available by spending little energy, instead of spending much of its limited energy by moving widely in searching good quality food and commuting to get water as speculated by Jachmann (1988). In Nilgiri Biosphere Reserve, in contrast to clans, bulls exhibited larger ranges during dry season. The reason for this might be that both the bulls used extreme ends of their annual ranges during the dry season. The central part of home range was used mostly for travelling between second wet and dry season ranges with relatively scattered use. During this period, they were also found to use small streams with very limited water supply relatively for longer period than the clans did. Bulls being alone can afford to withstand low water availability and sometimes even do without it unlike the females that cannot do the same as they live in larger groups and with dependent calves. Therefore, water availability may not be influencing bull's movement as much as it does clan's movement in the study area.

Apart from water, the other factor that influenced the dry season movement of elephants in the study area was forest fire which used to be very severe in the deciduous forest once in every 4 or 5 years due to very high accumulation of litter biomass from tall grass and with teak leaf fall. In years of severe forest fires, elephants during the dry season moved to their second wet season ranges and remained until mid-dry season as forest fire wipes off all the food resources available at the ground level. During wet season, rainfall influenced the movements of elephants by providing many temporary water sources and by favouring a luxurious growth of grass. Generally, the seasonal movement of elephants from the first wet season to second wet season range took place during the beginning or mid-October, and elephants remained in the second wet season ranges until the beginning of dry season. Unusually, in years when there was delay in the onset of second monsoon, elephants returned within a few days from their second wet season range to the first wet season range and waited for 2–3 weeks until the onset of monsoon and fresh growth of grass. However, such movements were restricted within the individual home range of the clans and bulls, and no wandering took place outside the home range due to rainfall. These findings clearly reveal the magnitude of rainfall influence on the wet season movements of elephants.

All the study animals (except *Wendy*) presently showed strong fidelity to their home ranges, a phenomenon recorded earlier by Baskaran and Desai (1996) and Baskaran (1998) in Asian elephants and Viljoen (1989a) in African elephants. In tropical forest, resource distribution and abundance vary within a habitat between seasons and thus within the home range too if home range lies in different habitats. It implies that the entire home range would not be always uniform regarding resource distribution and abundance. Parts of the home range would be with abundant resources at different times of the year depending on the season. This would mean that elephants use different parts of their home range during different seasons depending on changes in resource availability. Resources in a given habitat do not normally change between years. Therefore, the seasonal range would also remain stable, unless disturbed by drastic changes. If a long-lived species like elephant confines its movements within a small area (home range) and uses this area year after year, it could acquire knowledge about the resource distribution and abundance that vary spatially and temporally, especially in a heterogeneous habitat. Such knowledge about the resource distribution and abundance would certainly help to optimize their resource use and therefore enhance their reproductive success. But if the individual or group keeps shifting the home range from year to year, any familiarity gained in the previous year would be of little or no use in the new area. This means that the use of resources may be largely a matter of chance, which is not the best way of resource use for a long-lived species. As mentioned earlier, for an optimum use strategy to use the same home range repeatedly over the years and particular area (within the home range) every year during the same season, elephants should have strong fidelity to their home range and seasonal ranges. Thus, fidelity shown by elephants to their ranges could be a strategy adopted for optimal use of resources.

15.4.1 Range Overlap and Spacing

Home ranges of the clans *Hariny* and *Priyanka* overlapped each other, but the home range of the clan *Wendy* did not overlap much, as its range was near the periphery of the ranges of the other two clans. Similarly, little overlap of the home range between bulls in the present study could also be due to the fact that these bulls were basically from two different areas. The bulls' ranges overlapped little with those of clans in the present study, probably because of the absence of *musth* range in males. The breeding bulls have a strategy to build up the body condition during the non-*musth* time with limited movements resulting smaller non-*musth* range and range widely during *musth* period, in search of oestrous females (Joshua and Johnsingh 1995). It is obvious that overlap in home ranges among clans and between clans and bulls is more, and the degree of overlap varies widely depending on the location of home ranges. That is a clan or bull will overlap more with another clan or bull whose home range exists in the same area rather than with that of another clan or bull at the periphery of its home range. The degree of overlap may also be a function of elephant density and availability of resources as suggested in deer by Baker (1978). The overlap of home ranges is also determined by the spacing of essential resources most restricted in their distribution (Altman 1974).

The seasonal range overlap between the clans was more during the dry season followed by the second wet season and far less during the first wet season. A similar pattern of overlap was also observed between the clans and bulls. The limited availability of perennial water sources and the restriction of elephant movements around them during dry season could be the reason for greater overlap in the dry season range among the study animals. On the other hand, in the wet seasons, as elephants disperse over the temporary water areas, there was less overlap. The present finding of high overlap between clans in the dry season range differs from the findings of McKay (1990) who stated that home ranges of herds tended to overlap more extensively during the rainy season than during dry season in the Gal Oya National Park, Sri Lanka. Bull's seasonal ranges did not overlap each other in any season as they were basically from different areas. Joshua and Johnsingh (1995) estimated the seasonal range overlap between clan and bull and showed that the overlap was high during winter (14 km^2) compared to summer (9 km^2) and monsoon (7 km^2), being attributed to the *musth* period of the bull.

Although, the differences in the sizes of seasonal ranges of bulls can be due to variations in the habitat quality and environmental conditions of their ranges, differences between the clans may not be so because the spatial distribution and extensive overlap of home ranges of *Hariny* and *Priyanka* showed that these two clans operated in the same area. These two clans, with almost the same herd size, had an overlapping area of 466 km^2 which was used by them mostly in the same season. Hence, it can be reasonably expected that the two clans have a similar range use pattern as environmental factors in an area act on all the clans uniformly. However, the findings show that these two clans show more variations in seasonal range sizes and in overlap (within clan between seasons), which could be as a result of hierarchy and spacing. It has been suggested that in mammals, females are concerned with obtaining food while males compete for mates (Greenwood 1980, 1983; Dobson 1982). There were many occasions in which non-collared adult bulls were observed to feed within the vicinity of the bull *Admiral,* when males were not in *musth*. During the course of observation, no competition between bulls for a mate was observed, as the breeding bulls were extremely low in the population. However, Eisenberg et al. (1971), McKay (1990) and Desai (per. comm.) observed competition between bulls for mates. The observed tolerance among bulls in the feeding grounds may not be true when it comes to mating requirements.

The present study has answered, how a given space may be used by individuals of two different clans (*Hariny* and *Priyanka*), whose range overlap extensively. The spatial distribution of home range and its percentage overlap among the two clans *Hariny* and *Priyanka* suggest that these clans were sharing the same space within the population range. Both the clans, in the same season, used an overlapping area of 466 km^2 that constituted 83% and 70% home ranges of former and latter clans, respectively. However, these two clans were observed to encounter each other only once during the study period, and such observations suggest that normally, the clans space out themselves and mostly avoid each other, despite extensive spatial overlap. A recent study (De Villiers and Kok 1997) on African elephants observed such avoidance behaviour between females in core areas.

15.4.2 Determinants and Consequences of Interclan Encounter

Spacing has often been discussed in the context of resource defence in different animals (Zahavi 1971; Gill and Wolf 1975; Carpenter and MacMillen 1976; Simon and Middendorf 1976). Temporal partitioning of overlapping territories as a strategy to avoid interference competition has been documented for a lizard population (Simon and Middendorf 1976). Encounters among neighbours with extensive range overlap, which resulted in spacing, have been documented in some animals like chipmunks (Getty 1981). Among Asian elephants, the behaviour observed in the interclan encounter (*Hariny* and *Priyanka*) in a feeding ground suggested the existence of hierarchy and resource defence among clans. In the present study area, Desai (per. comm.) also had observed resource defence by clans and subtle aggression when two clans met at common water holes and feeding grounds. Studies on the behaviour of African elephants (Laws and Parker 1968; Laws 1969) suggest that territorial mechanism may operate at family or clan level. Douglas-Hamilton (1972) stated that, in competitive situations, attacks do occur, and such attacks may be within the family units, between family units of the same kin group or between apparently unrelated groups. If the resource is distributed unpredictably both in space and time, defence may be a costly strategy because there is no guarantee that a defended patch can provide sufficient resources. But defending a resource which is within the vicinity is not as costly as defending an entire range. For mammals, to defend foraging areas may be costly (Brown and Orians 1970). So elephant clans possibly defend resources within their immediate vicinity. The agonistic behaviour of dominant clan over the subdominant for the resource could be an important reason for encounter avoidance.

Many studies often cite the avoidance of agonistic encounters as the main factor promoting spacing (Recher and Recher 1969; King 1973; Tingay 1974; Young 1989). So clans space them self in such a way that they will not use a given space at the same time. This means a given space or patch can only be used by one clan at one time. So the overlapping clans in a population seem to use a given space on a rotational basis, based on their hierarchical position in the population, with dominant clan using a patch in optimal time and subdominant one in suboptimal time as shown by Baskaran (1998). Such a hierarchy and spacing among the overlapping clans also influence the seasonal habitat use pattern, resulting in dominant clans having prior access to good quality habitat and food (Baskaran 1998), thereby show better survival and reproductive success.

15.5 Management Recommendations

Small patches of revenue lands exist in Sigur and Singara Reserve Forests and these are encroached upon gradually, as clans and bulls show strong fidelity to their annual and seasonal ranges; further loss in habitats would increase human–elephant conflict. Therefore, the unoccupied revenue lands within the forest areas should be transferred from the revenue department as part of the Mudumalai Wildlife Sanctuary. The habitat in Nilgiri Biosphere Reserve is considered an optimal one for elephants in the whole of Asia. When elephant clans show a mean home range of over 600 km^2 in this optimal habitat, any other elephant reserve less than this size (600 km^2) may not therefore be viable. As overlap between clans varied from 30% to 100% (Desai et al. unpublished data), a minimum area of 900 km^2 is essential to provide sufficient space for overlapping clans. Therefore, elephant reserves less than this size should be enlarged wherever possible.

Water is the major limiting factor for elephants during the dry season. Elephants concentrate around perennial water sources in all the habitats during the dry season. At present, there are many places (viz. Doddakatti, Imparhallah and Onnaretty in Mudumalai Wildlife Sanctuary and Chemmanallah and Maddur in Bandipur Tiger Reserve), which are without perennial water sources. Creating water holes in these areas will help to spread out the elephant distribution in the dry season. Ben-Shahar 1993 reported that elephants in Africa cause more impact on vegetation around perennial water source than away from it. Studies in Africa and Asia clearly show that elephants concentrate around perennial water areas during the dry season. Therefore, providing more water sources in areas, where it is lacking, would reduce the dry season impact of elephant on vegetation.

Acknowledgements We are thankful to United States Fish and Wildlife Service for funding the study and the State Forest Department Tamil Nadu, Karnataka and Kerala for permitting us to undertake the study in their states. We extend our sincere thanks to late Mr. JC Daniel, the principal investigator of the project and former director of Bombay Natural History Society, Mumbai, for his consistent support. Last but not the least, we thank our field assistants Messers Krishnan, Chennan (late), Bomman, Kethan, Mathan and Maran and jeep drivers Mr. Sirul Kumar and Gopal for their wonderful cooperation during the study period.

References

Altman S (1974) Baboons, space, time and energy. Am Zool. 26: 185–206

Baker RR (1978) Evolutionary ecology of animal migration. Holmes and Meier, New York. 1012 p

Barnes RFW (1982) Elephant feeding behaviour in Ruaha National Park, Tanzania. Afr J Ecol 20:123–136

Baskaran N (1998) Ranging and resource utilization by Asian elephants (Elephas maximus Linnaeus) in Nilgiri Biosphere Reserve, South India. Ph.D. thesis. Tiruchirappalli, Bharathidasan University

Baskaran N (2013) An overview of Asian elephant (*Elephas maximus*) in Western Ghats and its implication for Western Ghats ecology. J Threatened Taxa 5(14):4854–4870

Baskaran N, Desai AA (1996) Ranging behaviour of the Asian elephant *(Elephas maximus)* in the Nilgiri biosphere reserve, South India. Gajah 15:41–57

Baskaran N, Balasubramanian M, Swaminathan S, Desai AA (1995) Home range of elephants and its implications for the management of Nilgiri biosphere reserve, India. In: Daniel JC, Datye HS (eds) A week with elephants. Proceedings of the International Seminar on the conservation of Asian elephant. Mudumalai Wildlife Sanctuary, June 1993. Bombay Natural History Society, Bombay, pp 296–313

Baskaran N, Balasubramanian M, Swaminathan S, Desai AA (2010) Feeding ecology of Asian elephant (*Elephas maximus* Linn.) in tropical forests of Nilgiri biosphere reserve, Southern India. J Bombay Nat Hist Soc 107(1):3–13

Baskaran N, Anbarasan U, Agoramoorthy G (2012) India's biodiversity hotspot under anthropogenic pressure – A case study of Nilgiri biosphere reserve. J Nat Conserv 20:56–61

Ben-Shahar R (1993) Pattern of elephant damage to vegetation in northern Botswana. Biol Conserv 65:249–256

Brown JL, Orians GH (1970) Spacing patterns in mobile animals. In: Johnston RF, Frank PW, Michener CD (eds) Annual ecology and systematics, vol 1, 239–257

Burt WH (1943) Territoriality and home range concept as applied to mammals. J Mammal 24:346–352

Carpenter FL, MacMillen RE (1976) Threshold models of feeding territoriality and test with a Hawaiian honeycreper. Science 194:639–641

Champion HG, Seth SK (1968) Revised survey of forest types in India. Manager of Publication, New Delhi, p 404

Daniel JC, Desai AA, Sivaganesan N, Datye HS, Rameshkumar S, Baskaran N, Balasubramanian M, Swaminathan S (1995) Ecology of the Asian elephant, Final report 1987–1994. Bombay Natural History Society, Bombay

De Villiers PA, Kok A (1997) Home range association and related aspects of elephants in the eastern Transvall Lowveld. Afr J Ecol 35:224–236

Desai AA (1991) The home range of elephants and its implications for the management of the Mudumalai wildlife sanctuary, Tamil Nadu. J Bombay Nat Hist Soc 88:145–156
Desai AA, Baskaran N (1996) Impact of human activities on the ranging behaviour of elephants in the Nilgiri biosphere reserve, South India. J Bombay Nat Hist Soc 93(3):559–569
Desai AA, Johnsingh AJT (1995) Social organization and reproductive strategy of the male Asian elephant (Elephas maximus) in Nilgiri biosphere reserve, South India. In: Daniel JC, Datye HS (eds) A week with elephants. Proceedings of the International Seminar on the conservation of Asian elephant. Mudumalai Wildlife Sanctuary, June 1993. Bombay Natural History Society, Bombay, p 532
Dobson FS (1982) Competition for mates and predominant juvenile male dispersal in mammals. Anim Behav 30:1183–1193
Douglas-Hamilton I (1972) On the ecology and behaviour of African elephant. Unpublished D. Phil. thesis, University of Oxford
Douglas-Hamilton I, Douglas-Hamilton O (1975) Among the elephants. Collins and Harvill Press, London, p 285
Dunham KM (1986) Movements of elephant cows in un-flooded middle Zambezi valley, Zimbabwe. Afr J Ecol 24:287–291
Easa PS (1988) Movement pattern of Asiatic elephant, (*Elephas maximus*). In: Parambikulam wildlife sanctuary, Kerala. Final report of the research project Wild 06/84, Kerala Forest Research Institute, Peechi
Eisenberg JF, McKay GM, Jainudeen MR (1971) Reproductive behaviour of the Asiatic elephant (*Elephas maximus maximus* L.) Behaviour 38:193–225
Getty T (1981) Territorial behaviour of eastern chipmunks (*Tamias striatus*): encounter avoidance and spatial time sharing. Ecology 62(4):915–921
Gill FB, Wolf LR (1975) Economics of feeding territoriality in the golden-winged sunbird. Ecology 56:333–345
Greenwood PJ (1980) Mating system, philopatry and dispersal in birds and mammals. Anim Behav 28:1140–1162
Greenwood PJ (1983) Mating system and evolutionary consequences of dispersal. In: Swingland IR, Greenwood PJ (eds) The ecology of animal movement. Clarendon Press, Oxford, pp 116–131
Jachmann H (1988) Numbers, distribution and movements of the Nazinga elephants. Pachyderm 10:16–22
Jennrich RI, Turner FB (1969) Measurement of non-circular home range. J Theor Biol 22:227–237
Joshua J, Johnsingh AJT (1995) Ranging pattern of elephants in Rajaji National Park : implication for reserve design. In: Daniel JC, Datye HS (eds) A week with elephants, Proceedings of the International Seminar on the conservation of Asian elephant. Mudumalai Wildlife Sanctuary, June 1993. Bombay Natural History Society, Bombay, pp 256–260
King JA (1973) The ecology of aggression behaviour. Annu Rev Ecol Syst 3:117–138
Laws RM (1969) Aspects of reproduction in African elephant *Loxodonta africana*. J Reprod Fertil 6. (suppl.:193–217
Laws RM, Parker ISC (1968) Recent studies on elephant populations in East Africa. Symp Zool Soc Lond 21:319–359
Leuthold W (1977) Spatial organization and strategy of habitat utilization of elephants in Tsavo National Park, Kenya. Z Säugetierkd 42:358–379
Leuthold W, Sale JB (1973) Movements and patterns of habitat utilization of elephants in Tsavo National Park, Kenya. E Afr Wildl J 11:369–384
Lindstedt S, Miller BJ, Buskirk SW (1986) Home range, time, and body size in mammals. Ecology 67(2):413–418
McKay GM (1990) Behavioural ecology of the Asiatic elephant in southern Ceylon, Smithsonian contribution to zoology, vol 125. Smithsonian Institution press, Washington, DC, pp 1–113
Moss CJ (1988) Elephant memories: thirteen years in the life an elephant family. William Morrow and Co, New York, p 336
Olivier RCD (1978) On the ecology of Asian elephant. Unpublished Ph. D. thesis, University of Cambridge

Project Elephant (2007) Census in project elephant reserves in India: http://envfor.nic.in/pe/elephant 2007.pdf. Accessed on 12 Mar 2009
Recher HF, Recher JA (1969) Some aspects of the ecology of migrant shorebirds. II. Aggression. Wilson Bull 81:140–154
Rodgers DH, Elder WH (1977) Movements of elephants in Luangwa Valley Zambia. J Wildl Manag 41(1):56–62
Seton ET (1909) Life histories of northern animals, vol I. Charles Scribner's Sons, New York, p 673
Simon CA, Middendorf GA (1976) Resource partitioning among an iguanid lizard: temporal and micro habitat aspects. Ecology 57:1317–1320
Sukumar R (1989) The Asian elephants ecology and management. Cambridge University press, Great Briton, p 251
Tchamba MN, Bauer H, De-Iongh HH (1995) Application of VHF-radio and satellite telemetry in norther Cameroon. Afr J Ecol 33:335–346
Thouless CR (1995) Long distance movement of elephants in the northern Kenya. Afr J Ecol 33:321–334
Thouless CR (1996) Home range and social organization of female elephants in the northern Kenya. Afr J Ecol 34:284–297
Thouless CR, Dyer A (1992) Radio-tracking of elephants in Likipia District, Kenya. Pachyderm 15:34–39
Tingay A (1974) Aggression in black swan. Emu 74:35–38
Viljoen PJ (1989a) Spatial distribution and movements of elephants *(Loxodonta africana)* in northern Namib desert region of the Kaokoveld, South West Africa/Namibia. J Zool (Lond) 219:1–19
Viljoen PJ (1989b) Habitat selection and preferred food plants of a desert-dwelling elephant population in the northern Namib desert, South West Africa/Namibia. Afr J Ecol 27:227–240
Young AD (1989) Spacing behaviour of visual-feeding and tactile-feeding shorebirds in mixed-species groups. Can J Zool 67:2026–2028
Zahavi N (1971) The social behaviour of white wagtail *Motacilla alba alba* wintering in Israel. Ibis 113:203–211

The Role of Elephants in the Forest Ecosystem and Its Conservation Problems in Southern India

16

Balasundaram Ramakrishnan, M. Ilakkia, S. Karthick, and A. Veeramani

Abstract

The planet earth is inhabited by diverse array of living organisms such as microorganisms, plants, animals and human beings which collectively constitute the biodiversity. Each and every element of the living component of the system has its own role, either positive or negative, to play as a system component. So preservation and conservation of living organisms, whether they are tiny or large, become immense important in playing beneficial role in maintaining biodiversity. Mega-herbivorous animal such as elephant has major impact on the terrestrial ecosystems in which they live and thus on the animals that depend on these habitats. Elephant can be referred as "keystone species" because it facilitates feeding by other herbivores that disperse seeds and supports large assemblages of invertebrates, such as dung beetles, and lower plants such as algae and fungi apart from enriching soil nutrients through dung piles. These algae and fungi are preferred nutrient plants for some reptiles such as monitor lizard and star tortoise in the semiarid tropical forests. Dung beetle accumulation attracts many insectivorous birds. Dung deposition into water holes is being benefited to the Pisces and amphibians. Seed dispersal through alimentary canal induces germination and survival capacity of the seedlings to maintain the forest heterogeneity. Elephant also does some of the silvicultural practices such as creation of paths in

B. Ramakrishnan (✉) · M. Ilakkia · S. Karthick
Department of Zoology & Wildlife Biology, Mammalogy and Forest Ecology Lab, Government Arts College, The Nilgiris, Tamil Nadu, India
e-mail: bio.bramki@gmail.com

A. Veeramani
PG and Research Department of Zoology, Government Arts College (Autonomous), Kumbakonam 612002, Tamil Nadu, India
e-mail: wildveera@gmail.com

C. Sivaperuman, K. Venkataraman (eds.), *Indian Hotspots*,
https://doi.org/10.1007/978-981-10-6605-4_16

dense forest, maintenance of grazing lawns and height of the trees and thinning in thick vegetation cover to keep the sustainable utility of the forest. Identification of subsoil water and natural salt licks through elephants' strong sense is also shared by the other animals especially the herbivores for which intake of minerals from the natural soil is most important for many physiological activities. The pachyderm is under severe threat due to various conservation problems such as loss of habitat, habitat quality and corridors, reduction of home range, population increase, impact of developmental activities, human-elephant conflict issues and poaching for ivory. Among the factors, some of them may be responsible for major proportions, and some of them involve less proportion. But these are the reasons listed as conservation problems for the long-run conservation of Asian elephant especially in Southern India.

Keywords
Asian elephant · Habitat loss · Fragmentation · Keystone species

16.1 Introduction

The planet earth is inhabited by diverse array of living organisms such as microorganisms, plants and animals, including human beings, which collectively constitute the "biodiversity". The living world, biosphere, is organized in a systematic way in which the living entities are influenced by nonliving entities, i.e. the soil and environment. In this system the nonliving parameters (abiotic) are interacting with one another and also interact with living components (biotic) of the system. Thus, the diverse arrays of biotic components are interdependent and interacting with one another, so as to sustain their own existence and the habitat and environment in which they live.

Thus it is obvious that each and every element of the living component of a system has its own role, either positive or negative, to play as a system component. If there is any disturbance or any adverse effect, it will lead to a sequence of repercussion and sometimes ends up in dire consequences. There had been major climatic changes, which led to geological, topographical and geographical alterations of land and extinction of many species, which include organisms ranging from tiny microorganisms to large animals such as dinosaurs. So, preservation and conservation of living organisms, whether they are tiny or large, become imminent, as they have to play beneficial role to welfare of the human beings either directly or indirectly. There are many living organisms that evolved over the earth before the existence of man, and they have originated before several millions of years. Elephant is one of such large animals which had originated before the history of man and exists as a component and "keystone species" in the forests, especially in the tropical regions of the world. These large mammals are found only in Asia and Africa. In India, elephants are not only a living component in the forest ecosystems but also found place in the social, cultural, ethical and religious activities of man. As an inhabitant of the forests in the tropical region, they play a vital role in the forest management, regeneration and maintenance of the equilibrium between the living components.

The Asian elephant (*Elephas maximus*) is endangered due to habitat degradation, loss and fragmentation of habitat and poaching for ivory. Rapid economic growth and increasing aspirations of the growing human population have had numerous adverse impacts, of which development projects such as dams, mines, railways and highways, industries and expansion of agriculture have taken a toll of elephant habitats leading to increase in human-elephant conflicts over most parts of their range. Further, ivory poaching especially in South India has resulted in a skewed sex ratio that in some population it is reduced to 1 adult male for every 100 adult females. The skewed ratio in favour of females is a clear indication that poaching has had a very adverse impact on the elephant population. The continued habitat loss, degradation and fragmentation due to loss of corridors resulted in isolation of populations interfering with the genetic connectivity between populations. Nowadays, the long-run conservation of Asian elephant becomes a difficult task for the managers. This paper describes the role of elephants as well as various conservation problems that are pertaining on elephant conservation in a whole.

16.2 Methods

This documentation is part of University Grants Commission's Major Research Project F.No. 42-594/2013 (SR). These observations were done during the last 2 years of fieldwork of the project. Focal animal and scan sampling methods were attempted whenever we encountered elephants to document the role of elephants (Altmann 1974). Foot survey and field observation were attempted to quantify the conservation problems.

16.3 Results and Discussion

16.3.1 The Role of Elephants in Forest Ecosystem

16.3.2 When Elephant Feeds

The general mode of feeding habit of elephants is foraging of the vegetation and at times through grazing the grassland and depredation of the crops in the fields (Figs. 16.1 and 16.2). Breaking the branches of the tree species and debarking and uprooting of the trees are some of the common behaviours of elephants during foraging. The elephants feed only on the barks and twigs of the fallen trees, and the remaining parts of the trees become the feed for other animals such as Indian gaur; sambar, spotted and barking deer; blackbuck; and other herbivores. It has been stated that certain essential mineral elements required for the herbivores are found available in the trees of the forests, which are inhabited by the elephants.

Fig. 16.1 Grassland habitat

Fig. 16.2 Herd of elephant

Pruning and thinning are some of the forest management practices adopted since a long time. These processes make the sunlight to reach up to the surface of the ground in thick forest areas. So that seeds on the grounds germinate and facilitate the regeneration and recruitment. Nevertheless these practices are currently abandoned by the forest department. However, natural way of thinning, pruning and regeneration of plants is being ingeniously carried out by the elephants without causing any undesirable effect to the forest elements.

It is also highlighted that breaking the branches of trees in the forests due to the absence of foraging plants during a particular season favours the development of new buds (auxiliary buds) into number of branches during the subsequent favourable seasons. This also helps to maintain the height of the trees at a particular level. In other words, the elephants help to maintain the height alteration of the trees in the forests, so that continuity and closeness of the canopy of the forest cover are established. This facilitates continuous availability of food for the elephants and other animals throughout the year.

Elephants graze over the grassland during a particular season. Generally they graze the tall grasses such as *Saccharum* sp., *Dichanthium* sp., *Cenchrus* sp. and *Themida* sp. (Easa 1988). These grasses are unpalatable to other herbivores because the leaves of these grasses developed during that particular season become thick, silicified and scarified and loose the tenderness. Generally such grasses are preferred by elephants and fed on them. The remaining lower portion of the grass species produces tender leaves during the subsequent seasons, which become preferred food species for other herbivores in the habitat. Thus elephants play a vital role, both directly and indirectly, in maintaining the food chain of the ecosystem and also the climax state of the community.

16.3.3 When Elephant Walks

The nature mobility of elephants as herds and their heavy weight make their routes visible. The routes are also being frequently used by other herbivores. In due course of time, their routes become countryside roads. Generally these routes lead to natural water sources such as reservoirs, ponds, streams or rivers. Most of the roads in the forest areas had been actually made by the frequent movements of elephants before the human settlement. Such routes are also being used by other animals, which help them to find and satisfy their water requirements. Sometimes they guide those who lost their routes, baffled and wandering in the thick and eerie forests.

Being a tropical country, the most prevalent types of vegetation include thorny bushes, scrub jungles and deciduous vegetation with open canopy. Interestingly, the species richness is remarkably high, and hence there is rich biodiversity, which also includes some rare species of plants and animals. It is amazing to note that there is the coexistence of tiger and elephants in such types of vegetation. It may be news that the presence of elephants is obligatory in a scrub jungle for the existence of tigers. There is prey-predator relationship in such forests, in which the role of elephants cannot be ignored.

For example, the herbivores such as black-naped hares, spotted and sambar deer and antelopes such as blackbuck and four-horned antelope are the prey species for the tigers. The populations of the prey and predator species are homeostatically controlled. The prey species protect themselves using the bushes as hiding places, and at the same time, they have to move fast so as to escape from the attack of the predator. Hence both of the activities of the predator and prey species, i.e. hunting and escaping, respectively, need free passage, which has been made and facilitated by the elephants when they walk in dense forests. The flies, beetles, butterflies, bugs and grasshoppers which fly out of the grasses as the elephants walk along the grassland are made prey for insectivorous birds (myna, cattle egret, etc.).

16.3.4 When Elephant Defecates

Generally an adult elephant defecates in an average of 16 times a day (Figs. 16.3 and 16.4). The fresh dung is basically ball shaped and warm and contains undigested remains of fibrous plant material mixed with organic matter and intestinal juices. The dung is useful to other living organisms in many ways either directly or indirectly. It is interesting to note that the freshly landed dung attracts butterflies. Elephant biologists have categorically established the relationship between the

Fig. 16.3 Elephant dung

Fig. 16.4 Elephant dung

butterflies and elephants, especially with the dung of elephants. In the cool early morning hours, the butterflies visit the freshly dropped dung and enjoy the warmth. It has been reported by entomologists that the elephant's dung has certain specific minerals, which are ingested by the male butterflies, and also it has been well established that these minerals are essential for their reproduction.

There is a clear tropic relationship between the organisms, which are attracted towards the dung. Many beetles and flies are attracted and visit the fresh moist dung before it gets dried up. It becomes the centre of attraction for the birds such as peafowl, grey jungle fowl, myna, Indian robin, etc. They visit that microhabitat and feed on the beetles and flies. Reptiles, such as *Calotes*, skinks, *Varanus*, etc., are also attracted towards the dung of the elephants as it provides their prey in the form of insects and beetles.

The dung of the elephants is also a very good substrate for the growth of mushrooms (Fig. 16.5). Mushrooms are the source of protein and medicine. These are the organisms that are mediating the degradation of lignocellulosic organic materials. There are categories of mushrooms, which are edible and delicious to eat. Hence the technology of mushroom cultivation becomes a rural and women entrepreneurship programme. In this technology, agricultural wastes such as paddy straw are commonly used as substrate for mushroom cultivation. The mushroom production is fundamentally based on the phenomenon of solid-state fermentation of paddy straw substrate. In mushroom culture the mycelium of the mushroom completely established over the paddy straw substrate is partially decomposed under suitable moisture and temperature and ultimately produces the fruit bodies. The food of the elephants includes mostly the different varieties of grasses, which have a well-developed fibre in their tissue. As these fibrous plant materials pass through the digestive system of the elephant, they undergo digestive processes, and the remains come out as dung. Casing of the substrates with mud is one of the procedures for mushroom production. The dung is very much similar to the cased substrate used for mushroom cultivation as the dung contains undigested fibrous plant materials, suitably compacted and encased with dung materials, and also provides with suitable moisture and temperature. All these parameters of the dung facilitate the colonization of the dung by the fungi and produce the fruit bodies of the mushroom. These naturally grown mushrooms are collected and eaten as foods by the natives of

Fig. 16.5 Mushroom growth in the elephant dung

the forest and also become the food for the star tortoise and sloth bears. In the development or succession of an ecosystem, either primary or secondary, arrival of the propagating organs (seeds or fruits) of the plants of the habitat is essential. This process is called migration. Migration of the seeds and fruits takes place through dispersal. It takes place through various agencies such as wind, water and animals. In the dispersal of the seeds and fruits, especially in the case of tree species, elephants play a very important role. In elephant, the process of digestion of engulfed plant materials is completed in 46–72 h. We know well that the elephants walk a distance of about 15–20 km in a day. Thus it is obvious that the elephants cross a distance of 50 km before the defecation of the feed that has been taken up by the elephants in a place. But during the process of digestion, not all the plant materials are completely digested. It has been pointed out that only 40–45% of the feed gets digested and the remaining part of the feed contains seeds and fruits of varieties of plant species and the fibrous materials. The seed and fruits that come out of the gut are viable and in a condition of ready to germinate. Thus these propagating organs of the plants travel along with the elephants to a far off place and get dispersed along with the dung.

Mere dispersal of the seeds and fruits does not ensure the successful establishment of the plant species in that habitat (Fig. 16.6). The migrated seeds/fruits have to germinate, grow, establish and colonize, the process which we ecologically call ecesis. Not the seeds of all the plant species readily germinate, when they get detached from the mother plant. They are in an inactive state, which we call seed dormancy, i.e. the seeds have to undergo a period of rest, which varies depending on the species. So the dormancy of the seeds is to be broken in order to make then to germinate readily. There are several methods of breaking of dormancy, which include scarification, stratification and using light and temperature. Among them scarification is the method of softening of the seed coat by mechanical or chemical treatment.

In the method of chemical scarification, the seed coats are treated with strong mineral acids or other chemicals. As the seeds and fruits of the plants pass through the alimentary tract, they are treated with digressive juice, enzymes and other

Fig. 16.6 Seed dispersal through elephant dung

organic substances, which act as agents of chemical scarification. Hence, they readily germinate when they reach the surface of the ground along with the dung. These substances not only break the dormancy of seeds/fruits but also enhance their germination and other growth potentialities. The dried and partially decomposed dung is being eaten up by the termites. It is well known that termites are the prey species for some of the reptiles and birds, and ultimately the dung enhances the fertility of the soil.

16.3.5 Identification of Nature Salt Licks

Minerals are the basic requirements for the growth and development. Minerals enter into different trophic levels through food chain. Plants obtain the minerals from the soil as soil solution through root system and translocated to different metabolic pathways and locked up in different components of the plants. There are different plant species, which obviously depend upon their different minerals and other nutrients for their growth. These minerals in the plant tissues satisfy the mineral requirements of the herbivores and thus enter into the food chain. In the absence of such plant species, the herbivores satisfy their requirement through some other sources or means (Fig. 16.7).

Elephants as herbivore satisfy their mineral requirement directly through the plants. Due to unavailability or scarcity of the mineral requirements from the plants to elephants, they need to obtain their requirement directly from the soil. The process is called "salt licking". It is obviously known that the elephants have very good sense of smell. With the help of this potential power, they detect the availability of the mineral salts under the soil. They hit the surface of the soil with the help of their

Fig. 16.7 Nature salt licks of elephant

Fig. 16.8 Elephant drinking subsoil water

forelegs and tusks and dig out the soil and eat, and thus they satisfy their mineral requirements. These mineral sources (salt licks) are also used by other herbivores such as gaur, sambar and spotted deer, etc. Thus the elephant as a large mammal in the biodiversity-rich tropical forests plays a vital role holistically in the management and sustainability of biodiversity and ecosystem as a whole.

16.3.6 Identification of Subsoil Water

During the summer all the water sources get dried up (Fig. 16.8). Since elephants are being large animals, their water requirement is also higher than that of other animals. Hence they are in a compulsion of finding out the availability of groundwater. The elephant with the help of its unique organ, the trunk, is capable of detecting the nearest water table from the surface of the earth. The trunk in made up of more than a lakh of muscle fibres. The elephants gently tap the surface of the earth with the trunk and create vibration. The fibres of the muscle are capable of sensing these vibrations and detect the level of water table. They make small ditches in the regions where the water table is very near from the surface of the earth, prepare mud and smear it over their body. Thus, they overcome the high temperature of scorching sun during the midday of the summer season. These ditches have continuous flow of water from springs, which become water source for other animals after the elephants left that place. Thus during the summer most of the animals in the forest use the tracks of the elephants in order to get their water requirements satisfied, which has been thoroughly established by the elephant biologists both in Asian and African elephants.

Fig. 16.9 Poached elephant for ivory

16.3.7 Conservation Problems

Conservation issues can be divided into two distinct categories: (1) activities that affect elephants directly such as hunting/poaching and capturing and (2) developmental activities and human activities leading either to the loss of elephant habitat or its qualitative degradation.

16.3.8 Poaching for Ivory

Ivory poaching is widespread in Asian countries, where substantial proportions of male elephants or tuskers have been particularly affected in southern India (Sukumar 1989a, b) (Fig. 16.9). This region also has the largest regional concentrations of elephants in Asia (Sukumar and Santiapillai 1996). Given the very high proportion of tusked male elephants here, the southern Indian region was also home to perhaps the largest numbers of tuskers until recent times when this situation changed (Ramakrishnan et al. 1998). An estimated 100 male elephants were killed in the country every year by ivory poachers between 1980 and 1986. This has seriously affected the male and female ratio in many areas. As we mentioned above, ivory poaching has led to an adult male to female ratio of 1:25 as was observed in the Mudumalai Tiger Reserve, Nilgiris (Arivazhagan 2005). It was a clear indication that poaching has had a very adverse impact on the population and at this stage, even the loss of one male was a very serious concern. While poaching has depleted the male population, the female population has been increasing over several decades. This is leading to local overabundance of elephants, especially in protected areas, and can have adverse impacts on vegetation and therefore on habitat quality and also increase in the conflict with surrounding human population. Baskaran et al. (1995) described that the elephants are having an adverse impact on their preferred food tree species; this problem is yet to be assessed and its implications to be understood.

16.3.9 Loss of Habitat

The most common cause of endangerment is habitat loss. Plants and animals need space to live and energy provided by food, just as humans do. As human population and consumption increase, wildlife habitat is converted to agriculture form lands, houses and highways. Forests are cut down for building materials, fuel and paper. Prairies and forestlands are turned into cropland and grazing land for our livestock. The major conservation issue today remains the exploitation of the elephant habitat, leading to qualitative degradation and fragmentation of the habitat. Biotic pressures from such a large population have led to the loss of much of the village and private forests. Irrespective of the suitability of the area for permanent agriculture, degradation of such lands into wastelands happens due to defective agricultural practices and absence of appropriate soil conservation measures. This has further increased the dependence of the local people and their cattle on the forests and is ultimately accelerated throughout the elephant ranges in the country. The continued loss, degradation and fragmentation of the elephant habitat reduce the elephant's ranges and lead to human-elephant conflicts. Further loss would be very adverse impacts on elephant populations. Apart from habitat loss and degradation, conversion of natural forests to commercial timber and monoculture plantations, etc. has affected the elephant habitat seriously. Constructions of dams, hydroelectric projects, railways, roads and reservoirs and establishment of human settlements in flat areas and in the valleys have also fragmented the elephant habitat and disrupted habitual migration patterns by fragmenting the habitat (Fig. 16.10).

16.3.10 Reduction of Home Range Size

Within the species, the home range varies widely between areas and within the area between the sexes and individuals (Fig. 16.11). These variations have been attributed to different factors. Variations in the home range size of elephants have been found to be related to habitat quality, both in Asia and Africa. Olivier (1978) found the home ranges of the Asian elephant to be larger in the primary forests and smaller in the secondary forests. Baskaran (1998) stated that the clans had home range sizes that varied from 562 to 800 km^2 with a mean size of 677 km^2. Olivier (1978), using radiotelemetry, estimated the home range size as 166.9 and 59.27 km^2 for two herds in Malayan rainforests. Sukumar (1985, 1989a, b), by resighting method, estimated minimum home range sizes of 105 and 115 km^2 for the two clans using 14 and 15 locations, respectively, in Sathyamangalam Forest Division, South India. Easa and Jayaraman (1998), using resighting method, reported the home range size of two

Fig. 16.10 Herd of elephant in the human habitation

Fig. 16.11 Adult elephant with calf

clans (family unit) in Parambikulam Wildlife Sanctuary (South India) to be 124.3 and 156.6 km^2 based on 226 and 200 locations, respectively. Desai (1991) estimated that the home ranges of three clans as 232, 265.6 and 112 km^2 based on 257, 60 and 56 resightings, respectively, in Mudumalai Wildlife Sanctuary and Bandipur National Park. Daniels et al. (1995) reported a home range of 224 km^2 using 14 locations for 1 clan in a large degraded part of the Hosur and Dharmapuri forest divisions of Tamil Nadu, South India. Joshua and Johnsingh (1995), using radiotelemetry, estimated the home range size of a clan as 30 km^2 based on 277 locations in Rajaji National Park, North India.

Bulls, during musth period, are said to move extensively in search of oestrous females (Desai and Johnsingh 1995; Joshua and Johnsingh 1995). Barnes (1982) stated that African elephant bulls travelled long distances in search of oestrous females during mating season. Sukumar (1989a) speculated that in more diverse region in terms of vegetation, the home range could be less for elephants to meet out their seasonal requirements within a relatively small area. Possibly, diversity of habitat may not be the only factor influencing the range size of elephants. The corridors are narrow in size, might disrupt significantly on the home range of elephants and would change the normal behaviour of the elephants.

16.3.11 Population Increase

It is notable that conservation and management of endangered species in the wild require adequate knowledge of their distribution and population size. Ecological parameters such as population estimation and sex ratios are greater relevance to survival of elephants in the long run, considering its multidimensional conservation issues. It is highly warranted to document current population and its structure to deal the conservation issues pertaining to elephants. The scope of the study was endorsed with the support of the Project Elephant, Ministry of Environment and Forests, Government of India. Of late, estimation of elephant numbers was calculated based on block count in several parts of its ranges across the country. The forest department has fine-tuned the method after eliminating the several field constraints and found suitable and convenient with the available force with them besides talented experts, NGOs and volunteers. Training modules on census techniques were provided by the forest department and representatives from reputed research institutions. This has gained a momentum to execute the census techniques fairly in a better manner with adequate training by the field staff. The line transect method developed by Burnham et al. (1980) has been used successfully for estimating elephant densities in Asia and Africa through direct counting (Varman and Sukumar 1995; Karanth and Sunquist 1992; Baskaran and Desai 2000) in areas with high animal density. The line transect method has also been used to estimate densities through enumeration of indirect evidence, e.g. dung piles (Barnes and Jenson 1987; Dawson 1990) in areas with low elephant density and poor visibility.

16.3.12 Loss of Corridors

Over a few decades, the vast elephant ranges have been fragmented and precariously connected by narrow corridors due to expansion of human habitations and agriculture, indiscriminate growth of various development activities, severe biotic

Fig. 16.12 Elephants moving in the tea estate

pressures, etc. "Forest corridors" can be defined as the "narrow strip of forests connecting two habitats that facilitate major functions such as exchange of genes between populations, dispersal, provide access to variety of seasonal foraging grounds, and prevention of faunal collapse" (Saunders and Reberia 1991). Elephant corridors have received lot of attention in conservation and are widely used in devising conservation strategies, especially in recent years, with the reduction of contiguous habitats into islands (Menon et al. 2005) (Fig. 16.12). The most important function of the corridor is to prevent wild animals from getting isolated in small pocket-like islands. The process of habitat fragmentation has been going on ever since man started agriculture. But this problem has, of late, become much more acute due to mounting pressure on land.

What should be the optimum size of a corridor? The length will naturally depend on the distance to be connected. In doing so, it may become necessary to take a circuitous route connecting existing jungles. There cannot be any hard and fast rule on the width. An elephant corridor requires a minimum width approximately 500 m to approximately 2 km. As much as wider is better for the movement of elephants. But limitations such as the lay of the land, the types of country and the practical consideration such as causing least disturbance to people who are likely to be affected by the provisions of corridors have to be taken into account in determining the width. The need of the corridors for the Asian elephants has been reported by many studies; the significance of the corridors to elephants in terms of ecological aspects as well as conflict issues had been studied in detail only recently by

Fig. 16.13 Domestic animals grazing in the natural habitat

Ramakrishnan (2008). Ramakrishnan and Ramkumar (2007) have documented micro- and macro-level ground realities in the crucial elephant corridors that would help the managers to prepare specific management plans for securing corridors especially in Coimbatore, Sathyamangalam, Nilgiris and Anamalai elephant ranges on behalf of the Wildlife Trust of India. Elephants use corridors as part of their ranging between habitats. The quality of elephant corridors is mostly affected by various anthropogenic pressures and development activities. These pressures on the corridors reduce the resource availability to elephants, which in turn leads to crop raiding and other human-elephant conflicts at the forest fringes. Conversion of elephant corridors into estates, buildings and crop cultivation sites results in crop damage and human causalities (Santiapillai 1987; Sukumar 1990; Easa and Sankar 1999).

16.3.13 Loss of Habitat Quality

Human settlements and their activities within the elephant's habitat have a detrimental impact on elephants directly by rendering the surrounding habitat unusable to elephants (Desai and Baskaran 1996) (Fig. 16.13). Sukumar (1985) has also cautioned that the elephant habitat is deteriorating due to high density of livestock in Sathyamangalam Forest Division. Ramakrishnan (2008) documented that intense grazing pressures seriously affected on the vegetation composition of forest corridors and degrade the habitat quality extensively in the Nilgiri Biosphere Reserve.

Fig. 16.14 Construction of building and developmental activities in the elephant habitat

He has stated that most of the corridors showed low densities of regeneration and recruitment classes of native tree species and major proportion of barren ground due to overgrazing by livestock. Ramakrishnan et al. (1997) recorded that there were three important plant species, namely, *Commiphora berryi*, *Acacia planifrons* and *Acacia leucophloea* were highly preferred by local people for fuel wood collection as well as preferred by elephants for browsing during dry season in Sathyamangalam and Coimbatore forest divisions. In the same study, it was documented the similar trend in non-timber forest produce collection by local people on *Terminalia chebula*, *Phyllanthus emblica* and wild mango. This kind of anthropogenic pressures sometimes may not affect in huge areas where elephants have space and resources, but this would definitely affect elephants when they move through narrow corridors.

16.3.14 Impact of Developmental Activities

During the past decades, the developmental activities were not much seen in Southern India especially along the corridors in and around Coimbatore, Nilgiri North and Sathyamangalam forest divisions (Fig. 16.14). After the 1990s, the indiscriminate growth of various developmental activities such as establishment of ashrams, amusement parks and educational institutions, change in agricultural practice and expanding agriculture was on the rise along the fringes of the corridors, thus resulted ever-increasing human-elephant conflict issues in the Southern India.

Fig. 16.15 Human-elephant conflict

16.3.15 Human-Elephant Conflict (HEC) Issues

The HEC is one of the biggest challenges for successful conservation of Asian elephants (*Elephas maximus*) in India today. Humans are increasingly competing with large mammals for space and other resources across the world (Madhusudan 2003). This competition for food, water and space between humans and elephants increases the conflict issues (Sukumar 1991; Hoare and du Toit 1999; Smith and Kasiki 1999; Sitati et al. 2003; Kumar 2006) such as crop raiding, damage to property and injury and death of humans, elephants and livestock (Nath and Sukumar 1998; Karanth and Madhusudan 2002; Hoare 2000; Sitati et al. 2003).

Large-scale conversion of forest areas due to increasing human pressure is the main reason behind for increasing trend of human-elephant conflicts in the plains of the Nilgiri Biosphere Reserve (Fig. 16.15). When the interactions between elephants and human beings become very close, undoubtedly there would be a conflict between man and elephant. Elephants cause crop depredation, and the attacks on human beings lead to injuries, severe wounds and ultimately to death. Besides, the elephants also cause damage to human properties.

There are some proximate causes that influenced the crop-raiding strategy of elephants in India. Sukumar and Gadgil (1988) stated that the elephants prefer feeding on crops because of greater nutritive content and palatability of crops. Though the elephants live in a variety of landscapes available in India, it has been well studied and conclusively reported that large contiguous areas are either surrounded by crop fields (Sukumar 1991; Balasubramaniyan et al. 1995), very degraded areas with other agricultural encroachments (Datye and Bhagwat 1995) or fragmented landscape with a mosaic of crop fields. Patches of forests (A.C. Williams and

Fig. 16.16 Human casualty due to elephant

A.J.T. Johnsingh, Wildlife Institute of India. Unpublished Report) are also the influencing factors for crop depredation by elephants. Blair et al. (1979) reported that the increased cultivated area and human movement in to the elephant habitats are also responsible for crop depredation by elephants. Another possible crop-raiding strategies adopted by elephants are that they are, especially males, related to social organization and the "high risk, high gain" to increase their fitness (Sukumar 1991). The pattern of crop raiding by elephants and immediate reasons behind on it might vary, but several of the above factors may play a significant role under particular circumstances. Not all elephants in a population raid crops (Balasubramaniyan et al. 1995). Elephants annually damage crops worth from a few thousand dollars to millions of dollars (A.C. Williams and A.J.T. Johnsingh, Wildlife Institute of India. Unpublished Report; Blair et al. 1979; Sukumar 1989a, b).

Every year more than 100 human beings and 40–50 elephants are killed during crop raiding in India (Johnsingh and Panwar 1992; Menon, Asian Elephant Conservation Center, Unpublished Report).

16.3.16 Human Deaths and Injury

A variety of direct interactions by elephants leading to human injury and killing have been widely reported from Africa and Asia (Fig. 16.16). Thouless (1994) opined that people's injury and killing by elephant in Lakipia District, Northern Kenya, have increased in the last few years. An exceptionally high kill and injury of human being in 1992 might have been partly due to draught conditions, resulted with elephant staying in well-watered southern areas close to human settlements (Thouless 1994). India experiences approximately 175–200 cases of human deaths caused by elephants annually (Bist 2002). Sukumar (1989a, b) reported killing of 30–50 people every year in southern India which has largest elephant habitat and population. Nath and Sukumar (1998) recorded an average of six human deaths by

Fig. 16.17 Elephant died due injury

elephants per year in Kodagu District of Karnataka. Dey et al. (1991) reported that the human deaths are ranging between 28 and 59 persons every year, from fragmented habitats of Northwest Bengal between 1980 and 1990. It has also been reported by Barua and Bist (1995) from the same region that the killing was at an average of 47 people per year since 1981. Datye and Bhagwat (1995) reported a total of 208 human deaths between 1980 and 1991 from South Bihar (134) and Southwest Bengal (74) through pocketed elephants on a fragmented landscape. Williams and Johnsingh (1996a, b) recorded a total death and injury of 115 humans from three districts of Garo hills, Meghalaya, between 1984 and 1995. Ramakrishnan (2008) reported that 21 numbers of human deaths are recorded in and around Coimbatore Forest Division between 1994 and 2005.

16.3.17 Elephant Deaths and Injury

Like humans, elephants do suffer due to the negative interactions inflicted on them through injury and killing (Fig. 16.17). Several kinds of equipment, devices and chemicals are being used for such purposes. Killings were done by gunshots or electrocution. Ramakrishnan and Durairasu (2005) reported that the crude electric fences fixed by unprofessional people have resulted in death of many elephants in and around the Coimbatore Forest Division, and most of them were the adult males either in early or prime puberty. Conflict incidences have also been reported from Sri Lanka by several authors through gunshots (McKay 1973; Santiapillai 1994; Santiapillai and De Silva 1994). Each year, nearly 100 elephants are killed in

Fig. 16.18 Elephant feeding in agricultural habitat

conflict-related instances in India (MoEF 2010). Ramakrishnan (2008) reported that 40 number of elephant deaths recorded in and around Coimbatore Forest Division between 1994 and 2005.

16.3.18 Crop Damage

Crop damage by elephants is also very serious in southern, central and northeastern India. Although adult male elephants are usually the most notorious raiders, considerable damage may also be caused by family herds (Fig. 16.18). Crop depredation by elephant is a critical problem among the human-elephant conflict issues in India. Fragmentation of habitat leading to trapping of elephants in isolated patches with cultivation all around is mentioned as the factor responsible for crop raiding in South India (Sukumar 1990). Further, factors such as degradation of habitat, competition for water, movement pattern and palatability and nutritive value of crops also led to crop depredation (Sukumar 1985, 1989a, b, 1990; Balasubramaniyan et al. 1995). Crop raiding and man-wildlife conflict have been documented in Kerala by Veeramani and Jayson (1995) and Veeramani et al. (1996). Gopinathan (1990) has also mentioned the crop-raiding problem in Wayanad Wildlife Sanctuary. Crop raiding and economic loss due to elephants were reported from Bihar by Mishra (1971) and Datye and Bhagwat (1993). Similar studies were also conducted in the Western Ghats, especially in Nilgiris, by Sukumar (1990) and Balasubramaniyan et al. (1995) and Eastern Ghats by Rameshkumar and Sathyanarayna (1993). Ramakrishnan (2008) recorded that non-preferable crops such as chilly, graphs, tomato, turmeric and marigold were high crop economic benefit to the farmers in elephant ranges of Sathyamangalam and Coimbatore forest divisions in Southern

Fig. 16.19 Damaged hut by wild elephant

India. Bist (2002) reported that about 10 lakh hectares of crop lost due to elephant depredation in India. MoEF (2010) report mentioned that nearly 500,000 families were affected by crop damage caused by elephants in India.

16.3.19 Assert Damage

The elephant is one of the most conflict-prone wildlife species in India, causing large-scale damage not only to crops and human lives, also to assert of humans (Fig. 16.19). Agriculture exploitation of groundwater by various other developmental activities also makes the entire area dry in a short span of time. In the long run, it may lead to early onset of drought in areas. The increasing incidences of recent human-elephant conflicts are due to total loss of inaccessibility to potential food patches and perennial water sources (in private areas particularly), and especially in summer and disappearance of critical microhabitats for elephants. These factors force the animals to explore to new areas for migration. There has been proliferation of construction activities such as educational institutions, research institutions, industries, amusement parks, a zoological park (proposed) and horticultural gardens over a decade. In construction of buildings, noise pollution also disturbs elephants. Developmental activities slowly arrest the traditional migratory path of elephants and thereby lead to prevention of gene exchange between populations. Bist (2002) reported that 15,000 houses were damaged annually by elephants in India. The amount spent on control measures and ex gratia payment towards human-elephant conflict runs to Rs 15 crores annually each year.

Fig. 16.20 Elephant died due train accident

16.3.20 Train and Road Traffic

The increasing trend of vehicle transport during day hours considerably affects the diurnal movement of elephants between habitats. In recent years the vehicular traffic is so congested and non-stop even during night hours (Fig. 16.20). This acts as a permanent barrier for the migrating elephants, which escalates human-elephant conflicts in adjacent areas. Although vehicle collision does not occur as frequent as train collision with elephants, there is every possibility that other small- to medium-sized animals may die due to accident. Singh (2001) stated that 17 elephants have died in train accidents in the last 14 years on the Haridwar-Rishikesh line in Chilla-Motichur Corridor in India. Therefore laying of roads and railway tracks in between narrow corridors is a serious impediment to the animals especially to migrating animals. In the period of 6 years, 118 elephants have died in railway accidents in 9 states in India.

16.4 Conclusion

Indeed we have lost a lot from the nature. Nevertheless we have a lot yet to be preserved and conserved in nature. Elephants are large mammals and from an important link in the energy flow system of the ecosystem. In the ecosystem they act as facilitator of flow of energy and matter, and also they play a vital role in the forest management practices. Hence, it is not an exaggeration to state the elephant as a "keystone species". Such an important species is being threatened due to habitat loss, fragmentation due to loss of corridors and human-induced disturbances. The incidences of human-elephant conflicts are also in the increasing trend. So, it is high time to think seriously about the events that lead to the human-elephant conflicts and to identify suitable mitigating measures for overcoming such problems. It is also emphasized that there is a need of suitable policies, planning and management for preserving and conserving the existing habitats and corridors as measures to conserve elephants for their long-run existence in the living planet.

Acknowledgements This paper is part of the University Grants Commission's Major Research Project F.No. 42-594/2013 (SR). We thank the UGC for providing fund the Major Research Project. Our sincere gratitude to the Principal Chief Conservator of Forests (Head of Forest Force), Tamil Nadu, for issuing us permission to enter into forest areas of the Tamil Nadu part of the Nilgiri Biosphere Reserve. Our special thanks to Mr. I. Anwardeen, IFS., the Conservator of Forests, Coimbatore Circle, for providing constant support to our research works. We record our sincere thanks to the District Forest Officers of Coimbatore, Nilgiri North, Nilgiri South and Gudalur forest divisions. We thank our college principal and head of the department for giving constant encouragement to our team. We thank our lab students Mr. A. Samson and Ms. A. Chitheena for helping us in various ways in the field as well as in lab.

References

Altmann J (1974) Observational study of behaviour: sampling methods. Behaviour 49:227–267

Arivazhagan C (2005) Population dynamics of Asian elephant (*Elephas maximus*) in Southern India. Unpublished Ph.D. thesis submitted to Bharathidasan University, Trichy

Balasubramaniyan M, Baskaran N, Swaminathan S, Desai AA (1995) Crop raiding by Asian elephant (*Elephas maximus)* in the Nilgiri Biosphere Reserve, South India. In: Daniel JC, Datye HS (eds) Proceedings of the international seminar on the conservation of Asian elephants. Bombay Natural History Society, pp 350–367

Barnes RF (1982) Elephant feeding behaviour in Ruaha National Park. Tanzania Afr J Ecol 20:123–136

Barnes RFW, Jenson KL (1987) How to count elephants in forest. IUCN African elephant & rhino specialist group. Tech Bull I:1–6

Barua P, Bist SS (1995) Changing patterns in the distribution and movement of wild elephants in north Bengal. In: Daniel JC, Datye HS (eds) A week with elephants. Bombay Natural History Society/Oxford University Press, Bombay, pp 66–84

Baskaran N (1998) Ranging and resource utilization by Asian elephant (*Elephas maximus Linnaeus*) in Nilgiri biosphere reserve, South India. Ph.D thesis submitted to the Bharathidasan University, Tiruchirappalli, Tamil Nadu, India

Baskaran N, Desai AA (2000) Elephant population estimation in Mudumalai wildlife sanctuary and National Park. Final report 1999–2000. Bombay Natural History Society, Bombay

Baskaran N, Balasubramanian M, Swaminathan S, Desai AA (1995) Home range of elephants in the Nilgiri biosphere reserve, south India. In: Daniel JC, Datye H (eds) Proceedings of the international seminar on the conservation of the Asian elephant. Bombay Natural History Society, Bombay

Bist SS (2002) An overview of elephant conservation in India. Indian Forester 128:121–136

Blair JAS, Boon GG, Noor NM (1979) Conservation or cultivation. The confrontation between the Asian elephant and land development in peninsular Malaysia. Land Dev Dig 2:27–59

Burnham KP, Anderson DR, Laake LL (1980) Estimation of density from line transect sampling of biological populations. Wildl Monogr 72:1–202

Daniels RJR, Gadgil M, Joshi NV (1995) Impact of human extraction on tropical humid forests in the western Ghats in Uttara Kannada, South India. J Appl Ecol 32:866–874

Datye HS, Bhagwat AM (1993) Estimation of crop damage and the economic losses caused by elephants and its implications in the management of the elephants, a week with elephants. In:

Proceedings of the international seminar on Asian elephants. Bombay Natural History Society, pp 375–389
Datye HS, Bhagwat AM (1995) Estimation of crop damage and the economic loss caused by elephants and its implications in the management of elephants. In: Daniel JC, Datye HS (eds) Proceedings of the international seminar on the conservation of Asian elephants. Bombay Natural History Society, Mumbai, pp 375–379
Dawson S (1990) A model to estimate density of Asian elephants (*Elephas maximus*) in forest habitats. M. Sc. thesis, University of Oxford, 157 p
Desai AA (1991) The home range of elephants and its implications for management of the Mudumalai wildlife sanctuary, Tamil Nadu. J Bombay Nat Hist Soc 88:145–156
Desai AA, Baskaran N (1996) Impact of human activities on the ranging behaviour of elephants in the Nilgiri biosphere reserve, South India. J Bombay Nat Hist Soc 93:559–569
Desai AA, Johnsingh AJT (1995) Social organization and reproductive strategy of the male Asian elephant (*Elephas maximus*) in Nilgiri biosphere reserve, South India. In: Daniel JC, Datye HS (eds) A week with elephants. Proceedings of the international seminar on the conservation of Asian elephant. Mudumalai Wildlife Sanctuary, June 1993. Bombay Natural History Society, Bombay, p 532
Dey A, Srivastava R, Gupta VK (1991) Robust designs – a review and bibliography. Cahiers du CERO 33(1–2):51–62
Easa PS (1988) Movement pattern of Asiatic elephant *Elephas maximus* in Parambikulam Wildlife Sanctuary, Kerala. Kerala Forest Research Institute, research report no. 54
Easa PS, Jayaraman K (1998) Population estimation of major mammals in the forests of Kerala: 1997. Kerala Forest Research Institute, Peechi
Easa PS, Sankar S (1999) Study on man-wildlife interaction in Waianae wildlife sanctuary, Kerala. KFRI Research report No.16. Kerala Forest Research Institute, Peechi
Gadgil S (1988) Recent advances in monsoon research with particular reference to the Indian monsoon. Aust Meteorol Mag 36:193–204
Gopinathan V (1990) Crop damage by elephants in Wayanad. In: Proceedings of the elephant symposium, Kerala Forest Department (Wildlife Wing), Thiruvananthapuram
Hoare RE, du Toit JT (1999) Coexistence between people and elephants in African savannas. Conserv Biol 13:633–639
Hoare R (2000) African elephants and humans in conflict: the outlook for co-existence. Oryx 34(01):34
Johnsingh AJT, Panwar HS (1992) Elephant conservation in India-problems and prospects. Theriological Congress
Joshua J, Johnsingh AJT (1995) Ranging pattern of elephants in Rajaji National Park: implication for reserve design. In: Daniel JC, Datye HS (eds) A week with elephants. Proceedings of the international seminar on the conservation of Asian elephant. Mudumalai Wildlife Sanctuary, June 1993. Bombay Natural History Society, Bombay, pp 256–260
Karanth KU, Madhusudan MD (2002) Mitigating human wildlife conflicts in southern Kerala. Indian Forester 22:949–953
Karanth KU, Sunquist ME (1992) Population structure, density and biomass of large herbivores in the tropical forests of Nagarahole, India. J Trop Ecol 8:21–35
Kumar MA (2006) Effect of habitat fragmentation on Asian elephant (*Elephas maximus*) ecology and behaviour patterns in a conflict-prone plantation landscape of the Anamalai hills, Western Ghats, India. Final report submitted to Rufford Maurice Laing Foundation, UK
Madhusudan MD (2003) Uneasy neighbours: human resource use and large mammal conservation in the tropical forests of Karnataka, south India. PhD thesis, National Institute of Advanced Studies/Manipal Academy of Higher Education, India
McKay GM (1973) The ecology and behaviour of the Asiatic elephant in south eastern Ceylon. Smithson Contrib Zool 125:1–113
Menon V, Tiwari SK, Easa PS, Sukumar R (2005) Right of passage: elephant corridors of India. Conservation Reference Series 3. Wildlife Trust of India, New Delhi

Mishra J (1971) An assessment of annual damage to crops by elephants in Palamau District, Bihar. Bombay Nat Hist Soc 68:307–310
MoEF (2010) Project elephant report. Ministry of environment and forests. Government of India, New Delhi
Nath CD, Sukumar R (1998) Elephant-human conflict in Kodagu, southern India: distribution patterns, people's perceptions and mitigation methods. Asian Elephant Conservation Centre, Bangalore
Olivier RDC (1978) Distribution and status of the Asian elephant. Oryx 14:379–424
Ramakrishnan B (2008) Significance of corridors to Asian elephant (*Elephas maximus*) and strategies for their conservation and management in the Nilgiri Biosphere Reserve, Tamil Nadu, South India. Unpublished Ph.D thesis submitted to the Bharathidasan University, Trichy, India
Ramakrishnan B, Durairasu P (2005) Human-elephant conflict mitigating measures in the Coimbatore Forest Division, Tamil Nadu, India. OSAI Technical Report No. 3, Coimbatore
Ramakrishnan B, K Ramkumar (2007) Land acquisition perspectives of vital elephant corridors in the Coimbatore and Sathyamangalam Forest Divisions, Tamil Nadu, South India. Report submitted to Wildlife Trust of India, New Delhi
Ramakrishnan B, Sivaganesan N, Srivastava RK (1997) Human interference and its impact on the elephant corridors in Sathyamangalam and Coimbatore Forest Divisions, Tamil Nadu, Southern India. Indian J For 20(1):8–19
Ramakrishnan U, Santosh JA, Ramakrishnan U, Sukumar R (1998) The population and conservation status of Asian elephant in Periyar Tiger Reserve, Southern India. Curr Sci 74:110–113
Rameshkumar S, Sathyanarayna MC (1993) Crop raiding patterns in hosur and Dharmapuri forest divisions Dharmapuri district Tamil Nadu. Abstracts, International seminar on the conservation of the Asian elephant, Mudumalai Wildlife Sanctuary, June 1993. Bombay Natural History Society, Bombay, p 29
Santiapillai C (1987) Action plan for Asian elephants conservation – a country by country analysis-a complication. World wide Fund for Nature, Indonesia
Santiapillai C (1994) Elephant mortality in Sri Lanka. Gajah 72:48–54
Santiapillai C, De Silva M (1994) An action plan for the conservation and management of elephant (*Elephas maximus*) in Sri Lanka. Gajah 13:1–24
Saunders DA, Reberia CP (1991) Values of corridors to avian populations in a fragmented landscape. In: Saunders DA, Hobb RJ (eds) Nature conservation: the role of corridors. Survey Beathy and Sons, Clipping Norton, pp 221–240
Singh P (2001) The construction of rail roads, highways and canals are playing havoc with migratory patterns, segregating elephants populations and other wildlife. Apart from accidental death, inbreeding and shrinking genetic pools, man-animal conflict is also on the raise. Sanctuary Asia, August
Sitati NW, Walpole MJ, Smith RJ, Leader-Williams N (2003) Predicting spatial aspects of human-elephant conflict. J Appl Ecol 40:667–677
Smith R, Kasiki S (1999) A spatial analysis of human-elephant conflict in the Tsavo ecosystem, IUCN African elephant specialist group report. Submitted to world wildlife fund for nature, Nepal, Nature Conservation Foundation
Sukumar R (1985) Ecology of the Asian elephants (*Elephas maximus*) and its interaction with man in South India. Ph.D. Thesis. Indian Institute of Science, Bangalore
Sukumar R (1989a) Ecology of the Asian elephant in southern India. I movement and habitat utilization patterns. J Trop Ecol 5:1–18
Sukumar R (1989b. The Asian elephants ecology and management, Cambridge University Press, Great Britain, 251 p
Sukumar R (1990) Ecology of the Asian elephants in Southern India. Part: II feeding habitats and crop raiding patterns. J Trop Ecol 6:33–53
Sukumar R (1991) The management of large mammals in relation to male strategies and conflict with people. Biol Conserv 55:93–102
Sukumar R, Gadgil M (1988) Male-female differences in foraging on crops by Asian elephants. Anim Behav 36(4):1233–1235

Sukumar R, Santiapillai C (1996) *Elephas maximus*: status and distribution. In: Shoshani J, Tassy P (eds) The proboscidea: evolution and palaeoecology of elephants and their relatives. Oxford University Press, Oxford, pp 327–331

Thouless CR (1994) Conflict between humans and elephants on private land in northern Kenya. Oryx 28:119–127

Varman KS, Sukumar R (1995) The line transect method for estimating densities of large mammals in a tropical deciduous forest: an evaluation of models and field experiments. J Biosci 20:273–287

Veeramani A, Jayson EA (1995) A survey of crop damage by wild animals in Kerala. Indian Forester 22:949–953

Veeramani A, Jayson EA, Easa PS (1996) Man-wildlife conflict: cattle lifting and human casualties in Kerala. Indian Forester 122:897–902

Williams AC, Johnsingh AJT (1996a) Threatened elephant corridors in Garo hills, North east India. Gajah 16:61–68

Williams AC, AJT Johnsingh (1996b) Status survey of elephants (*Elephas maximus*), their habitats and an assessment of elephant – human conflict in Garo hills, Meghalaya. Final Report, Wildlife Institute of India

Influence of Ranging and Hierarchy on the Habitat Use Pattern by Asian Elephant (*Elephas maximus*) in the Tropical Forests of Southern India

17

Nagarajan Baskaran, R. Kanakasabai, and Ajay A. Desai

Abstract

In tropical forests, resource quality and quantity change across habitats due to spatiotemporal varition in environmental conditions. Wide ranging species like elephants that adapted to live in different habitats have been documented to use different habitats in various seasons. Such movement generally has been attributed to environmental factors and its resultant variance in resource quality and quantity, assuming that all habitats are free to all the elephants to use optimally in any season. However, Asian elephant clans in Southern India have been documented to show hierarchy and spacing; besides environmental factors, there appears a behavioural factor in the use of habitat types among individual clans. This chapter presents the first quantitative information on elephants as to how ranging and spacing influence on use of habitat types and their preference by individual clan and bull, by monitoring the movement patterns of three clans and two bulls, ranging in a large contiguous habitats with high-density population of Asian elephants in Nilgiri Biosphere Reserve, Southern India, between 1991 and

N. Baskaran (✉)
Bombay Natural History Society, Hornbill House, S.B. Singh Road, Mumbai 400023, Maharashtra, India

Department of Zoology and Wildlife Biology, A.V.C. College (Autonomous), Mannampandal, Mayiladuthurai 609305, Tamil Nadu, India
e-mail: nagarajan.baskaran@gmail.com

R. Kanakasabai
Department of Zoology and Wildlife Biology, A.V.C. College (Autonomous), Mannampandal, Mayiladuthurai 609305, Tamil Nadu, India

A. A. Desai
Bombay Natural History Society, Hornbill House, S.B. Singh Road, Mumbai 400023, Maharashtra, India

84, BC Camp, Belgaum 590001, Karnataka, India

C. Sivaperuman, K. Venkataraman (eds.), *Indian Hotspots*,
https://doi.org/10.1007/978-981-10-6605-4_17

1994. The results on habitat use pattern show that the number of habitat types used by clans and bulls varied among clans, bulls and between them. Such variations are attributed to spatial location of home range and home range fidelity. Therefore, though many habitat types were available in adjoining areas of their home, not all of them were used by all the clans and bulls. While some clans and bulls have used all types of habitat available in a population range, others have just restricted to one or two habitats round the year. Further, among overlapping clans, the intensity of use of various habitats and their habitat preferences were significantly different, and these have been related to the observed hierarchy in space use and its resultant spacing among overlapping clans. Therefore, it is revealed that apart from influence of environmental factors, hierarchy, a behavioural (social) factor, also plays an important role in the strategy of habitat use. The interclan encounter discussed in the Chap. 15 also goes in support of this view. Thus, the elephant clans are not free to use all the habitats of their choice and appeared to exhibit a hierarchy-based habitat use, which seems to fit with the Fretwell's 'ideal-despotic distribution' model of habitat selection rather than 'ideal-free distribution' as assumed by earlier studies. Such a hierarchy-based habitat use pattern among the elephant clans could result in dominant clans having prior access to good-quality habitat and food, thereby showing better survival and reproductive success.

Keywords
Asian elephant · Habitat use · Hierarchy · Spacing

17.1 Introduction

In tropical forests, resource quality and quantity are not uniform throughout the year in all the habitats due to abiotic factors, which cause the animals to move from one habitat to another during different seasons. The living elephants besides their wide ranging nature are also highly adaptable to live in wide verities of habitats ranging from tropical rain forests to dry thorn forest or semiarid desert (Olivier 1978a, b; Viljoen 1989a, b). Habitat use pattern has been the major focus of many studies on elephants widely in Asia (McKay 1973; Olivier 1978a; Sukumar 1985, 1989a, b; Sivaganesan 1991; Daniel et al. 1995) and in Africa (Wing and Buss 1970; Leuthold and Sale 1973; Caughley and Goddard 1975; Western 1975; Williamson 1975; Eltringham 1977; Leuthold 1977; Short 1983; Jachmann 1988; Viljoen 1989b; Dublin 1996). Most of these studies in Asia and Africa have looked at population or subpopulation levels and generally attributed the seasonal movements and habitat use pattern of elephants to environmental factors and their influence on food quality and quantity. These studies have outlined a pattern of elephant movements in various habitats during different seasons. In the absence of a mass migration of elephants from one habitat to another, all the individuals in a population did not seem to follow any one pattern. Studies in Africa (Leuthold and Sale 1973; Dunham

1986; Viljoen 1989a; Tchamba et al. 1995; Thouless 1996) and in Asia (Baskaran et al. 1995; Daniel et al. 1995; Baskaran and Desai 1996) have shown that individual clans and bulls have specific home ranges and do not move randomly. Further, there is hierarchy and spacing among the clans in the use of space (Baskaran 1998). Therefore, besides environmental factors, there appears a behavioural factor in the use of habitat types among individual clans (Baskaran 1998). This paper presents the first quantitative information on how ranging and spacing, the resultant of hierarchy, influence on use of habitats and their preference by individual clans and bulls in Nilgiri Biosphere Reserve, Southern India.

17.2 Methods

17.2.1 Study Area

Nilgiri Biosphere Reserve (76° 0′ E and 77° 15′ E and 12° 15′ N and 10° 45′ N), spread over an area of 5520 km^2, is situated at the junction of three southern states—Tamil Nadu, Karnataka and Kerala. It has an undulating terrain with an average elevation of 1000 m above MSL. Rivers such as Nugu, Moyar and Bhavani and most of their tributaries are perennial and drain the area. The Nilgiri Biosphere Reserve (NBR) has a diverse climate due to its varied reliefs and topography. The temperature ranges from 7 °C in December to 37 °C in April and receives rainfall both from the southwest (May to August) and northeast (September to December) monsoons. The mean annual rainfall varies from 600 (in the eastern side) to 2000 mm (in the western side). The dry season is from January to April. Corresponding to the gradient in rainfall, the vegetation varies from southern tropical thorn forest in the east to moist deciduous forest in the west with dry deciduous forest in between the two forest types (Champion and Seth 1968). The reserve forms part of Brahmagiri, Nilgiri-Eastern Ghats landscape, which is the largest contiguous habitat (13,058 km^2) available for Asian elephant anywhere in Asia (Baskaran 2013). NBR is also known for remarkable faunal diversity and is well known for supporting the largest population of Asian elephants with an estimated population of 5750 individuals (Project Elephant 2007; Baskaran 2013; Baskaran et al. 2010) and relatively less disturbed population (Desai and Baskaran 1996). Overgrazing by domestic cattle and fire wood collection are serious problems in the eastern fringes of NBR (Baskaran et al. 2012).

17.2.2 Seasonal Habitat Use

The study was carried out using radiotelemetry to monitor the movement of individual clans and bulls to different habitats in relation to season. This method gives very precise information on how different clans and bulls use various habitat types. The habitat use pattern of individual animals within their home range has been extensively studied in several species of mammals; roe deer (Cederlund 1983), buffalo (Funston et al. 1994), sloth bears (Joshi et al. 1995) and birds barred owl (Nicholls and Warner 1972). The study area habitats were classified into four major types, viz. moist deciduous, dry deciduous, mixed deciduous and dry thorn forests. The proportion of time spent in each habitat was estimated based on the number of locations which fell in different habitat types. As there was a time interval between two locations and since these intervals were not uniform, the locations were weighted (depending on the time interval between two locations) to represent the time spent in a particular habitat type in a given season. Habitat utilization pattern was studied for two clans (viz. *Hariny* and *Priyanka*) and two bulls (*Salim Ali* and *Admiral*). As the third clan (*Wendy*) primarily used the dry thorn forest (except for a short period of less than a month in the first year) and because of small sample size, no detailed analysis was possible for the dataset on this clan.

17.2.3 Habitat Preference

Habitat preference in relation to seasonal and annual use of habitat was estimated by availability (area of habitat types) and utilization (time spent in each habitat) method. Since the clans and bulls show very strong fidelity to their annual and seasonal ranges (Viljoen 1989a; Daniel et al. 1995; Baskaran and Desai 1996; Baskaran 1998) and do not move randomly within the entire elephant range in a given area, only habitat types available within the home range were considered as available to each clan and bull. Time spent in each habitat was calculated based on weighted locations of each collared elephant clan and bull within each habitat. Proportion of various habitat types available within the home range of individual clans and bulls were estimated using the Spatial Ecology Analysis System (SEAS) programme developed by John Carey, Wisconsin University, USA. Preference rate was calculated as described by Neu et al. (1974) and Byers (1984) by using PREFER, a computer programme developed by Prasad and Gupta (Wildlife Institute of India, Dehra Dun).

17.3 Results

17.3.1 Habitat Use Pattern

The number of habitat types and their extent varied among clans, bulls and also between them. *Hariny* used only three habitat types (dry deciduous, mixed deciduous and dry thorn forest), while the *Priyanka* clan used all the four types (Table 17.1). Similarly, *Admiral* used all the four habitat types, while *Salim Ali* used only three. Most of *Hariny* and *Priyanka* home ranges consisted of dry deciduous forest (75% and 59%, respectively) followed by the dry thorn forest (21% each). But the *Priyanka* clan also had 17% of its range in the moist deciduous forest. *Admiral* utilized moist deciduous (32%), dry deciduous (27%) and dry thorn forest (26%) habitats, while mixed deciduous forest formed only 14% of its range. On the other

Table 17.1 Percentage of various habitat types available within the home range of individual elephant clans and bulls in Nilgiri Biosphere Reserve, Southern India

	Identification of elephants				
Habitat types available	Hariny	Priyanka	Wendy	Admiral	Salim Ali
Dry deciduous forest	74.9	59.1	–	27.3	86.7
Moist deciduous forest	–	17.2	–	32.3	9.7
Mixed deciduous forest	3.7	2.6	1.42	14.2	3.6
Thorn forest	21.4	21.1	98.58	26.2	–

Table 17.2 Habitat utilization pattern shown by elephant clans and bulls in different seasons in Nilgiri Biosphere Reserve, Southern India

Habitat types available	Identification of elephants			
	Hariny	Priyanka	Admiral	Salim Ali
Dry season				
Dry deciduous forest	91.19	52.09	0.97	88.32
Moist deciduous forest	–	0.23	17.36	3.78
Mixed deciduous forest	3.26	0.52	6.34	7.9
Thorn forest	5.55	47.15	75	–
First wet season				
Dry deciduous forest	96.39	85.81	23.75	92.71
Moist deciduous forest	–	13.94	75.39	7.29
Mixed deciduous forest	2.94	–	0.86	–
Thorn forest	0.67	0.24	–	–
Second wet season				
Dry deciduous forest	74.94	38.41	11	61.03
Moist deciduous forest	–	1.33	20.12	38.97
Mixed deciduous forest	6.54	1.31	–	–
Thorn forest	18.52	58.95	68.87	–

hand *Salim Ali's* home range mostly consisted of dry deciduous forest (87%). Whereas *Wendy* used 11 km^2 of mixed deciduous forest in the first year, but this clan shifted its home range during the second year to a new area consisting only of dry thorn forest, which it used till the end of the study. Therefore, a detailed analysis of seasonal habitat use was not possible for this clan. Overall the results show that not all habitat types are available within the home range of all clans and bulls, and their extent also varied considerably among study subjects.

17.3.2 Seasonal Habitat Use

Seasonal habitat use pattern also varied among clans, bulls and between them in all the seasons (Table 17.2). For example, *Hariny* clan used mostly the dry deciduous forest in the dry season (91%). In contrast, the *Priyanka* clan spent almost equal time in dry deciduous (52%) and dry thorn forest (47%) areas. The *Hariny* clan did not use moist deciduous forest though this habitat was available adjoining to its home range. *Admiral* mostly used the dry thorn forest (75%) during the dry season, whereas *Salim Ali* depended mostly on the dry deciduous forest (88%). A similar variation was also observed both during first and second wet seasons (Table 17.2) indicating that some clans and bulls never used some habitats though they are in the adjoining areas of their home ranges, as seen in the case clan *Hariny*, *Wendy* and *Salim Ali*, which never used the moist deciduous, dry deciduous and the dry thorn forests, respectively, during the entire period of this study. Secondly, the intensity of use of a given habitat also varied among the clans, bulls and between them in each season.

17.3.3 Habitat Preference

Overall habitat preference shown to different habitat types was not uniform among clans and bulls (Table 17.3). The *Hariny* clan showed overall preference to the dry deciduous forest and used the mixed deciduous forest in relation to its availability and dry thorn forest less than the expected. In contrast, the *Priyanka* clan showed a significant preference to the dry thorn forest but used the dry deciduous forest in proportion to its availability. Similarly, clans also varied in their seasonal habitat preferences (Table 17.4). In the dry season, the *Hariny* clan showed preference to the dry deciduous forest. In contrast, the *Priyanka* clan showed preference to the dry thorn forest. During the first wet season also, the *Hariny* clan used the dry deciduous forest more intensively with a strong preference. But the mixed deciduous forest was used in proportion to its availability, and the dry thorn forest was used significantly less than expected. Similarly, the *Priyanka* clan showed significant preference to the dry deciduous forest, and its use of mixed deciduous and dry thorn forests was significantly less than expected. During the second wet season, both clans used the dry deciduous, mixed deciduous and dry thorn forests. The *Hariny*

Table 17.3 Overall habitat preference shown by elephant clan in Nilgiri Biosphere Reserve, Southern India

Habitat type	Proportion of habitats available	Habitat usage		Confidence limit (90%)	
		Observed	Expected	Lower	Upper
Hariny					
Dry deciduous	0.749	879	759	0.841	0.892 M
Mixed deciduous	0.037	45	38	0.029	0.060 P
Dry thorn	0.214	90	217	0.067	0.110 L
Priyanka					
Dry deciduous	0.591	674	686	0.545	0.617 P
Moist deciduous	0.172	57	200	0.033	0.065 L
Mixed deciduous	0.026	7	30	0.000	0.012 L
Dry thorn	0.211	422	245	0.328	0.399 M

L = used less than the expected, *M* = used more than the expected and *P* = used in proportion to availability

Table 17.4 Seasonal habitat preference shown by elephant clan in Nilgiri Biosphere Reserve, Southern India

Habitat types	Proportion of habitats available	Habitat usage		Confidence limit (90%)	
		Observed	Expected	Lower	Upper
Dry season					
Hariny					
Dry deciduous	0.749	325	266.577	0.876	0.948 M
Mixed deciduous	0.037	12	13.169	0.010	0.055 P
Dry thorn	0.214	20	76.165	0.027	0.085 L
Priyanka					
Dry deciduous	0.591	217	246.435	0.460	0.582 L
Moist deciduous	0.172	1	71.725	0.000	0.008 L
Mixed deciduous	0.026	2	10.841	0.000	0.014 L
Dry thorn	0.211	197	87.983	0.410	0.533 M
First wet season					
Hariny					
Dry deciduous	0.749	281	218.251	0.938	0.990 M
Mixed deciduous	0.037	9	10.781	0.006	0.053 P
Dry thorn	0.214	2	62.357	0.000	0.018 L
Priyanka					
Dry deciduous	0.591	311	214.462	0.812	0.904 M
Moist deciduous	0.172	51	62.415	0.094	0.185 P
Mixed deciduous	0.026	0	9.435	0.000	0.000 L
Dry thorn	0.211	1	76.568	0.000	0.009 L
Second wet season					
Hariny					
Dry deciduous	0.749	273	273.001	0.695	0.804 P
Mixed deciduous	0.037	24	13.486	0.034	0.096 P
Dry thorn	0.214	68	78.001	0.137	0.234 P
Priyanka					
Dry deciduous	0.591	146	224.125	0.322	0.447 L
Moist deciduous	0.172	5	65.228	0.000	0.028 L
Mixed deciduous	0.026	5	9.860	0.000	0.028 P
Dry thorn	0.211	224	80.018	0.526	0.653 M

L = used less than the expected, *M* = used more than the expected and *P* = used in proportion to availability

clan showed no significant preference for any habitat type and used all of them equally in proportion to their availability. On the other hand, the *Priyanka* clan preferred the dry thorn forest significantly more than expected, and its use of dry deciduous forest was significantly less than expected. The mixed deciduous forest was used in proportion to its availability.

Overall, the bull *Admiral* showed preference to moist deciduous and dry thorn forest habitats (Table 17.5), while *Salim Ali* showed significant preference only to moist deciduous. During the dry and second wet seasons, *Admiral* showed preference for the dry thorn forest, using this habitat more intensively than

Table 17.5 Overall habitat preference shown by elephant bull in Nilgiri Biosphere Reserve, Southern India

Habitat type	Proportion of habitats available	Habitat usage		Confidence limit (90%)	
		Observed	Expected	Lower	Upper
Admiral					
Dry deciduous	0.273	76	175.266	0.087	0.150 L
Moist deciduous	0.323	252	207.366	0.344	0.441 M
Mixed deciduous	0.142	17	91.164	0.011	0.042 L
Dry thorn	0.262	297	168.204	0.413	0.512 M
Salim Ali					
Dry deciduous	0.867	336	358.592	0.762	0.854 L
Moist deciduous	0.097	68	40.352	0.120	0.207 M
Mixed deciduous	0.036	12	14.976	0.009	0.048 P

L = used less than the expected, *M* = used more than the expected and *P* = used in proportion to availability

expected (Table 17.6). But during the first wet season, it preferred the moist deciduous forest. In contrast, *Salim Ali* did not show preference to any habitat during dry season but preferred the dry and moist deciduous forests during the first wet and second wet seasons, respectively, indicating use of habitats and preference were not uniform among individual clan and bulls in the population.

17.4 Discussion

17.4.1 Seasonal Habitat Use and Habitat Preference

Seasonal use of habitats by elephants has been a favourite subject of study both in Asia (Eisenberg and Lockhart 1972; McKay 1973; Santiapillai et al. 1984; Sukumar 1985, 1989a, b; Sivaganesan 1991; Daniel et al. 1995) and Africa (Buss 1961; Buechner et al. 1963; Wing and Buss 1970; Leuthold and Sale 1973; Caughley and Goddard 1975; Western 1975; Williamson 1975; Eltringham 1977; Leuthold 1977; Short 1983; Jachmann 1988; Viljoen 1989b; Dublin 1996). These studies, looking at either the entire population or a subpopulation level, have generally attributed the seasonal movements of elephants to seasonal changes in environmental factors (rainfall, temperature) with corresponding changes in food quality (fresh grass, browse and their nutrient composition) and quantity (biomass availability of grass and browse) and water and shade availability. The differential use of habitats was examined using the densities without due consideration to their home range and social organization. In the present study, the seasonal movements of elephants to various habitats were found to be influenced by the seasonal climatic changes and corresponding changes in food and water availability. However, such factors influenced the seasonal movements of individual clans and bulls only within their home ranges, and no individuals moved randomly in response to climatic factors. Therefore, there are elephants in all the habitats in all the seasons as stated by

Table 17.6 Seasonal habitat preference shown by elephant bull in Nilgiri Biosphere Reserve, Southern India

Habitat type	Proportion of habitats available	Habitat usage		Confidence limit (90%)	
		Observed	Expected	Lower	Upper
Dry season					
Admiral					
Dry deciduous	0.273	2	64.676	0.000	0.026 L
Moist deciduous	0.323	41	76.522	0.112	0.235 L
Mixed deciduous	0.142	15	33.641	0.023	0.103 L
Dry thorn	0.262	178	62.070	0.683	0.823 M
Salim Ali					
Dry deciduous	0.867	138	135.252	0.822	0.945 P
Moist deciduous	0.097	6	15.132	0.001	0.074 L
Mixed deciduous	0.036	12	5.616	0.027	0.131 P
First wet season					
Admiral					
Dry deciduous	0.273	55	63.568	0.168	0.307 P
Moist deciduous	0.323	176	75.211	0.683	0.824 M
Mixed deciduous	0.142	2	33.065	0.000	0.024 L
Dry thorn	0.262	0	61.007	0.000	0.000 L
Salim Ali					
Dry deciduous	0.867	114	106.390	0.871	0.983 M
Moist deciduous	0.097	9	11.903	0.017	0.129 P
Mixed deciduous	0.036	0	4.418	0.000	0.000 L
Second wet season					
Admiral					
Dry deciduous	0.273	19	47.163	0.050	0.169 L
Moist deciduous	0.323	35	55.801	0.125	0.278 L
Mixed deciduous	0.142	0	24.532	0.000	0.000 L
Dry thorn	0.262	119	45.263	0.601	0.777 M
Salim Ali					
Dry deciduous	0.867	84	118.727	0.511	0.710 L
Moist deciduous	0.097	53	13.283	0.290	0.489 M
Mixed deciduous	0.036	0	4.930	0.000	0.000 L

L = used less than expected, *M* = used more than expected and *P* = used in proportion to availability

Eltringham (1977) that elephant migration was not complete because some elephants were found in grassland and in forest at all times of the year. The reason for the presence of elephants in all the habitats during all the seasons does not support the general reasons (environmental factors) attributed by earlier studies and likely be the influence of ranging and mechanism of spacing among elephants, and this is the first study to consider the behavioural component (ranging and social organization) of the elephant to understand adequately the ecology of habitat use by the species.

17.4.2 Influence of Home Range on Habitat Use

Since the present study clans and bulls showed a strong fidelity to their home ranges, seasonal movements were restricted to the habitat types that were within the home ranges. The habitat utilization pattern revealed remarkable variations in diversity of habitat types available to different clans and bulls, which utilized them on available basis. For example, the home ranges of the clan *Hariny* and the bull *Salim Ali* had three habitat types within their ranges, while *Priyanka* and *Admiral* had four habitat types. On the other hand, the *Wendy* clan's range consisted of only two habitat types during 1991 before its range shifted and only one during 1992–1995. Though the moist deciduous and the dry thorn forests were available adjoining the home ranges to *Hariny* and *Salim Ali*, respectively, they were not utilized. Similarly, *Wendy* did not use the dry deciduous forest throughout the first year (1991) from the second year onwards; the mixed deciduous forest was also not used. But the clan *Priyanka* used all four habitat types. Home range fidelity has been reported to influence habitat use in many deer species. Black-tailed deer starved on their traditional summer ranges instead of moving a few km to recently burnt area, where abundant forage was available (Dasmann and Taber 1956). Verme (1973) also noted that deer in Northern Michigan were not using what appeared to be good summer ranges. Nelson (1979) stated that available data on deer suggest that most of the deer prefer familiar ground rather than good habitat. White-tailed deer continued to use their traditional home ranges in spite of disturbances in their migration route (Nelson and Mech 1981). Although many studies on habitat use of elephants have been carried out, the influence of home range fidelity on habitat use has not been reported earlier in Asia. However in Africa, Viljoen (1989a) reported that individual elephants in northern Namib desert showed strong fidelity to home ranges and thus subjected themselves to seasonal ranges within their individual home ranges irrespective of higher rainfall or river floods in adjacent areas outside their home ranges. He also pointed out that home range fidelity is the main cause for not occupying vacant areas with overabundance of food. The present results are also consistent with his findings. Therefore, the spatial location of home range can rule out access to some habitat types. This means no matter how many habitat types are available in the adjoining areas or how good the adjoining habitats are, individual clans and bulls use only what is available within their range, thereby putting constraints on their options for habitat types. Thus, the spatial location of the home ranges in a given area seems to influence the degree of fitness of each clan and bull.

17.4.3 Influence of Spacing on Habitat Use

There were differences between clans and bulls in the use of available habitats. For example, though moist deciduous and dry deciduous forests were available within the ranges of the two bulls, the bull *Admiral* preferred the moist deciduous forest during the first wet season, while *Salim Ali* preferred the dry deciduous forest. This difference between bulls could be attributed to climatic conditions, availability of microhabitats and plant communities in their ranges which differed spatially. However, the same reasons could not be attributed to the spatially overlapping clans (*Hariny* and *Priyanka*) that provided a unique chance to examine the influence of spacing on habitat utilization. It is apparent from the results that the proportion of the dry thorn forest within the home ranges of *Hariny* and *Priyanka* were similar, while that of dry deciduous forest were different. However, as both clans mostly used the overlapping area during dry and second wet seasons, the habitat types available within these overlapping areas were accessible to both clans. The overlapping areas of 466 km^2 consisted mainly of dry deciduous and dry thorn forest habitats. Therefore, any environmental factor that may influence on the habitat use of elephants should act on both the clans uniformly. Moreover, in any given time (season), two different habitats cannot be similar in their resource (food, water and shelter) quality and quantity; definitely one could be superior to the other. Therefore, it is reasonable to expect both the clans to show similar strategy of habitat use pattern, but the observed results showed remarkable differences in the use of habitat types. For example, *Hariny* used dry deciduous forest far more than the dry thorn forest (91% and 6%, respectively) during the dry season, while the *Priyanka* clan used the dry deciduous (52%) and the dry thorn forest (47%) almost equally in the same season. Similarly, during the second wet season also, the *Hariny* clan mainly utilized the dry deciduous forest, while the dry thorn forest was used much less, whereas the *Priyanka* clan used the dry thorn forest more intensively than the dry deciduous forest. The results on the habitat preference also show that the above-mentioned habitat uses differ significantly from the expected use. As such, differences in the habitat use strategy among the overlapping clans appear to a consequence of social hierarchy and spacing, apart from the influence of environmental factors. Such information is lacking in the field of elephant ecology, but there is some documentation of the effect of hierarchy in the use of habitat types in other taxa. In oystercatchers, many of the subdominant (second and third wintering) birds leave the mussel beds (an optimal habitat) and move to mud flats in autumn when adults return from their breeding area (Goss-Custard et al. 1982). Boats and Goss-Custard (1992) stated that there is a possibility of subdominant oystercatchers being unable to feed at an adequate rate on mussel beds, where the density of competitors is very high, and instead may feed on other prey in open flats, where competitors can be much less. Similarly, the American redstarts furnish another possible example of the role of intraspecific competition in habitat selection. First-year males arrive later than the adult-plumaged birds; they do not compete successfully with older birds for favoured sites (Ficken and Ficken 1967; Morse 1973) and are frequently aggressively displaced by the latter (Morse 1985). Svardson (1949) suggested that increasing density may be associated with an increase in the range of habitat occupied by a species because intraspecific competition forces some individuals to occupy marginal habitats. Clutton-Brook (1982) recorded that threat display in red deer hinds was more and intense among animals other than matrilineal relatives. Such threats often removed the recipient from a particular feeding site due to competition in a preferred feeding site. The displacement of subordinates by the dominant one from good feeding patches has been shown to exist in the well-defined social groups of

African buffalo and red deer due to social hierarchies (Sinclair 1992). Geist (1981) stated that resource defence by mule deer is facultative. In elephants, based on each clan's position in the social hierarchy, they use a habitat in a given season without overlapping in time, with other clans. Thus, the dominant clans use the optimal habitat, while the subdominant ones use suboptimal habitat. There is thus a significant difference in the habitat use pattern among the overlapping clans. The interclan encounter discussed in the Chap. 15 and in Baskaran (1998) also supports this view. Thus, the elephant clans are not free to use the habitat of their choice and appeared to exhibit a hierarchy-based habitat use; this pattern seems to fit with the Fretwell's (1972) 'ideal-despotic distribution' model of habitat selection and not 'ideal-free distribution' as assumed by earlier studies. Such a hierarchy-based habitat use pattern among the elephant clans results in dominant clans having priority access to good-quality habitat and food (Baskaran 1998), thereby showing better survival and reproductive success.

Acknowledgements We are thankful to United States Fish and Wildlife Service for funding the study and the State Forest Department Tamil Nadu, Karnataka and Kerala for permitting us to undertake the study in their states. We extend our sincere thanks to late Mr. J. C. Daniel, the principal investigator of the project and former director of Bombay Natural History Society, Mumbai, for his consistent support. Last but not least, we thank our field assistants Messers Krishnan, Chennan (Late), Bomman, Kethan, Mathan and Maran and jeep drivers Mr. Sirul Kumar and Gopal for their wonderful cooperation during the study period.

References

Baskaran N (1998) Ranging and resource utilization by Asian elephants (*Elephas maximus* Linnaeus) in Nilgiri Biosphere Reserve, South India. Ph.D. Thesis, Bharathidasan University, Tiruchirappalli, Tamil Nadu

Baskaran N (2013) An overview of Asian elephant (*Elephas maximus*) in Western Ghats and its implication for Western Ghats Ecology. J Threatened Taxa 5(14):4854–4870

Baskaran N, Desai AA (1996) Ranging behaviour of the Asian elephant *(Elephas maximus)* in the Nilgiri Biosphere Reserve, South India. Gajah 15:41–57

Baskaran N, Balasubramanian M, Swaminathan S, Desai AA (1995) Home range of elephants and its implications for the management of Nilgiri Biosphere Reserve, India. In: Daniel JC, Datye HS (eds) A week with elephants. Proceedings of the International Seminar on the conservation of Asian elephant. Mudumalai Wildlife Sanctuary, June 1993. Bombay Natural History Society, Bombay, pp 296–313

Baskaran N, Balasubramanian M, Swaminathan S, Desai AA (2010) Feeding ecology of Asian elephant (*Elephas maximus* Linn.) in tropical forests of Nilgiri Biosphere Reserve, Southern India. J Bombay Nat Hist Soc 107(1):3–13

Baskaran N, Anbarasan U, Agoramoorthy G (2012) India's biodiversity hotspot under anthropogenic pressure – a case study of Nilgiri Biosphere Reserve. J Nat Conserv 20:56–61

Boates JS, Goss-Custard JD (1992) Foraging behaviour of oystercatchers *Haematopus ostralegus* specializing on different species of prey. Can J Zool 70:2398–2404

Buechner HK, Buss IO, Longhurst WM, Brooka AC (1963) Number and migration of elephants in Murchinson Falls National Park. Uganda J Wildl Manage 27:36–53

Buss IO (1961) Some observation of food habits and behaviour of African elephant. J Wildl Manag 25(2):131–148

Byers CR, Steinhorst RK, Krauman PK (1984) Clarification of a technique for analysis of utilization and availability data. J Wildl Manag 48:1050–1053

Caughley G, Goddard J (1975) Abundance and distribution of elephants in Luangwa Valley. Zambia E Afr Wildl J 13:39–48

Cederlund G (1983) Home range dynamics and habitat selection by Roe deer in Boreal area in central Sweden. Acta Theriol 28(30):443–460

Champion HG, Seth SK (1968) Revised survey of forest types in India. Manager of Publication, New Delhi, 404p

Clutton-Brook TH, Guinness FE, Albon SD (1982) Red deer behaviour and ecology of the two sexes. The University of Chicago. Edinburgh University Press, Edinburgh, 378p

Daniel JC, Desai AA, Sivaganesan N, Datye HS, Rameshkumar S, Baskaran N, Balasubramanian M, Swaminathan S (1995) Ecology of the Asian elephant. Final Report 1987–1994. Bombay Natural History Society, Bombay

Dasmann RF, Taber RD (1956) Behaviour of Columbian black-tailed deer with reference to population ecology. J Mammal 32(2):143–163

Desai AA, Baskaran N (1996) Impact of human activities on the ranging behaviour of elephants in the Nilgiri Biosphere Reserve, South India. J Bombay Nat Hist Soc 93(3):559–569

Dublin HT (1996) Elephants of the Masai Mara, Kenya: seasonal habitat selection and group size patterns. Pachyderm 22:25–35

Dunham KM (1986) Movements of elephant cows in un-flooded middle Zambezi valley, Zimbabwe. Afr J Ecol 24:287–291

Eisenberg JE, Lockhart MC (1972) An ecological reconnaissance of Wilpattu National Park. Ceylon Smithson Contrib Zool 101:1–118

Eltringham SK (1977) Number and distribution of elephant *Loxodonta africana* in the Rwenzori National Park and Chambura Game Reserve. Uganda E Afr J Wildl 15:19–39

Ficken MS, Ficken RW (1967) Age-specific differences in the breeding behaviour of the American redstarts. Wilson Bull 79:188–199

Fretwell FD (1972) Populations in a seasonal environment. Monogr Popul Biol 5:1–220. Princeton University Press, Princeton

Funston PJ, Skinner JD, Dott HM (1994) Seasonal variation in movement patterns, home range and habitat selection of buffaloes in semi-arid habitat. Afr J Ecol 32:100–114

Geist V (1981) Behaviour: adaptive strategies in mule deer. In: Wallmo OC (ed) Mule deer and black-tailed deer of North America. Lincoln Press, University of Nebraska, Lincoln

Goss-Custard JD, AleVdit Durell SE, McGrorty S, Reading CJ (1982) Use of mussel, *Mytilus edulis* beds by Oystercatchers *Haematopus ostralegus* according to age and population size. J Anim Ecol 51:543–554

Jachmann H (1988) Numbers, distribution and movements of the Nazinga elephants. Pachyderm 10:16–22

Joshi AR, Garshelis DL, Smith JLD (1995) Home rages of sloth bear in Nepal: implication for conservation. J Wildl Manag 59(2):204–214

Leuthold W (1977) Spatial organization and strategy of habitat utilization of elephants in Tsavo National Park, Kenya. Z Säugetierkd 42:358–379

Leuthold W, Sale JB (1973) Movements and patterns of habitat utilization of elephants in Tsavo National Park. Kenya E Afr Wildl J 11:369–384

McKay GM (1973) Behavioural ecology of the Asiatic elephant in southern Ceylon. Smithson Contrib Zool 125:1–113

Morse DH (1973) The foraging of small population of Yellow Warblers and American Redstarts. Ecology 54:346–355

Morse DH (1985) Habitat selection in North American Parulid Warblers. In: Cody ML (ed) Habitat selection in birds. Academic, Orlando, pp 131–157

Nelson ME (1979) Home range location of the White-tailed deer. USDA. For. Serv. Res. Pap. NC-173. Nor. Cent. For. Expt. Sta., St. Paul

Nelson ME, Mech LD (1981) Deer social organization and wolf predation in northeastern Minnesota. Wildl Monogr 77:1–48

Nichollas TH, Warner DW (1972) Barred owl habitat use as determined by radio-telemetry. J Wildl Manag 36:213–224
Nue CW, Byers CR, Peek JM (1974) A technique for analysis of utilization-availability data. J Wildl Manag 38:541–545
Olivier RCD (1978a) On the ecology of Asian elephant. Unpublished Ph. D. thesis, University of Cambridge
Olivier RCD (1978b) Distribution and status of Asian elephant. Oryx 14:379–424
Project Elephant (2007) Census in project elephant reserves in India: http://envfor.nic.in/pe/elephant 2007.pdf. Accessed 12 Mar 2009
Santiapillai C, Chambers MR, Ishwaran N (1984) Aspect of the ecology of the Asian elephant (*Elephas maximus*) L. in the Ruhuna National Park, Sri Lanka. Biol Conserv 29:47–61
Short JC (1983) Density and seasonal movements of elephant (*Loxodonta africana cyclotis*, Matschie) in Bia Natioal Park, Ghana. Afr J Ecol 21:175–184
Sinclair ARE (1992) Do large mammals disperse like small mammals? In: Stenseth NC, Lidicker WZ Jr (eds) Animal dispersal. Chapman and Hall, London, pp 229–242
Sivaganesan N (1991) The ecology of the Asian elephant in Mudumalai Wildlife Sanctuary, with special reference to habitat utilization. Unpublished Ph. D. thesis, Bharathidasan University, Tiruchirappalli
Sukumar R (1985). Ecology of Asian elephant *(Elephas maximus)* and its interaction with man in south India. Ph. D. thesis, Indian Institute of Science, Bangalore, India
Sukumar R (1989a) Ecology of Asian elephant in south India. I. Movement and habitat utilization patterns. J Trop Ecol 5:1–18
Sukumar R (1989b) The Asian elephants ecology and management. Cambridge University press, Great Briton, 251p
Svardson G (1949) Competition and habitat selection in birds. Oikos 1:157–174
Tchamba MN, Bauer H, De-Iongh HH (1995) Application of VHF-radio and satellite telemetry in northern Cameroon. Afr J Ecol 33:335–346
Thouless CR (1996) Home range and social organization of female elephants in the northern Kenya. Afr J Ecol 34:284–297
Verme LJ (1973) Movement of white-tailed deer in upper Michigan. J Wildl Manag 37(4):545–552
Viljoen PJ (1989a) Spatial distribution and movements of elephants (*Loxodonta africana*) in northern Namib desert region of the Kaokoveld, South West Africa/Namibia. J Zool Lond 219:1–19
Viljoen PJ (1989b) Habitat selection and preferred food plants of a desert-dwelling elephant population in the northern Namib desert, South West Africa/Namibia. Afr J Ecol 27:227–240
Western D (1975) Water availability and its influence on the structure and dynamics of a Savannah large mammals community. E Afr Wildl J 13:265–286
Williamson BR (1975) Seasonal distribution of elephants in Wankie National Park. Arnoldia Rhod 7(11):1–16
Wing LD, Buss IO (1970) Elephants ant forests. Wildl Monogr 19:92p

Human-Elephant (*Elephas maximus*) Conflict in Southern India: Spatial Movement Pattern of Asian Elephants Outside Reserved Forests of Coimbatore Forest Division

18

Kalyanasundaram Ramkumar, Balasundaram Ramakrishnan, and R. Saravanamuthu

Abstract

This study revealed that a total of 422 crop fields were visited by elephants across the six forest ranges of Coimbatore Forest Division during the study period. The number of crop fields which are affected by elephants was high in Boluvampatti range (116) followed by Coimbatore (107), PN Palayam (105), Mettupalayam (64) and Sirumugai (25). Very less number of affected crop fields was recorded in Karamadai range (5). The highest number of crop fields was affected in 0–0.5 km category from forest boundary in five forest ranges, i.e. Coimbatore, PN Palayam, Mettupalayam, Sirumugai and Karamadai. On the contrary, in Boluvampatti, the highest number of crop fields was affected in 0.5–1 km category. Frequency of crop fields affected in different distance categories from forest boundary (0–0.5 km, 0.5–1 km, 1–1.5 km, 1.5–2 km, 2–2.5 km) and within each forest range was significant ($F = 5.98$, $P = 0.002$). Frequency of crop fields affected in different distance categories from forest boundary (0–0.5 km, 0.5–1 km, 1–1.5 km, 1.5–2 km and 2–2.5 km) versus between forest ranges was also significant ($F = 5.36$, $P = 0.002$). As overall in Coimbatore Forest Division, crop fields located close to forest (0–0.5 km) were highly (35.55 %) affected by elephant visits. An overall intensity of elephant visits into the crop fields revealed

K. Ramkumar (✉)
Wildlife Trust of India, Noida, Uttar Pradesh, India
e-mail: ramkumar@wti.org.in

B. Ramakrishnan
Department of Zoology & Wildlife Biology, Mammalogy and Forest Ecology Lab, Government Arts College, Udhagamandalam, Tamil Nadu, India

R. Saravanamuthu
Department of Botany, A.V.C. College, Mannampandal, Tamil Nadu, India

C. Sivaperuman, K. Venkataraman (eds.), *Indian Hotspots*,
https://doi.org/10.1007/978-981-10-6605-4_18

that low intensity was traced in 40% of crop fields followed by medium (38%), and high intensity was noticed in 22 % of the crop fields. Statistical test revealed that frequency of crop fields affected in different distance categories (0–0.5 km, 0.5–1 km, 1–1.5 km,1.5–2 km, 2–2.5 km and 2.5–3 km) from forest and within various intensities of elephant visits (1–10, 11–20 and 21–30) was significant ($F = 8.707$, $P = 0.002$).

Keywords
Asian elephant · Conflict · India · Spatial movement · Tropical forest

18.1 Introduction

"Elephant on house visit scares Kovai", screamed the newspaper headlines on 18 March 2013. A wild tusker entered a thickly populated residential area and caused havoc. Ramanathan and his family trembled in fear as another elephant broke open the iron gate of their house and entered the premises at Press Enclave in Kovaipudur at the crack of dawn. This is a scenario in Coimbatore district since 2002. Sukumar (1985a) pointed out that elephant incursion into human habitation is not something unusual or new. It has been going on ever since man took to agriculture within elephant habitat. The Gaja sastra that can be traced back to the fifth or sixth century BC narrates that wild elephants invaded the kingdom of Anga and caused considerable damage, a reference no doubt to crop raiding.

Many studies have quantified conflict to be intense in the proximity of a forest area (Graham 1973; Sukumar 1989; Kiiru 1995; Nath and Sukumar 1998; Osborn 1998; Naughton et al. 1999; Nyhus et al. 2000; Talukdar et al. 2006; Lahkar et al. 2007; Riddle 2007; Daniel et al. 2008). It has also been found that conflict intensified farther away from the forest boundary (Smith and Kasiki 2000; Kumar et al. 2004), while still others found no correlation at all (Osborn 1998; Hoare 1999; Di Fonzo 2007). One African study found that conflict levels were significantly lower in areas bordering the forest but suggested that there was anecdotal evidence of successful mitigation strategies being practiced by local farmers as well as the wildlife authorities (Smith and Kasiki 2000). In Nepal, sites with good forests along edge habitats suffered less conflict (WWF 2008). Studies on the influence of refuges and corridors on the distribution and intensity of conflict outside protected areas are the need of the hour (Sitati et al. 2003). The spatial distribution of damage was highly skewed even within this high-risk zone closest to the forest, some fields were totally destroyed, whereas other areas nearby were left untouched (Naughton et al. 1999; Naughton-Treves and Treves 2005; Sitati et al. 2005; Lahkar et al. 2007; Parker et al. 2007).

In India, a distributional analysis of damage along the periphery of forests is yet to be done. In fragmented forests, the longer boundary shared with cultivation increases the likelihood of elephant visits (Nath and Sukumar 1998; Smith and Kasiki 2000; Sukumar 2003; Madhusudan 2003) and worsens the damage (Daniel et al. 2008). In South Africa elephants moving into new territory indicates an increase in population numbers (Grant et al. 2008); in India it appears to be also driven by large-scale disturbances to the habitat (Sarma and Easa 2006), extreme drought (Sukumar 1995), severe poaching or when habitat within a home range is lost or highly degraded (IUCN 2006).

The elephant movements in this division are mostly restricted to very narrow paths of the foothills of the large mountains naturally near the human habitations. However, depredation is higher when compared to other largely populated elephant habitats. In Coimbatore Forest Division, elephants are moving into the human habitation and crop fields located even more than 5 kilometres away from the forest boundary, during the past decade. However, there is no scientific study and detailed information available on this perception. Thus, this study was initiated to assess the spatial movement pattern of elephants outside reserved forests of Coimbatore Forest Division.

18.2 Methods

18.2.1 Study Area

The Coimbatore Forest Division covers an area of 694 km^2 and is situated in the south-east of the Nilgiri Biosphere Reserve (NBR). Also Coimbatore Forest Division falls under the Elephant Reserve No. 8, in which Nilambur-Silent Valley of Kerala forming the major portion of the tract. The Coimbatore Forest Division is also part of Nilgiris and Eastern Ghats Landscape which is holding single largest Asian elephant population in the world. This forest division has six ranges, namely, Sirumugai, Mettupalayam, Karamadai, PN Palayam, Coimbatore and Boluvampatti. The area lies between latitude 10°51′ and 11°27′ and longitude 76° 39′ and 77° 4′.

This forest division has a wide range of altitude from 450 to 1450 m. The Pillur slopes are the steepest; a shear drop is observed as the ground falls from 450 to 1530 m suddenly. The Melur slopes, Hulical Drug and Nellithurai forests are on the lower hill mountains. The Boluvampatti hill elevation ranges from 450 to 530 m. Above 530 m the ground rises sharply to the crest of the hill range to the north, west and south; the maximum elevation is 1986 m on the Velliangiri peak. Apart from these sloping hillocks, this reserve has Velliangadu Valley, Naicken Palayam Valley, Thadagam Valley, Boluvampatti Valley and the Walayar Valley. The Naicken Palayam rises sharply from 460 to 1614 m on the Nadukondanboli forming a tri-junction point for the entire three valleys. Innumerable little streams originate from Coimbatore Forests and drain the plateau. This network of streams resolves itself into Bhavani and Noyyal rivers. The vegetation types range from tropical thorn forest at the foothills to evergreen forest, in relation to terrain, altitude and rainfall. The study was carried out in the villages located all along the foot hills of forest ranges such as Boluvampatti, PN Palayam, Sirumugai, Coimbatore, Mettupalayam and Karamadai (Map 18.1).

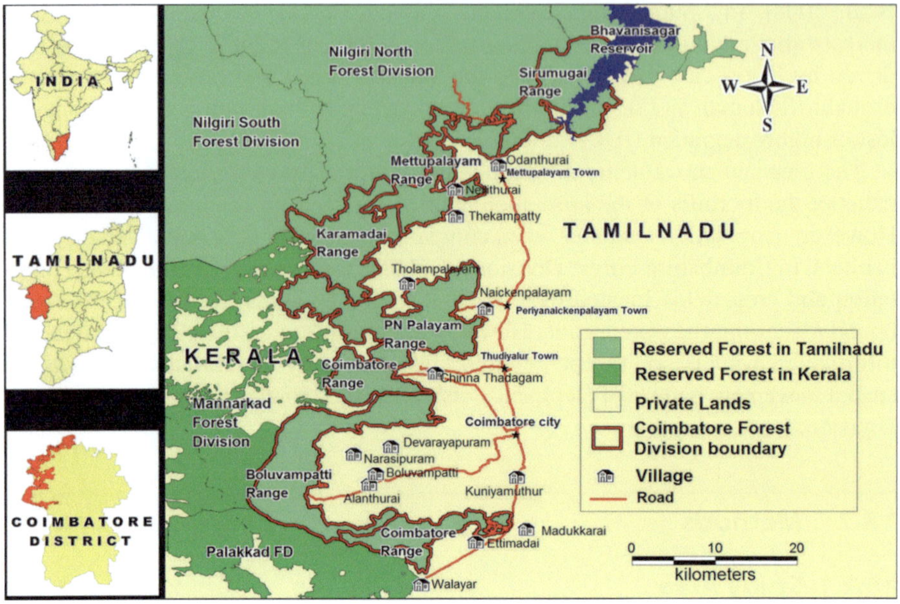

Map 18.1 Study area map

18.2.2 Assessment of Elephant Movement Outside Reserved Forests

A survey was made in the villages located up to 5 km from the forest boundary of six forest ranges of the division at an interval of a fortnight for 1 year. In each survey, enquiry was made with forest range offices, section offices, beat offices and villagers on elephants' recent visits/stray incidences into the farm lands. Based on those information, all elephant-strayed places were visited and recorded the geo coordinates by Global Positioning System (GPS).

The geo coordinates of elephant-visited crop fields/places were depicted in the Survey of India 1:50,000 digital maps. Also buffer layer is created at an interval of every 0.5 km up to 5 km from forest boundary. The geo coordinates of elephant-visited crop fields were counted in relation to distance category and forest ranges. Significant differences between frequency of crop fields affected in distance categories from forest boundary and forest ranges were determined using the two-way ANOVA test. Distance categories such as 0–0.5 km, 0.5–1 km, 1–1.5 km, 1.5–2 km and 2–2.5 km only were used to perform the analysis, and remaining categories were not considered due to nil comparative values.

18.2.3 Quantifying the Intensity of Elephant Visits Outside the Reserved Forests

A survey was made in the villages located up to 5 km from the forest boundary of six forest ranges of the division fortnight for 1 year. In each survey, enquiry was made with forest range offices, section offices, beat offices and villagers on elephants' recent visits/straying incidences in to the farm lands. Based on those information, the owner or caretaker of the elephant-visited crop fields was selected for the interview. A total of 422 persons were interviewed from the six forest ranges of the Coimbatore Forest Division.

Questionnaire method was followed to find out the intensity of elephant visits into the crop fields of different forest ranges in relation to distance from forest boundary. During the interviews, information such as the farmer's name and intensity of elephant visits were collected. Intensity of elephant visits was categorized as low (1–10 visits), medium (11–20 visits) and high (20–30 visits) based on replication of elephant visit to the same crop fields. Distance categories such as 0–0.5 km, 0.5–1 km, 1–1.5 km, 1.5–2 km, 2–2.5 km and 2.5 km only were used to perform the ANOVA, and the remaining categories were not considered due to nil comparative values.

18.2.4 Depicting the Spatial Movement Pattern of Elephants Outside of the Reserved Forests

The geo coordinates of elephant-visited crop fields and its intensity were depicted on the Landsat Mosaic-EarthSat 2003 image. GIS softwares were used to prepare the maps.

18.3 Results

18.3.1 Crop Fields Affected by Elephants in Different Forest Ranges

This study revealed that a total of 422 crop fields were visited by elephants across the six forest ranges of Coimbatore Forest Division during the study period. The number of crop fields which are affected by elephants was high in Boluvampatti range (116) followed by Coimbatore (107), PN Palayam (105), Mettupalayam (64) and Sirumugai (25) (Fig. 18.1). Very less number of affected crop fields was recorded in Karamadai range (5).

In the pattern of elephant depredation with respect to distance from forest boundary, the highest number of crop fields was affected in 0–0.5 km category within the forest range as Coimbatore (36), PN Palayam (33) and Mettupalayam (31), Sirumugai (17) and Karamadai (4) (Table 18.1). On the contrary, in Boluvampatti, the highest number of crop fields was affected in 0.5–1 km category. In Coimbatore

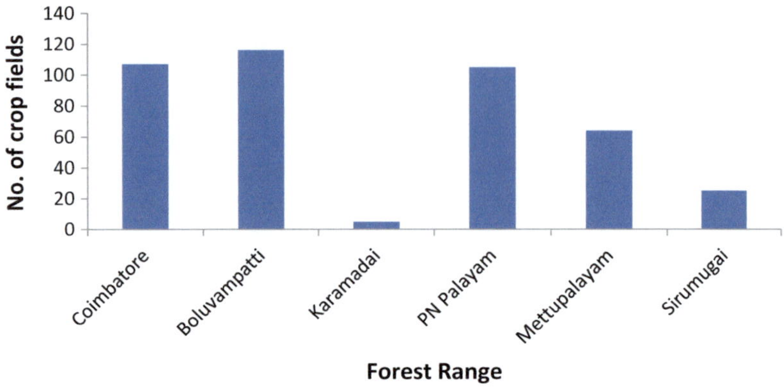

Fig. 18.1 Crop fields affected by elephants in different forest ranges

crop fields were affected in almost all category of distance interval except 3.5–4 km and 4–4.5 km. The longest elephant visit was recorded in two crop fields of Coimbatore with respect to all forest ranges. Apart from this elephant's longest visit from RF recorded, the distance visited in the remaining forest ranges was Boluvampatti (3–3.5 km), Sirumugai (2.5–3 km), PN Palayam (2–2.5 km), Mettupalayam (1.5–2 km) and Karamadai (0.5–1 km). This study revealed that elephants are visiting more than 5 km outside the forest areas of Coimbatore Forest Division due to various reasons.

Frequency of crop fields affected in different distance categories from forest boundary (0–0.5 km, 0.5–1 km, 1–1.5 km, 1.5–2 km, 2–2.5 km) and within each forest ranges was significant ($F = 5.98$, $P = 0.002$). Frequency of crop fields affected in different distance categories from forest boundary (0–0.5 km, 0.5–1 km, 1–1.5 km, 1.5–2 km and 2–2.5 km) versus between forest ranges was also significant. All the differences were significant ($F = 5.36$, $P = 0.002$).

18.3.2 Intensity of Elephant Visits vs Distance from Forest in Different Forest Ranges

Intensity of elephant visits was categorized as low (1–10 visits), medium (11–20 visits) and high (20–30 visits) based on replication of elephant visit to the same crop fields. Intensity of elephant visits with respect to location of crop fields from forest boundary in Coimbatore range revealed that crop fields which are located in 0–0.5 km attracted more elephants as 21–30 visits in 11.2% of crop fields, 11–20 visits in 10.3% of crop fields and 1–10 visits in 12.1% of crop fields (Fig. 18.2). Totally, 33.6 % of crop fields were affected by elephants in 0–0.5 km. It was interesting to note that 1–10 visits of elephant were recorded in 4.5–5 km and >5 km category in 0.9% and 1.9% of crop fields, respectively.

Table 18.1 Frequency of crop fields affected by elephants with relation to distance from forest boundary in different forest ranges

Distance from forest boundary (Km)	No. of crop fields affected by elephants in different forest ranges											
	Coimbatore		Boluvampatti		Karamadai		PN Palayam		Mettupalayam		Sirumugai	
	No. of crop fields	Relative (%)	No. of crop fields	Relative %	No. of crop fields	Relative (%)	No. of crop fields	Relative (%)	No. of crop fields	Relative (%)	No. of crop fields	Relative (%)
0–0.5	36	33.6	29	25.0	4	80	33	31.4	31	48.4	17	68
0.5–1	13	12.1	47	40.5	1	20	26	24.8	18	28.1	2	8
1–1.5	9	8.4	27	23.3	0	0	32	30.5	11	17.2	1	4
1.5–2	20	18.7	10	8.6	0	0	13	12.4	4	6.3	4	16
2–2.5	12	11.2	2	1.7	0	0	1	1.0	0	0.0	0	0
2.5–3	12	11.2	0	0.0	0	0	0	0.0	0	0.0	1	4
3–3.5	2	1.9	1	0.9	0	0	0	0.0	0	0.0	0	0
3.5–4	0	0.0	0	0.0	0	0	0	0.0	0	0.0	0	0
4–4.5	0	0.0	0	0.0	0	0	0	0.0	0	0.0	0	0
4.5–5	1	0.9	0	0.0	0	0	0	0.0	0	0.0	0	0
>5	2	1.9	0	0.0	0	0	0	0.0	0	0.0	0	0

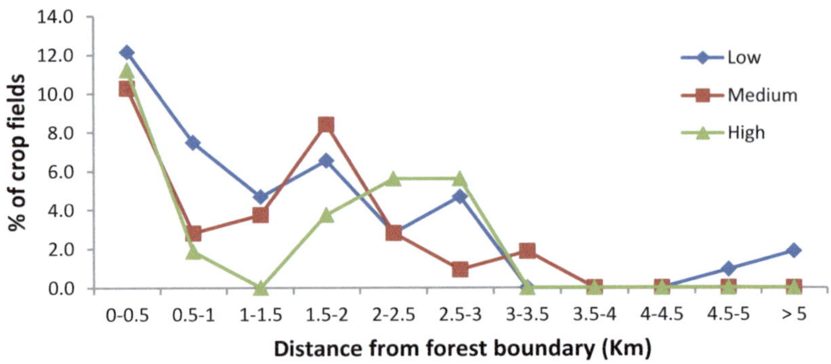

Fig. 18.2 Intensity of elephant visits into the crop fields adjoining to different forest ranges in Coimbatore Forest Division with relation to distance from forest boundary – Coimbatore range

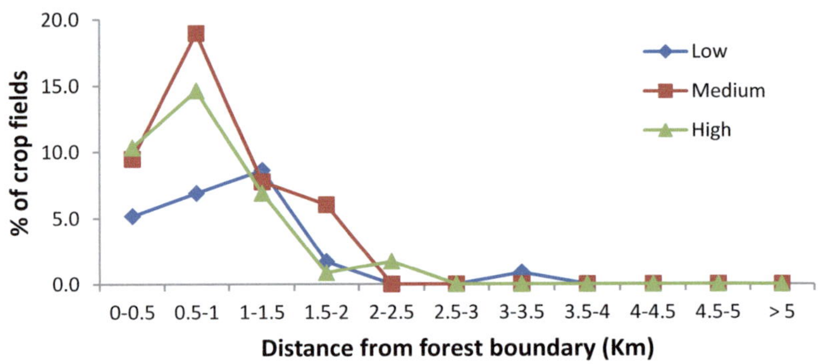

Fig. 18.3 Intensity of elephant visits into the crop fields adjoining to different forest ranges in Coimbatore Forest Division with relation to distance from forest boundary – Boluvampatti range

In Boluvampatti range, highest percentage of crop fields (19%) was affected in 0.5–1 km with 11–20 visits of elephant followed by 14.7% crop fields with 21–30 visits (Fig. 18.3). Crop fields located in the 0–0.5 km and 1–1.5 km category were also remarkably affected by elephant visits at different levels (Maps 18.2, 18.3, and 18.4). In PN Palayam, high intensity of elephant visits was recorded in 0–0.5 km category (4.8% of crop fields) followed by 1–1.5 km (2.9%), 0.5–1 km and 1.5–2 km (1.9% each) (Fig. 18.4). Medium intensity of elephant visits was noticed in 1–1.5 km category (18.1% crop fields) followed by 0.5–1 km (12.4%), 1.5–2 km (8.6%) and 0–0.5 km (6.7%). Among the crop fields, highest percentage (20% crop fields) was affected by low intensity of elephant visits in 0–0.5 km category. Elephant visits were not noticed after 2.5 km from forest boundary.

In Karamadai, elephant depredation was noticed only up to 1 km from forest boundary. Of which crop fields which are located in 0–0.5 km category were affected more by different levels of intensity of elephant visits as high (20% crop fields), medium (20% crop fields) and low (40% fields) (Fig. 18.5). In Mettupalayam,

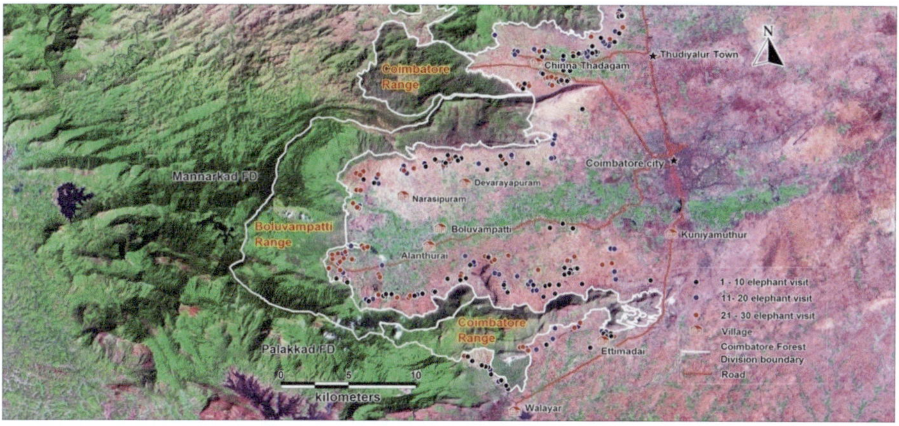

Map 18.2 Intensity of elephant visits in Boluvampatti and Coimbatore ranges

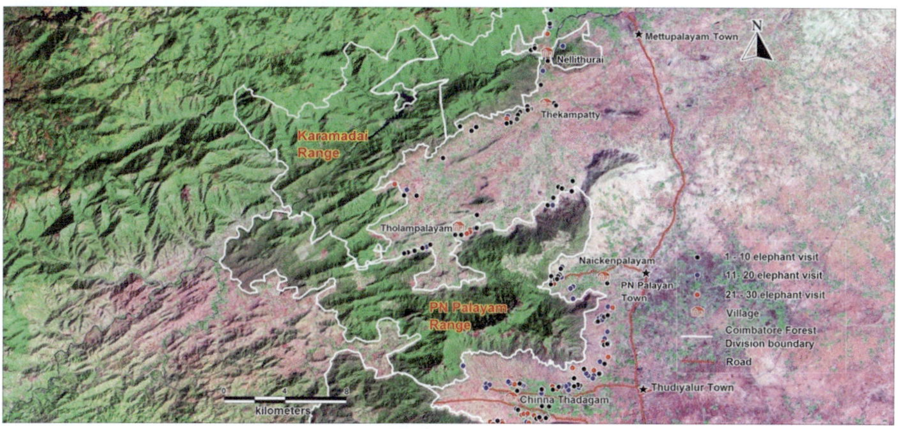

Map 18.3 Intensity of elephant visits in PN Palayam and Karamadai ranges

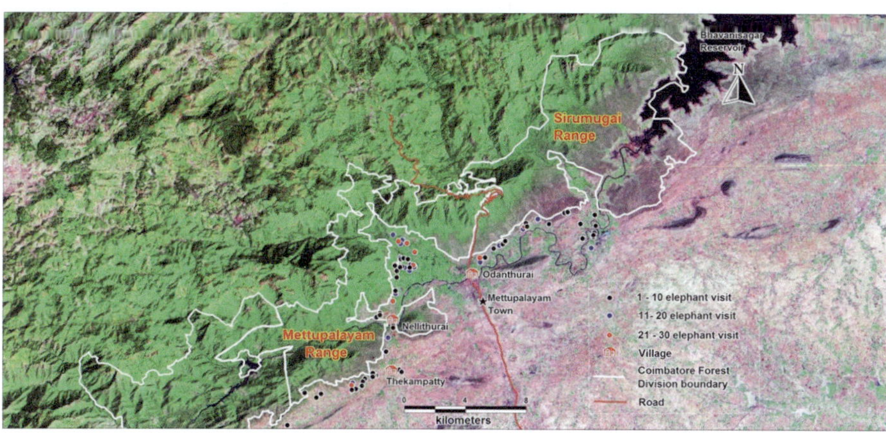

Map 18.4 Intensity of elephant visits in Mettupalayam and Sirumugai ranges

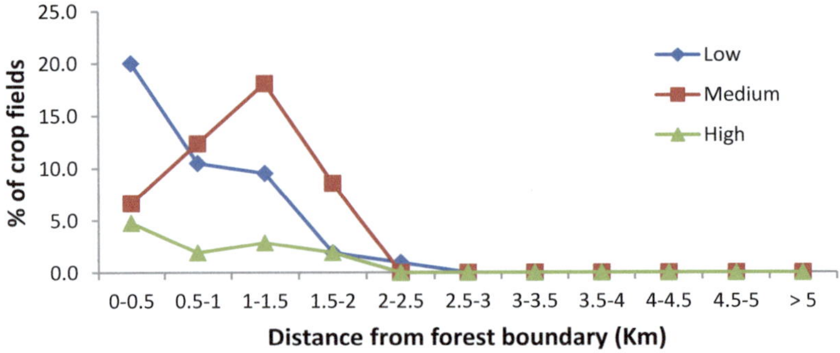

Fig. 18.4 Intensity of elephant visits into the crop fields adjoining to different forest ranges in Coimbatore Forest Division with relation to distance from forest boundary – PN Palayam range

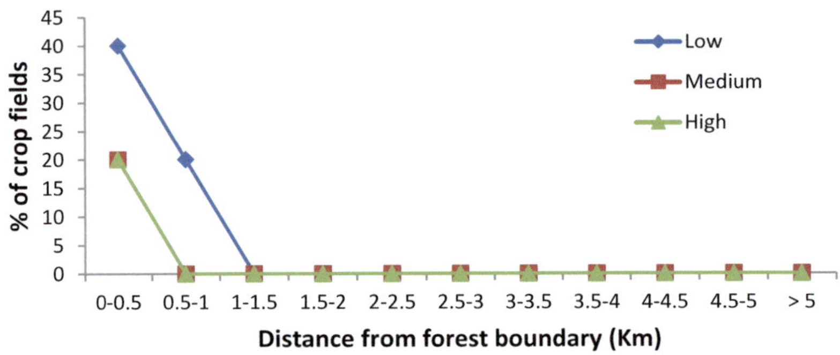

Fig. 18.5 Intensity of elephant visits into the crop fields adjoining to different forest ranges in Coimbatore Forest Division with relation to distance from forest boundary – Karamadai range

high proportion of crop fields was affected in 0–0.5 km category by different intensity of elephant visits as low (34% crop fields), medium (9.4% crop fields) and high (4.7% crop fields) (Fig. 18.6). High intensity of elephant visits was recorded up to 2 km from forest boundary. In Sirumugai, high proportion of crop fields was affected in 0–0.5 km category by different intensity of elephant visits as low (36% crop fields), medium (24%) and high (8%) (Fig. 18.7). Elephant visits were not noticed after 3 km from forest boundary.

18.3.3 Intensity of Elephant Visits vs Overall Distance from Forest

Crop fields which are located in 0–0.5 km from forest boundary were highly (35.55 %) affected by elephant visits (Fig. 18.8). The trend of elephant visit was decreased with respect to increase in the distance of crop fields from forest boundary up to 4.5 km. Below 1% of crop fields was visited by elephants 4.5–5 km and >5 km distance category.

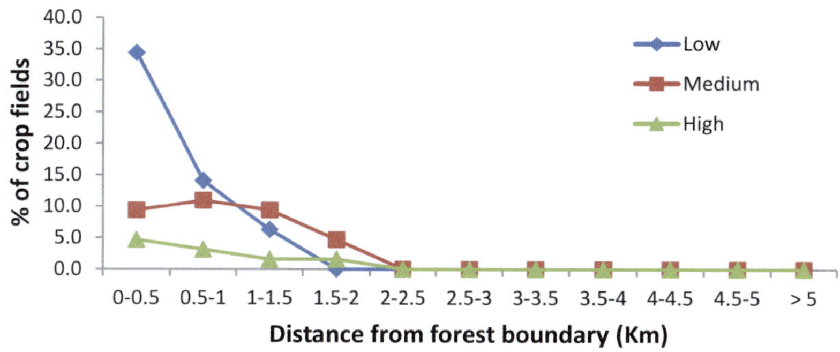

Fig. 18.6 Intensity of elephant visits into the crop fields adjoining to different forest ranges in Coimbatore Forest Division with relation to distance from forest boundary – Mettupalayam range

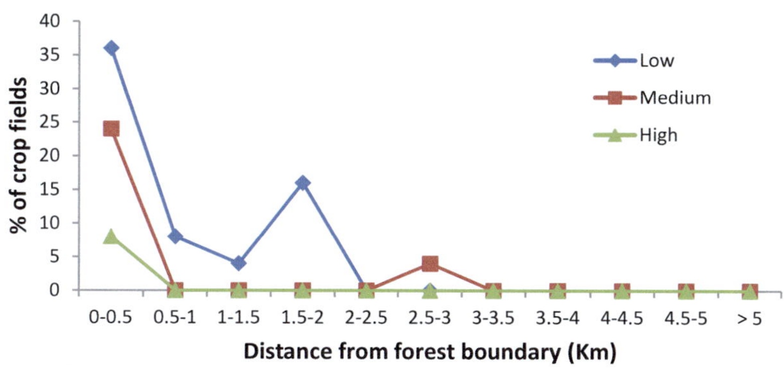

Fig. 18.7 Intensity of elephant visits into the crop fields adjoining to different forest ranges in Coimbatore Forest Division with relation to distance from forest boundary – Sirumugai range

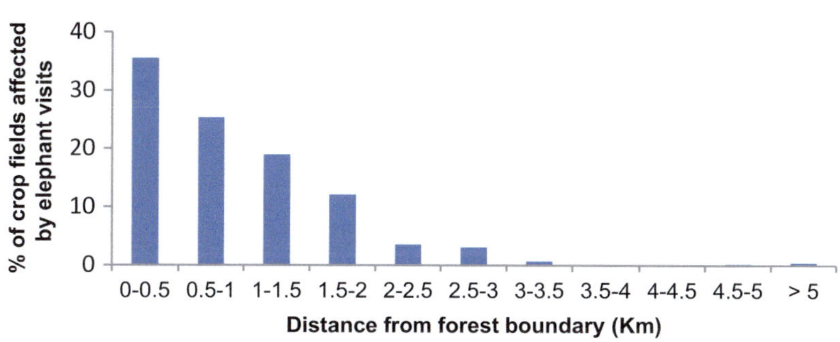

Fig. 18.8 Percentage of crop fields affected by elephant in relation to distance from forest boundary in Coimbatore Forest Division

Table 18.2 Intensity of elephant visits with relation to distance from forest boundary

	Intensity of elephant visits into the crop fields					
	Low (1–10 visits)		Medium (10–20 visits)		High (21–30 visits)	
Distance from forest boundary	No. of crop fields	Relative%	No. of crop fields	Relative%	No. of crop fields	Relative%
0–0.5	73	42.9	42	26.3	35	38.0
0.5–1	39	22.9	45	28.1	23	25.0
1–1.5	30	17.6	38	23.8	12	13.0
1.5–2	15	8.8	28	17.5	8	8.7
2–2.5	4	2.4	3	1.9	8	8.7
2.5–3	5	2.9	2	1.3	6	6.5
3–3.5	1	0.6	2	1.3	0	0.0
3.5–4	0	0.0	0	0.0	0	0.0
4–4.5	0	0.0	0	0.0	0	0.0
4.5–5	1	0.6	0	0.0	0	0.0
>5	2	1.2	0	0.0	0	0.0

Among the low intensity of elephant visits with respect to location of crop fields from forest boundary, high percentage was recorded in 0–0.5 km (42.9 %), followed by 0.5–1 km (22.9%) and 1–1.5 km (17.6%), and the lowest was recorded in 3.5–4 km and 4–4.5 km as 0% each (Table 18.2). Interestingly 0.6–1.2% of crop fields were affected by elephants in the 4.5–5 km and >5 km category, respectively. Within the medium intensity category, percentage of crop fields affected by elephants was recorded high in 0.5–1 km (28.1%), followed by 0–0.5 km (26.3%), 1–1.5 km (23.8%) and 1.5–2 km (17.5%), and the lowest was recorded in 2.5–3 km and 3–3.5 km as 1.3% each. Among the high intensity of elephant visits into crop fields, high percentage was recorded in 0.0.5 km (38%) followed by 0.5–1 km (25%) and 1–1.5 km (13%).

Statistical test revealed that frequency of crop fields affected in different distance categories (0–0.5 km, 0.5–1 km, 1–1.5 km,1.5–2 km, 2–2.5 km and 2.5–3 km) from forest and within various intensities of elephant visits (1–10, 11–20 and 21–30) was significantly different ($F = 8.707$, $P = 0.002$). Frequency of crop fields affected in different distance categories (0–0.5 km, 0.5–1 km, 1–1.5 km,1.5–2 km, 2–2.5 km and 2.5–3 km) from forest versus various intensities of elephant visits (1–10, 11–20 and 21–30) was not found significant ($F = 2.475$, $P = 0.133$). The overall intensity of elephant visits into the crop fields revealed that low intensity was 40% of crop fields followed by medium (38%), and high intensity was noticed 22 % of the crop fields (Fig. 18.9).

Among the crop fields which located in 0–0.5 km from forest boundary, intensity of elephant visit was low (48.7 % of crop fields) followed by medium (28% crop fields) and high (23% crop fields) (Fig. 18.10). Crop fields which are located in 0.5–1 km, 1–1.5 km and 1.5–2 km were highly prone for medium intensity of elephant visits as 42.1%, 47.5% and 54.9% of crop fields were affected, respectively. High intensity of elephant visits was recorded in 53.3% and 46.2 % of crop fields in 2.2.5 km and 2.5–3 km category, respectively.

Fig. 18.9 Intensity of elephant visits in the crop fields of Coimbatore Forest Division

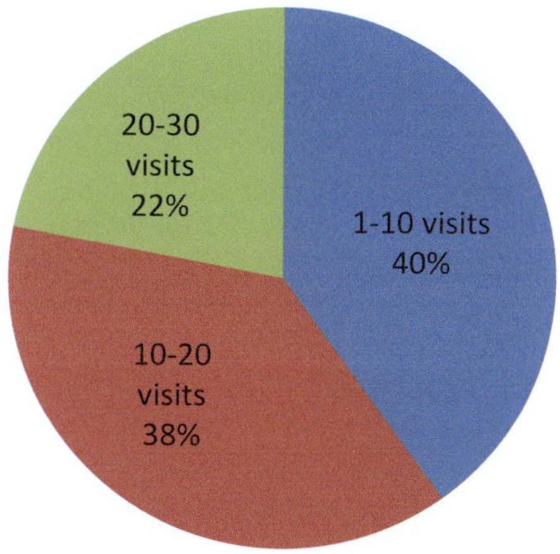

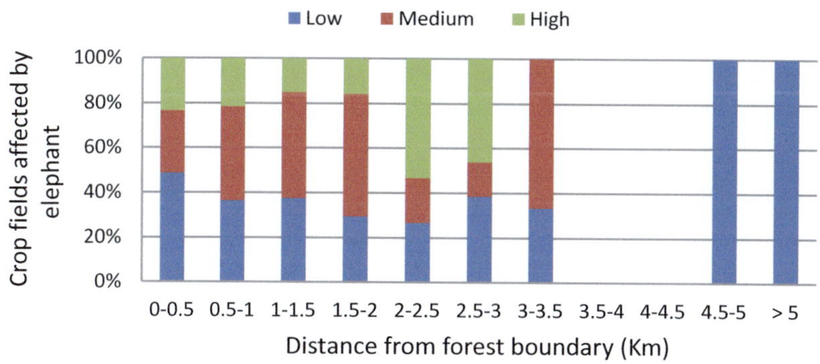

Fig. 18.10 Intensity of elephant visits into the crop fields with relation to distance from forest boundary

18.4 Discussion

This study has provided several interesting results with respect to elephant movement patterns outside forest areas in Coimbatore Forest Division. There appear to be significant differences between elephant visits outside forests in all forest ranges as well as between ranges. Crop fields located even 2 km away from forest boundary are highly prone to elephant visits. On few occasions, elephants even visited areas more than 5 km outside the forest areas of Coimbatore range during the study period. Before the study period (2007), it had been recorded that a herd of elephants had even strayed 25 km away from forests of Coimbatore forest division and intruded into the human habitation areas in Sulur – Kalangal. Mostly, elephants visit crop fields during night hours (after 11 PM) and stay in the crop fields till early morning 5 AM and then return to the forest again. If elephants failed to go back before sunrise due to some reasons, they will be virtually trapped outside forest areas for the entire day, surrounded by human crowd. During this time, elephants sometimes take refuge in some bushes available nearby and try to go back to forest late evening or midnight. At other times, they were driven by villagers from all directions, causing them to move further into human settlements, making it difficult for them to return to the forests. This is one of the major reasons for elephants straying for long distance into the private lands. Such incidences are happening occasionally in Coimbatore Forest Division.

Elephants have been recorded to similarly stray over long distances in other forest ranges also – Boluvampatti (3–3.5 km), Sirumugai (2.5–3 km), PN Palayam (2–2.5 km) and Mettupalayam (1.5–2 km). It clearly shows that elephants are straying outside RF for great distances across forest ranges of Coimbatore Forest Division. Interview with the villagers in Boluvampatti range revealed that elephants are mostly entering into crop fields located in long distance from forest boundary through the dry streams during night time. Sukumar (1989) pointed out that in a single night, a bull can travel up to 3–6 km through cultivated fields, while herds are eating crops opportunistically, venturing not more than 1 km from the forest boundary.

Seidensticker (1984) categorized movement of elephant herds across three different scales: (1) daily movement between feeding and water sources, (2) seasonal movement between dry and wet season forage areas within home ranges and (3) medium-term movements between sectors of the seasonal home range. Coimbatore Forest Division experiences all three scales of elephant movement described by Seidensticker; and this could be one of the major reasons for the intensity of elephant movement into the crop fields over long distance. In some areas, elephants are forced to move through human-dominated landscapes to water sources (Sukumar 1989) and/or different parts of the home range.

The overall history of human-elephant conflict in Coimbatore Forest Division revealed that elephant stray incidences started after developments of huge buildings along the forest foothills in 1995. Coimbatore Division never experienced such kind of human-elephant conflict incidences before the developmental activities took place along the elephant migratory routes. Balasubramanian et al. (1995) also pointed out that the Nilgiri Biosphere Reserve has very high elephant densities, and yet crop loss was considered low. Ramakrishnan (2008) pointed out that the indiscriminate growth in the construction of buildings in the forms of resorts, educational institutions, ashrams and amusement parks in the fringes of the corridors considerably affects the movement of elephants which becomes a cause for the human-elephant conflict. In Coimbatore Forest Division, about 22 buildings were established in fringes of narrow forest foothills during the period of 1995–2005, causing heavy restrictions on the movement of elephants. Sukumar (1989) pointed out that loss or fragmentation of habitat blocked the traditional routes and restricts elephants' access to forage, refuge areas, salt licks and water. Ramakrishnan (2008) assumed that these elephants get bounced back due to blockage of corridor and are brought into conflict with people. It is reported that they compensate for this loss by eating crops; bulls in particular may take advantage of the easy availability of crops and stored grain. Ramakrishnan and Durairasu (2005) pointed out that if hindrance to elephant movement exists, the animal will try to find an alternative route, which sometimes leads them into crop fields. Apart from that, many studies have found that elephant-human conflict is a symptom of inappropriate land use practices such as permanent human settlements and growing irrigated food crops adjacent to elephant range lands (Lahiri-Choudhury 1991; Johnsingh and Joshua 1994; Sivaganesan and Johnsingh 1995; Hoare 2001a; Treves et al. 2006; IUCN 2006; Fernando et al. 2008).

Crop fields were greatly affected in and around 0–0.5 km from forest boundary in all forest ranges. The intensity of elephant visits varied across the forest ranges. Intensity of elephant visits with respect to location of crop fields from forest boundary was varied across the forest ranges. Overall intensity of elephant visits in Coimbatore Forest Division revealed that even though elephants started straying in a small scale during the late 1990s due to various reasons, the present scenario has made the elephant herds to learn and practice and habituated to crop raiding. Sukumar (1985b, 1995) pointed out that eating crops may also be a learned behaviour. While calves may learn from the adults in the herd; young bulls that disperse from herds that generally do not eat crops also learn by associating with bulls (Osborn 1998). Sitati et al. (2003) pointed out that elephants may return to areas where they remember eating crops successfully in the past (Sitati et al. 2003). It has been reported that increasing human and elephant populations led to a corresponding increase in conflict (Smith and Kasiki 2000). Bell (1984) found elephant densities to be higher in valleys than plateaus, and therefore villages located in valleys suffered greater levels of crop loss. This is corroborated by our findings that more crop fields located in valleys such as Velliankadu, Naicken Palayam, Thadagam and Boluvampatti were highly affected by elephants.

A study also indicates that elephants do not stop using "high-risk" areas which may have been part of their home range in the past; instead they change their behaviour by using these areas only at night or walking faster than normal (Foley 2002) "four times faster" (Douglas-Hamilton et al. 2005). The shape and size of the forest also contributes to the intensity of conflict with large and square or circular forests faring better than long and thin ones (Sutton 1998; IUCN 2006; Boominathan et al. 2008). When elephants leave their natural habitat from Coimbatore Forest Division, it is assumed to be anomalous behaviour, and speculations on the causes include increasing elephant numbers, inadequate food resources in forests and blockage of corridor, among others. The reasons for human-elephant conflict in Coimbatore Forest Division could be due to man-made factors coupled with natural factors. The natural factors include:

1. The eastern boundary of the forest division shares its entire 350 km stretch with human habitations and farm lands.
2. The terrain is highly undulated and approximately 50% of the area is not conducive for elephant movement.
3. The four valleys and the villages located inside are surrounded by the forests from all except on one side, making them prone to elephant visits from any of these directions.

4. Alternately, about 25% of the forest areas are surrounded by villages in three directions, making elephant movement in any direction prone to contact with villages.
5. Water scarcity in foothill forests due to lack of retention as a result of the hilly terrain.

Since natural factors were not favourable across the division, elephants have used the foothill forests and adjoining private fallow lands for migration over decades.

The major man-made factors that influenced conflicts could be the land use changes in the forest fringe villages. During the mid-1990s, development activities started along the forest foothills. In the last 15 years, lands near foothills have become a hub for establishing new educational institutions, ashrams, factories and resorts due to availability of cheap fallow lands, which were used by elephants traditionally. Also cropping pattern along the forest fringes and further inside the villages was notably changed. In the recent years, loans provided to farmers and corresponding increase in the number of bore-wells made the farmers to grow more cash crops such as coconut; sugarcane around Coimbatore Forest Division may have attracted elephants further into the human habitation. Management strategies in this division should be aimed at regulating the land use changes in private lands at least 2 km away from forest boundary, habitat improvement in foothill forests and detailed research on factors of human-elephant conflict and new techniques on control measures. Private lands located at least 200 m from foothill forest should be freed from all sort of physical barriers. Water sources could be provided along every 5 km in the forest foothills during summer, to negate the need for elephants to move further inlands into human areas.

Acknowledgements The authors are highly grateful to Mr. K. Kalidasan, President, OSAI Environmental Organisation, Coimbatore, for the financial support and valuable suggestion for this study. The authors are also thankful to Mr. I. Anwardeen, IFS, District Forest Officer, Coimbatore Division, for the permission and all logistic support.

References

Balasubramanian M, Baskaran N, Swaminathan S, Desai AA (1995) Crop raiding by Asian elephant (*Elephas maximus*) in the Nilgiri biosphere reserve, South India. In: Daniel JC, Datye HS (eds) A week with elephants. Bombay Natural History Society, Oxford University Press, Bombay, pp 350–367

Bell RHV (1984) The man-animal interface: an assessment of crop damage and wildlife control. In: Bell RHV, McShane-Caluzi (eds) Conservation and wildlife management in Africa. US Peace Corps Seminar, Malawi

Boominathan D, Mohanraj N, Aziz T, Desai A (2008) Management of the Asian elephant in the Nilgiris and Eastern Ghats: human-elephant conflict in Somwarpet Subdivision (Madikeri Forest Division). WWF AREAS

Daniel JC, Desai A, Kumar A, Sakthivel C (2008) Evaluating population enumeration methods and human elephant conflict mitigation methods in Mudumalai Tiger Reserve, Tamil Nadu, India. Bombay Natural History Society, Mumbai

Di Fonzo MMI (2007) Determining correlates of human-elephant conflict reports within fringe villages of Kaziranga National Park, Assam. Masters thesis, University of London and Imperial College, London

Douglas-Hamilton I, Krink T, Vollrath F (2005) Movements and corridors of African elephants in relation to protected areas. Naturwissenschaften 92(4):158–163

Fernando P, Kumar MA, Williams AC, Wickramanayake E, Aziz T, Singh SM (2008) Review of human-elephant conflict mitigation measures practiced in South Asia. WWF-World Wide Fund for Nature, AREAS Technical Support Document submitted to World Bank

Foley LS (2002) The influence of environmental factors and human activity on elephant distribution in Tarangire National Park, Tanzania. Masters thesis, International Institute for Geo-Information Science and Earth Observation, The Netherlands

Graham A (1973) The gardeners of Eden. George Allen & Unwin, London

Grant CC, Bengis R, Balfour D, Peel M, Davies-Mostert W, Killian H, Little R, Smit I, Garai MH, Anthony B, Hartley P (2008) Controlling the distribution of elephant. In: Mennell KG, Scholes RJ (eds.) Assessment of South African elephant management 2007, 246–277

Hoare RE (1999) Determinants of human-elephant conflict in a land-use mosaic. J Appl Ecol 36:689–700

Hoare RE (2001) A decision support system for managing human-elephant conflict situations in Africa. http://data.iucn.org/themes/ssc/sgs/afesg/hec/pdfs/hecdssen.pdf#page=1. Accessed 15 Jan 2008

IUCN (2006) Asian elephant range states meeting, 24–26 January 2006, Kuala Lumpur, Malaysia: Report. Switzerland

Johnsingh AJT, Joshua J (1994) Conserving Rajaji and Corbett National Parks – the elephant as a flagship species. Oryx 28(2):135–140

Kiiru W (1995) The current status of human-elephant conflict in Kenya. Pachyderm 19:15–19

Kumar MA, Mudappa D, Raman TRS, Madhusudan MD (2004) The Elephant Hills: conservation of wild Asian elephants in a landscape of fragmented rainforests and plantations in the Anamalais, India. CERC Technical Report No. 10, Nature Conservation Foundation, Mysore, India

Lahiri-Choudhury DK (1991) Keeping wild elephants at bay. In: Eltringham SK (ed) The illustrated encyclopedia of elephants. Salamander Books, London, pp 166–169

Lahkar BP, Das JP, Nath NK, Dey S, Brahma N, Sarma PK (2007) A study of habitat utilization patterns of Asian elephants Elephas maximus and current status of human elephant conflict in Manas National Park within Chirang-Ripu Elephant Reserve, Assam. Aaranyak, Guwahati, India

Madhusudan MD (2003) Living amidst large wildlife: livestock and crop depredation by large mammals in the interior villages of Bhadra Tiger Reserve, South India. Environ Manag 31(4):466–475

Nath CD, Sukumar R (1998) Elephant-human conflict in Kodagu: southern India. Asian Elephant Research and Conservation Centre, Bangalore

Naughton L, Rose R, Treves A (1999) The social dimensions of human-elephant conflict in Africa: a literature review and case studies from Uganda and Cameroon. http://data.iucn.org/themes/ssc/sgs/afesg/hec/pdfs/hecugcarev.pdf. Accessed 2 Apr 2008

Naughton-Treves L, Treves A (2005) Socio-ecological factors shaping local support for wildlife: crop-raiding by elephants and other wildlife in Africa. In: Woodroffe R, Thirgood S, Rabinowitz A (eds) People and wildlife: conflict or coexistence. Cambridge University Press, Cambridge, pp 252–277

Nyhus PJ, Tilson R, Sumianto (2000) Crop-raiding elephants and conservation implications at Way Kambas National Park, Sumatra, Indonesia. Oryx 34(4):262–274

Osborn FV (1998) The ecology of crop-raiding elephants in Zimbabwe. Ph.D. thesis, Cambridge, UK

Parker GE, Osborn FV, Hoare RE, Niskanen LS (2007) Human-elephant conflict mitigation – a training course for community based approaches in Africa: Participant's Manual. http://data.iucn.org/themes/ssc/sgs/afesg/hec/pdfs/heccombappmen.pdf. Accessed 22 Jan 2008

Ramakrishnan B, Durairasu P (2005) Human-elephant mitigating measures in the coimbatore forest division, Tamil Nadu, India. OSAI Technical Report No.3, Coimbatore

Ramakrishnan B (2008) Status of wildlife corridors and their use by selected endangered mammals in the Nilgiri Biosphere Reserve, India. Unpublished PhD thesis Bharathidasan University, Tamil Nadu

Riddle H (2007) Elephant response Units (ERU). Gajah 26:47–53

Sarma UK, Easa PS (2006) Living with giants: understanding human-elephant conflict in Maharashtra and adjoining areas. Wildlife Trust of India, Delhi

Seidensticker J (1984) Managing elephant depredation in agricultural and forestry projects. A World Bank Technical Paper, The World Bank, Washington DC

Sitati N, Walpole MJ, Leader-Williams N (2005) Factors affecting susceptibility of farms to crop raiding by African elephants: using a predictive model to mitigate conflict. J Appl Ecol 42:1175–1182

Sitati NW, Walpole MJ, Smith RJ, Leader-Williams N (2003) Predicting spatial aspects of human-elephant conflict. J Appl Ecol 40:667–677

Sivaganesan N, Johnsingh AJT (1995) Food resources crucial to the wild elephants in Mudumalai wildlife sanctuary, South India. In: Daniel JC, Datye HS (eds) A week with elephants. Bombay Natural History Society/Oxford University Press, Bombay, pp 405–423

Smith RJ, Kasiki SM (2000) A spatial analysis of human-elephant conflict in the Tsavo ecosystem, Kenya. www.kent.ac.uk/anthropology/dice/resources/Smith_&_Kasiki_HEC_report.pdf. Accessed 17 Oct 2008

Sukumar R (1985a) Ecology and conservation of the Asian elephant in South India – with special reference to the Chamrajnagar and Sathyamangalam Forest Divisions. Report submitted to the AsESG/WWF – Project No. 3032

Sukumar R (1985b) Ecology of the Asian elephant (*Elephas maximus*) and its interaction with man in South India, vol 2. PhD thesis, Centre for Ecological Studies, Indian Institute of Science, Bangalore

Sukumar R (1989) The Asian elephant: ecology and management. Oxford University Press, Cambridge

Sukumar R (1995) Elephant raiders and rogues. Nat Hist 104(7):52–61

Sukumar R (2003) The living elephants: evolutionary ecology, behavior and conservation. Oxford University Press, New York

Sutton WR (1998) The cost of living with elephants in Namibia. www.agecon.ucdavis.edu/people/faculty/facultydocs/Jarvis/elephants/sutton.pdf. Accessed 19 Dec 2008

Talukdar BK, Boruah JK, Sarma P (2006) Multi-dimensional mitigations initiatives to human-elephant conflicts in Golaghat district and adjoining areas of Karbi Anglong, Assam, India. Proceedings, International Elephant Conservation & Research Symposium, Copenhagen Zoo, Denmark

Treves A, Wallace RB, Naughton-Treves L, Morales A (2006) Co-managing human wildlife conflicts: a review. Hum Dimens Wildl 11:383–396

WWF (2008) Common ground: Solutions for reducing the human, economic and conservation costs of human-wildlife conflict. http://assets.panda.org/download/hwc_final_web_2.pdf. Accessed 2 Aug 2008

Statistical Techniques for Estimating the Abundance of Asiatic Elephants Based on Dung Piles

19

M. Sivaram, K. K. Ramachandran, E. A. Jayson, and P. V. Nair

Abstract

In this chapter we presented the field methods and statistical models used for estimating elephant density and elephant population for different Elephant Reserves of Kerala. The Elephant Census was organized in the four Elephant Reserves of Kerala State during 2005 and 2007 by the Kerala State Forest and Wildlife Department.

Keywords

Line transect sampling · Dung decay rate · Dung defecation rate · Kerala · Western Ghats

19.1 Introduction

Information on elephant population in forests is essential for its effective management. Different methods have been used for the direct survey of elephants like total count, sample count, water hole count, and line transect sampling – direct sighting. These direct methods are usually more prone to sample error due to scattered occurrence of elephants, group behavior, and its vast home range (Jachmann 1991).

M. Sivaram (✉)
Southern Regional Station, ICAR-National Dairy Research Institute, Bangalore, India
e-mail: sivaram.ndri@gmail.com

K. K. Ramachandran · E. A. Jayson
Wildlife Biology Discipline, Kerala Forest Research Institute, Thrissur District, Kerala, India

P. V. Nair
GIS Discipline, Kerala Forest Research Institute, Thrissur District, Kerala, India

C. Sivaperuman, K. Venkataraman (eds.), *Indian Hotspots*,
https://doi.org/10.1007/978-981-10-6605-4_19

Further, direct sighting of elephants over vast area is practically problematic. However, elephants leave indirect evidence such as dung, which continues to be present in the area for a considerable time period. The estimation of elephant population through surveys of dung is practically an easy method and becoming popular.

Standing crop method and clearance plot method are the possible indirect methods for estimating elephant population. The standing crop method is based on the assumption that there is a stable relationship between the amount of dung present and the number of animals. This method requires one-time survey of dung, and dung count is corrected by defecation and decay rate. The clearance plot method involves clearing dung from marked plots at regular intervals, counting the droppings, and correcting the counts by the defecation rate (Staines and Ratcliffe 1987). Most of the studies in tropics fall under the framework of standing crop method. In order to convert the dung density into elephant density (number of elephants per unit area), the following formula is usually adopted:

$$\text{Estimated density of elephants, } \hat{D}_e = \frac{\text{Dung density}}{\text{Defecation rate}} \times \text{Dung decay rate}$$

Dung density (number of dung per unit area) is usually estimated through surveys of dung using quadrat sampling, strip transects sampling, or line transect sampling. Line transect sampling was followed in this study. Defecation rate (number of dung defecated per day per animal) can be estimated by monitoring captive elephants or by placing a known number of elephants in an enclosure previously cleared off dung and estimating the number of dung produced over a fixed time period. The dung decay rate is defined as the number of dung decayed per day and is expressed as the reciprocal of the estimated mean time to decay (Barnes and Barnes 1992).

Dung decay rate can be estimated by conducting experiments in which the fresh dung piles are located and marked and monitored over a period of time until they disappear. Most of the studies use decay rate for estimating elephant population with the assumption that the system is in a steady state throughout the period of decay experiments. The steady-state assumption states that the number of dung piles being deposited each day equals the number of dung piles disappeared each day, i.e., the number of dung piles per unit area remains constant from day to day (Barnes and Jenson 1987; Barnes and Barnes 1992). Further, dung decay rate has been estimated assuming an exponential rate of decay independent of age or by curve fitting over age-specific mortality of dung piles. However, these estimations have been found confounded with steady-state assumption and biases such as seasonal variation in decay rates. It is also seen that the decay rate (as well as defecation rate) is borrowed from other studies conducted in similar areas or even from distant places for the estimation of elephant population (Easa et al. 2002).

Laing et al. (2003) provided a robust methodology for estimating mean time to decay, which they termed it as "retrospective estimate" of the mean time to decay. In this method, fresh signs of the animal are located and marked on several dates in the lead-up to the survey, chosen so that the proportion of signs surviving from the earliest date to the survey is expected to be small, and to return to marked signs just once, at the time of the survey. Data on status of the signs are then binary, recording whether or not the signs survive to the date of the survey. This binary data are subjected to the logistic regression analysis, and mean time to decay is estimated. In this chapter, the dung survey methods and statistical models employed for the population estimation of wild elephants in the forests of Kerala State are presented.

19.2 Survey Methods

19.2.1 Organization of the Census Program

The Elephant Census was organized in the four Elephant Reserves of Kerala State under the direction and guidance of the Chief Conservator of Forests (Wildlife) (Fig. 19.1). The Field Director of the Periyar Tiger Reserve was the State Coordinator of the census. The Conservators of the Wildlife Wing were nominated as the Coordinators of the respective Regions. The actual census was carried out on 2 days. In the year 2005, the block count method was carried out on 5 May 2005 and dung survey using line transects sampling on 6 May 2005 (Sivaram et al. 2005). In the year 2007, the block count method was carried out on 7 May 2007 and dung survey using line transects sampling on 9 May 2007 (Sivaram et al. 2007).

A 1-day training program was organized for the selected forest officers (resource persons) at different places. The officials in the meeting were briefed on the field techniques to be followed in the census, the method of filling the proforma and also on the care to be taken while collecting the data. The doubts of the resource persons were cleared during the discussions that followed. The method of census and the procedures to be followed in the field for the success of the program were explained to the field staff in detail in the regional meetings of the forest officers convened by the respective coordinators.

The toposheets of the forest areas were taken to the Divisional Forest Offices/ Ranges, and the blocks were demarcated by the Forest Range Officers and the field staff. The copies of such maps were sent to the Divisional Forest Offices with instructions on laying transects in the selected sample blocks. Transects were laid by the forest officers in the selected blocks and marked with paint/colored biodegradable ribbons.

A proforma for the census was prepared and got printed along with the instructions to the participants. The materials such as field compass, measuring tapes, notebooks, and pencils were procured. The required number of kits containing these materials and the proforma were distributed to the offices of the Conservators. The Forest Range Officers later collected these items.

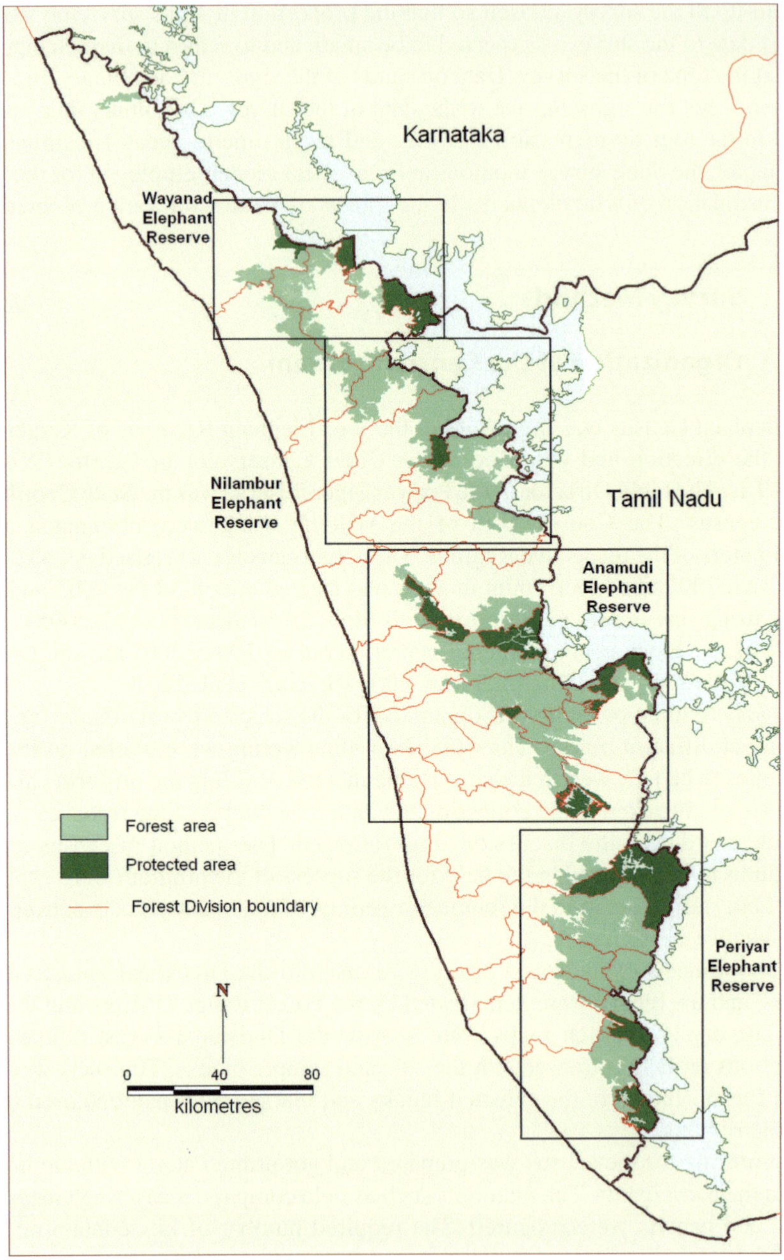

Fig. 19.1 Elephant Reserves of Kerala

Table 19.1 Total area of sampled blocks and number of blocks sampled

	Number of blocks sampled		Total area of sampled blocks (km^2)	
Elephant reserve	2005	2007	2005	2007
Wayanad ER	74	82	501.75	584.71
Nilambur ER	89	93	576.50	884.09
Anamudi ER	188	195	1163.69	1324.77
Periyar ER	166	213	1359.22	1862.02
Total	517	583	3601.16	4655.59

19.2.2 Sampling of Blocks

The total forest area of each Protected Area/Territorial Forest Division was divided into number of small blocks utilizing the Survey of India maps. A random sample of blocks was chosen in each Protected Area/Territorial Forest Division for the enumeration. The total number of blocks sampled was 517 in the year 2005 and 583 in the year 2007. The details of the total number of blocks sampled and area sampled in each Elephant Reserve are given in Table 19.1.

19.2.3 Line Transect Sampling: Elephant Dung Survey

19.2.3.1 Line Transect Sampling

Line transect sampling is one of the widely used scientific methods (Buckland et al. 2001). If the method is applied properly, it provides a viable technique to determine point estimates and measures of variance of animal density (in the present context dung density). In line transect sampling, the observer(s) perform a standardized survey along a series of lines, searching for objects of interest such as cluster of animals and dung piles. For each object detected, they record the perpendicular distance from the line to the object or radial distance from the observer to the object along with the angle of sighting. The main advantage with line transect sampling is that even without encountering all the objects of interest in the area, it is possible to develop an estimate of the total number of objects or their density through appropriate statistical analyses. However, the method presupposes adequate sample size in terms of sightings without which precise estimate of density cannot be obtained by this method. Burnham et al. (1980) recommend a minimum of 40 sightings of objects of interest for satisfactory estimation of the detection function in the area.

19.2.4 Line Transect Sampling: Elephant Dung Survey

The technique of line transect sampling was adopted in all the sampled blocks. In each sample block, transect of about 2 km length was laid by marking trees with paint or colored biodegradable ribbons. These transects were covered on foot

Table 19.2 Details of line transects laid in different Elephant Reserves of Kerala during the year 2005 and 2007

	Total number of transects		Total length of transect (km)		Total number of dung piles recorded	
Elephant reserve	2005	2007	2005	2007	2005	2007
Wayanad ER	75	82	148.0	163.15	1434	3557
Nilambur ER	88	93	179.8	183.01	718	1353
Anamudi ER	173	195	363.6	388.92	3286	4615
Periyar ER	166	214	331.8	420.95	1908	3528
Total	**502**	**584**	**1023.2**	**1156.03**	**7346**	**13,053**

recording the perpendicular distance to the geometric center of the elephant dung piles. The perpendicular distance was measured using a tape. The details of the number of transects sampled in each Elephant Reserve are given in Table 19.2.

19.2.5 Elephant Dung Decay Experiments

The dung decay experiments were conducted on a sample of fresh dung piles in all the four Elephant Reserves of Kerala representing different vegetation types following the retrospective method suggested in Laing et al. (2003). A number of field visits were made, adding fresh dung piles to the sample and also recording the state of the dung piles previously marked. A fresh dung pile is the one, which is 0–24 h old. The state of the dung pile was recorded as present (= 1) or absent (= 0) indicating the decay status of the dung piles. Present is defined as any stage where some dung material is still left. Absent is a stage where only traces (e.g., plant fiber remains, termite mounds, mud, etc.) are left and no dung material is present. Absent also includes "total disappearance" of dung pile (e.g., washing away in heavy rains).

Each dung pile was marked and numbered uniquely using one of the following methods.

1. During each visit, the previously marked dung piles were visited and their status noted.
2. If, however, a marker was missing and the marked dung pile could not be located accurately, it was excluded from the sample and fresh dung piles were marked during that visit.
3. During the last visit, the status (Presence/Absence) of all previously marked dung piles was noted. No fresh dung pile was marked on this visit.

The experiment was initiated about 105 days before the actual census. The field visits were made every fortnight as per the schedule prepared for each Forest Range, searching for fresh elephant dung in each vegetation type and marking the same for assessing its future status (present/absent). The details of the experiments were recorded using an observation form.

19.3 Statistical Models

19.3.1 Extent of Actual Elephant Habitat

Extent of actual elephant habitat is a crucial multiplication factor in extrapolating elephant population. Therefore, efforts were made to arrive at the actual extent of elephant habitat by consulting government notifications, published reports, and forest working plans. Apart from the forest areas, which are definitely devoid of elephants such as Thrissur Forest Division and Kumily Forest Range, the blocks that are devoid of elephants in various Forest Divisions, water bodies, and other enclosures were accounted for and the actual elephant habitat worked out. The details of the area devoid of elephant and the actual elephant habitat for various Elephant Reserves are presented in Table 19.3.

19.3.2 Line Transect Sampling: Dung Survey

In strip sampling, if strips of width 2ω and total length L are surveyed, an area of size $a = 2\omega L$ is censused. All n objects within the strips are enumerated, and estimated density is the number of objects (in our case dung piles) per unit area:

$$\hat{D} = \frac{n}{2\omega L}.$$

In line transect sampling, only a proportion of the objects (dung piles) in the area a surveyed is detected. Let this unknown proportion be P_a. If P_a can be estimated from the distance data, then we would estimate density by

$$\hat{D} = \frac{n}{2\omega L\hat{P}_a}$$

Table 19.3 Details of area devoid of elephant and actual area used for extrapolating elephant population in different Elephant Reserves of Kerala

Elephant reserve	Forest area (km²)	Effective forest area (km²)[a]	Area devoid of elephants (km²)	Actual area used for extrapolating elephant population (km²)
Wayanad ER	1200	1101.05	166.89	934.16
Nilambur ER	1419	1255.71	113.41	1142.30
Anamudi ER	3728	3365.92	548.47	2817.45
Periyar ER	3742	3411.73	385.32	3026.41
Total	**10,089**	**9134.41**	**1214.09**	**7920.32**

[a]Effective forest area was computed for different Elephant Reserves by allocating the total effective forest area of the state (9400 km²) in proportion to the total forest area of territorial forest divisions falling under the respective Elephant Reserve

where the unconditional probability of detecting an object in the strip is

$$P_a = \frac{\int_0^{\omega} g(x)\mathrm{dx}}{\omega}.$$

Substituting the estimator of P_a into $\hat{D}$ gives

$$\hat{D} = \frac{n}{2L\int_0^{\omega} \hat{g}(x)\mathrm{dx}}.$$

The probability density function of the perpendicular distance data, conditional on the object being detected is

$$f(x) = \frac{g(x)}{\int_0^{\omega} g(x)\mathrm{dx}}.$$

By assumption g(0), the probability of detection on the line is 1, so that the pdf, evaluated at zero distance, is

$$f(0) = \frac{1}{\int_0^{\omega} g(x)\mathrm{dx}}.$$

Therefore, the general estimator of density for line transect sampling is

$$\hat{D} = \frac{\mathrm{n}\hat{\mathrm{f}}(0)}{2L}. \tag{19.1}$$

The formula for $var(\hat{D})$ is given in Buckland et al. (2001). Some generally useful models of g(x) are given in Table 19.4. The series expansion added in the model is used to adjust the key function, using one or two more parameters, to improve the fit of the model to the distance data.

The perpendicular distances to dung piles formed the input data for the estimation of dung density. The density estimates were obtained using the formula (19.1) above. Univariate half normal distribution with the series expansion of simple polynomial/hermite polynomial was used as detection function for estimating the dung density. A 5% truncation of the largest perpendicular distance values was adopted to improve the precision of the density estimates. The software DISTANCE 5.0 Release 2 developed by Thomas et al. (2006) was used for all calculations. The dung density estimates were worked out at the habitat level and Elephant Reserve level by pooling the data appropriately.

19.3.3 Measuring Dung Decay Rate

19.3.3.1 Choice of Statistical Technique

Dung decay rate is the reciprocal of the average survival time of the dung piles. The average survival time is estimated by fitting appropriate mathematical model

Table 19.4 Models of the detection function used in line transect sampling

Key functions	Series expansion
Uniform, $1/\omega$	Cosine, $\sum_{m}^{j=1} a_j \cos(j\pi x/\omega)$
Uniform, $1/\omega$	Simple polynomial, $\sum_{m}^{j=1} a_j (x/\omega)^{2j}$
Half-normal, $\exp(-x^2/2\sigma^2)$	Cosine, $\sum_{m}^{j=2} a_j \cos(j\pi x/\omega)$
Half-normal, $\exp(-x^2/2\sigma^2)$	Hermit polynomial, $\sum_{m}^{j=2} a_j H_{2j}(x_s)$, where $x_s = x/\sigma$
Hazard-rate, $1-\exp(-(x/\sigma)^{-b})$	Cosine, $\sum_{m}^{j=2} a_j \cos(j\pi x/\omega)$
Hazard-rate, $1-\exp(-(x/\sigma)^{-b})$	Simple polynomial, $\sum_{m}^{j=1} a_j (x/\omega)^{2j}$

relating survival status of the dung piles with the age of the dung piles (i.e., duration of the dung piles up to which the dung piles survived). It was attempted to subject the data collected from dung decay experiments conducted in different Elephant Reserves to logistic regression model following Laing et al. (2003). However, it was observed from detailed analysis that the data collected from different Elephant Reserves were not found to follow the logistic regression model. Further, to apply logistic regression model, it was suggested that 90% of the indirect evidences followed up should be decayed by the end of the experiment (i.e., by the end of 105 days). In the dung decay experiments conducted in Kerala for the present elephant population estimation, such a situation was not seen. In Wayanad Elephant Reserve, a large number of dung piles were surviving even after 105 days. In other Elephant Reserves, most of the dung piles decayed well before 105 days (Sivaram et al. 2005). So, we had to resort to alternative statistical technique for analysis. We used Kaplan-Meier survivorship function (Hosmer and Lemeshow 1999) for estimating average survival time of the dung piles (referred to as time to decay).

19.3.3.2 Kaplan-Meir Survival Analysis

In studying the survival time of the dung piles using a follow-up study for a specified period of time, the primary outcome variable concerned is the number of days that the respective dung piles will survive. At the end of the study, it may be found that there will be dung piles that survived over the whole study time even if they have entered late and other dung piles that failed to follow up. The follow-up time is different for each dung piles as the entering time for each dung piles is different. The dung piles that failed to follow up may be ignored considering as missing data (since most of them are "survivors"). These dung piles, however, contain partial information on the survival time. Such observations are called censored observations. The censored observations include dung piles still alive at the end of study or dung piles lost to follow up or left study before the end or event not recorded properly. The Kaplan-Meier estimator of the survivorship function is free of models (nonparametric) and incorporates information from all of the observations available, both uncensored and censored, by considering survival to any point in time as a series of steps defined by the observed survival and censored times. This estimator is a product of a number of conditional probabilities resulting in an estimated survival function *S(t)* in the form of a step function. This estimator is also called as the product limit estimator.

The following is the general formulation of Kaplan-Meier estimator (Hosmer and Lemeshow 1999). Assume that we have a sample of n independent observations on dung piles denoted $(t_{(i)}, c_{(i)})$, $i = 1, 2, \ldots, n$ of the underlying survival time variable T and the censoring indicator variable C. Assume that among the n observations, there are $m \leq n$ recorded times of absence of dung pile. We denote the rank-ordered survival times as $t_{(1)} < t_{(2)} < \cdots < t_{(m)}$. Let the number of dung piles at risk of decaying at $t_{(i)}$ be denoted n_i, and the observed number of dung piles decayed be denoted d_i. The Kaplan-Meier estimator of the survivorship function at time t is obtained from the equation

$$\hat{S}(t) = \prod_{t_{(i)} \leq t} \frac{n_i - d_i}{n_i}$$

with the convention that

$$\hat{S}(t) = 1 \text{ if } t < t_{(1)}.$$

Using delta method, variance of the survivorship function is obtained as

$$\hat{\mathrm{Var}}\left(\hat{S}(t)\right) = \left(\hat{S}(t)^2\right) \sum_{t_{(i)} < t} \frac{d_i}{n_i (n_i - d_i)}.$$

In the analysis of survival time, the sample mean is not as important a measure of central tendency as in other settings due to the fact that censored survival time data are most often skewed to the right. The use of median time is the best option in such cases. The median survival time was used to work out the decay rate.

Median time is the second quartile (50th percentile), denoted by $\hat{t}_{50}$. The interpretation of this value is that we estimate that 50% will survive at least up to the time point $\hat{t}_{50}$.

$\hat{t}_{50} = \min\left\{t : \hat{S}(t) \leq 0.50\right\}$. In general, the estimate of the *pth* percentile is

$\hat{t}_p = \min\left\{t : \hat{S}(t) \leq (p/100)\right\}$. The estimator for the variance of the estimator of the *pth* percentile is

$$\hat{\text{Var}}\left(\hat{t}_p\right) = \frac{\hat{\text{Var}}\left(\hat{S}\left(\hat{t}_p\right)\right)}{\left[\hat{f}\left(\hat{t}_p\right)\right]^2}$$

where $\hat{f}\left(\hat{t}_p\right) = \dfrac{\hat{S}\left(\hat{u}_p\right) - \hat{S}\left(\hat{l}_p\right)}{\hat{l}_p - \hat{u}_p}$.

The values $\hat{u}_p$ and $\hat{l}_p$ are chosen such that $\hat{u}_p < \hat{t}_p < \hat{l}_p$ and are obtained from the equations shown below.

$$\hat{u}_p = \max\left\{t : \hat{S}(t) \geq (p/100) + 0.05\right\}$$

and $\hat{l}_p = \min\left\{t : \hat{S}(t) \leq (p/100) - 0.05\right\}$.

The end points of 100(1–α) percent confidence interval of $\hat{t}_p$ are

$$\hat{t}_p \pm z_{1-\alpha/2}\, \hat{\text{SE}}\left(\hat{t}_p\right)$$

where $\hat{\text{SE}}\left(\hat{t}_p\right) = \sqrt{\hat{\text{Var}}\left(\hat{t}_p\right)}$.

The assumptions to be made in using Kaplan-Meier estimator are:

- Censored individuals have the same prospect of survival as those, which continue to be followed. This cannot be tested for and can lead to a bias that artificially reduces *S(t)*.
- Survival prospects are the same for early as for late recruits to the study (can be tested for).
- The event studied (e.g., disappearance of dung) happens at the specified time. Late recording of the event studied will cause artificial inflation of *S(t)*.

19.3.3.3 Log-Rank Test

Log-rank function provides methods for comparing two or more survival curves where some of the observations may be censored and where the overall grouping may be stratified. The methods are nonparametric in the sense that they do not make assumptions about the nature or shape of the underlying distributions of survival estimates. The null hypothesis tested is that the risk of death/event is the same in all groups. In the case of absence of censorship, log-rank test reduces to Mann-Whitney test for two groups of survival times and Kruskal-Wallis test for more than two groups of survival times. The mathematical formulation for log-rank test for l factor levels to be compared over s stratum levels is as follows (Hosmer and Lemeshow 1999; SPSS 2006).

Let $n^{(s)}$ be the number of subjects in stratum s. Let $t_1^{(s)} < \cdots < t_{m_s}^{(s)}$ be the observed failure times (responses) and

$n_{\mathrm{li}}^{(s)}$ = the number of individuals in group l at risk just prior to $t_i^{(s)}$ in stratum s

$d_{\mathrm{li}}^{(s)}$ = number of deaths $t_i^{(s)}$ in group l

and

$$d_i^{(s)} = \sum_{g}^{l=1} d_{\mathrm{li}}^{(s)}$$

$$n_i^{(s)} = \sum_{g}^{l=1} n_{\mathrm{li}}^{(s)}.$$

Hence, the expected number of events in group l at time $t_i^{(s)}$ is given by

$$E_{\mathrm{li}}^{(s)} = \frac{d_i^{(s)} n_{\mathrm{li}}^{(s)}}{n_i^{(s)}}.$$

Define

$$U_s = \left(U_1^{(s)}, \ldots, U_{g-1}^{(s)}\right)'$$

with

$$U_l^{(s)} = \sum_{m_s}^{i=1} \left(d_{\mathrm{li}}^{(s)} - E_{\mathrm{li}}^{(s)}\right) \text{for } l = 1, \ldots, g-1.$$

Also, let V_s be a $(g - 1) \times (g - 1)$ covariance matrix with

$$V_{\mathrm{jl}}^{(s)} = \sum_{m_s}^{i=1} \frac{d_i^{(s)}\left(n_i^{(s)} - d_i^{(s)}\right)}{n_i^{(s)} - 1} \frac{n_{\mathrm{ji}}^{(s)}}{n_i^{(s)}} \left(\delta_{\mathrm{jl}} - \frac{n_{\mathrm{li}}^{(s)}}{n_i^{(s)}}\right) \text{for } j, l = 1, \ldots, g-1$$

where

$$\delta_{\mathrm{jl}} = \begin{cases} 1 \text{ if } j = 1 \\ 0 \text{ otherwise} \end{cases}.$$

Define

$$U = \sum_{p}^{s=1} U_s$$

and

$$V = \sum_{p}^{s=1} V_s .$$

The test statistic for the equality of the g survival functions is defined by

$$\chi^2 = U' V^{-1} U.$$

χ^2 has an asymptotic chi-square distribution with (g-1) degrees of freedom.

19.3.4 Estimating Elephant Population from Dung Density Estimates

The dung density of elephants was converted into animal density using the following formula:

$$\text{Elephant density}\left(\text{No.} / \text{sq km}\right) = \frac{\text{DD}}{\text{DR}} \times \text{DDR}$$

where DD = dung density, DR = defecation rate, and DDR = dung decay rate

The defecation rate of 16.33 per day, as obtained from wild elephants in Mudumalai by Watve (1992), was used in the above formula. As far as dung decay rate is concerned, dung decay rate obtained from dung decay experiments conducted in Wayanad Elephant Reserve in 2005 alone was relied upon. The elephant population was extrapolated for various Elephant Reserves by multiplying density estimates with their respective extent of elephant habitat.

19.4 Results

19.4.1 Elephant Dung Survey

19.4.2 Dung Density Estimates Based on Line Transect Survey

Table 19.5 shows the dung density estimates for different Elephant Reserves irrespective of vegetation type.

19.4.3 Dung Decay Rate

A total of 624 dung piles were followed up in Wayanad Elephant Reserve. Of these, 151 dung piles were followed up in evergreen forests, 235 dung piles in moist/dry deciduous forests, and 238 in plantations. About 50% of the total observations were

Table 19.5 Estimated elephant dung density in different Elephant Reserves of Kerala State

	Dung density (No. of dung piles/km^2)		% CV	
Elephant reserve	2005	2007	2005	2007
Wayanad ER	1512	2124	18.45	17.14
	(1050–2176)	(1514–2981)		
Nilambur ER	468	930	14.68	14.26
	(350–625)	(702–1232)		
Anamudi ER	1307	1423	9.99	7.81
	(1074–1590)	(1220–1660)		
Periyar ER	857	878	10.54	7.99
	(697–1055)	(750–1027)		

Figures given in the parenthesis are 95% lower and upper confidence limit
SE standard error, *LCL* 95% lower confidence limit, *UCL* 95% upper confidence limit

found censored in each of the forest types. The age distribution of the dung piles marked and followed up in Wayanad Elephant Reserve shows that most of the dung piles were still surviving for more than 105 days (Table 19.6). Kaplan-Meier product limit estimates of the survivorship function and their standard errors were calculated at 15 days of time in each of the habitat types and presented in Table 19.7. The Kaplan-Meier survivorship function shown in Fig. 19.2 depicts a decreasing staircase function, dropped at the values of the observed failure times and constant between observed failure times; it also shows that there were many dung piles surviving for a longer time in the study period. The minimum probability value of the survivorship function is not zero since the largest observed time (105 days) is a censored observation.

The estimated quartiles of the survival time distribution can also be obtained from the estimated survivorship function. The estimated median survival time for all the habitat types was found to be $\hat{t}_{50}$ =90 days. This means that 50%, i.e., half of the dung piles, were estimated to decay within 90 days (Table 19.8). As most of the censored observations were skewed to the right (Fig. 19.2), the sample mean was not taken as an important central tendency, and hence median was considered as an appropriate measure of central tendency. The estimated median time to decay is 90 days, and thus decay rate is 0.0111.

An inspection of the proportion of values that are censored and the pattern of censoring from the graph (Fig. 19.2) indicates that the censoring experience of the three habitat types was similar. Thus it appears that the assumption, which is necessary for the tests for equality of the survivorship function, seems to hold. The results of the log-rank test (Table 19.9) revealed that there was no significant difference between habitat types with respect to the survival pattern of the dung piles ($p > 0.05$).

Table 19.6 Distribution of dung piles across different age groups in different vegetation types of Wayanad Elephant Reserve

	Age groups (in days)																				
	15			30			45			60			75			90			105		
Vegetation type	N	0	1	N	0	1	N	0	1	N	0	1	N	0	1	N	0	1	N	0	1
Evergreen	10	0	10	12	7	5	16	12	4	23	17	6	37	25	12	24	14	10	29	1	28
Moist/dry deciduous	9	0	9	13	8	5	27	13	14	35	25	10	50	29	21	42	23	19	59	18	41
Plantation	13	0	13	17	5	12	30	15	15	33	16	17	55	48	7	38	25	13	52	16	36
Total	32	0	32	42	20	22	73	40	33	91	58	33	142	102	40	104	62	42	140	35	105

N Number of marked dung piles, *0* decayed, *1* not decayed (survived)

Table 19.7 Cumulative survival probability of the dung piles followed up in different vegetation types of Wayanad Elephant Reserve

Vegetation type	Time (in days)	Cumulative survival probability of the dung piles	Standard error	Cumulative number of dung piles disappeared	Number of dung piles remaining for follow-up
Evergreen	15	1.0000	–	0	151
	30	0.9504	0.0183	7	134
	45	0.8619	0.0294	19	117
	60	0.7323	0.0383	36	96
	75	0.5289	0.0443	61	65
	90	0.3892	0.0457	75	39
	105	0.3757	0.0460	76	28
Moist/dry deciduous	15	1.0000	–	0	235
	30	0.9646	0.0123	8	218
	45	0.9057	0.0196	21	200
	60	0.7840	0.0283	46	161
	75	0.6334	0.0340	75	122
	90	0.4892	0.0375	98	78
	105	0.3399	0.0391	116	41
Plantation	15	1.0000	–	0	238
	30	0.9778	0.0098	5	220
	45	0.9073	0.0198	20	193
	60	0.8257	0.0265	36	162
	75	0.5524	0.0368	84	97
	90	0.3989	0.0372	109	65
	105	0.2762	0.0363	125	36

19.4.4 Estimated Elephant Population Based on Dung Density

An attempt was made to estimate the elephant density and elephant population for different Elephant Reserves based on estimates of dung density, dung decay rate, and dung defecation rate. The pooled dung density estimates irrespective of vegetation types presented in Table 19.4 were used for estimating elephant density. The decay rate of 0.0102 per day from the experiments conducted in Wayanad Elephant Reserve in 2005 was used for both the census periods 2005 and 2007. With respect to defecation rate, 16.33 per day, as obtained from wild elephants in Mudumalai by Watve (1992), was used for estimating elephant density. The estimated elephant density and elephant population are presented in Table 19.10 for different Elephant Reserves.

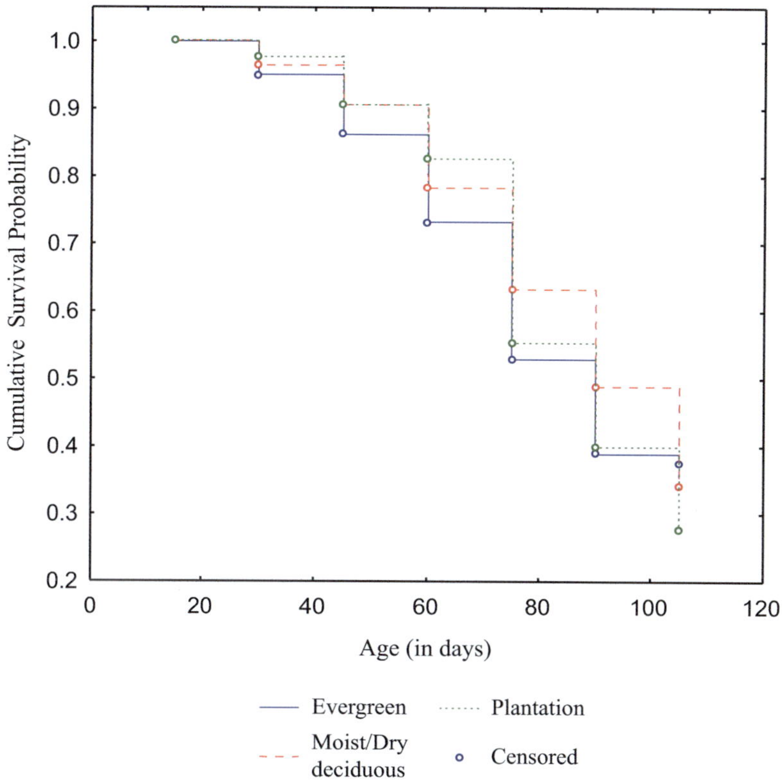

Fig. 19.2 Plots of estimated survivorship function for dung piles followed up in different vegetation types of Wayanad Elephant Reserve

Table 19.8 Kaplan-Meier estimates of survival time of the dung piles followed up in different vegetation types of Wayanad Elephant Reserve

Vegetation type	Mean survival time (SE)	95% confidence interval	Median survival time (SE)	95% confidence interval
Evergreen	82 (2)	78–86	90 (4)	82–98
Moist/dry deciduous	87 (2)	84–90	90 (4)	83–97
Plantation	85 (1)	82–88	90 (4)	83–97

SE standard error

Table 19.9 Log-rank statistic for testing the significance of the equality of the survivorship function in different vegetation types of Wayanad Elephant Reserve

Vegetation type	Evergreen	Moist/dry deciduous
Moist/dry deciduous	0.90	–
	($p > 0.05$)	
Plantation	0.00	1.46
	($p > 0.05$)	($p > 0.05$)

$p > 0.05$ indicates nonsignificant probability levels

Table 19.10 Estimated elephant density and elephant population in different Elephant Reserves based on dung density estimates

	Elephant density (no. of elephants/km^2)		No. of elephants	
Elephant reserve	2005	2007	2005	2007
Wayanad ER	0.9444	1.3271	882	1240
			(613–1270)	(884–1739)
Nilambur ER	0.2923	0.5807	334	663
			(250–446)	(501–879)
Anamudi ER	0.8161	0.8890	2299	2505
			(1889–2798)	(2148–2921)
Periyar ER	0.5354	0.5484	1620	1660
			(1317–1994)	(1418–1942)
Total			5135	6068
			(4069–6508)	(4950–7481)

Elephant density was estimated with the decay rate of 0.0102 and the defecation rate of 16.33
Figures given in the parenthesis are 95% lower and upper confidence limit

19.5 Discussion

Line transect sampling requires proper placement of transects at random and accurate measurement of distance to dung piles. Large-scale surveys such as the present ones require a good training on field methods and proper monitoring of field work to improve the quality of the data. The experience of the analysts has also implications in choosing the appropriate key function and series expansion for estimating the dung density. Stratification is an important means by which the precision of the estimates can be improved. In the surveys conducted so far, prior stratification could not be effected for lack of information. The GIS technology enables to have the survey design drawn in the maps along with relevant details such as habitat type. In future censuses, such maps should be made available well ahead of the survey. The actual extent of elephant habitat is crucial in extrapolating elephant population. Therefore, concerted efforts should be made to estimate the actual extent of elephant habitat in each of the Elephant Reserves.

In estimating elephant population based on dung piles, dung decay rate is the most sensitive factor as it appears in the numerator part of the correction formula. Therefore, it is important that the dung decay rate is measured as precisely as possible. There have been experiments conducted in the past with the assumption that the system is in steady state where the production rates are equal to the decay rates. With regard to the empirical technique used for determining decay rate, Barnes and Jenson (1987) fitted exponential decay rate equation with the assumption that the decay rate is independent of age. Later, Barnes and Barnes (1992) again estimated the dung decay rate using some other equations assuming constant and variable age-specific mortality following curve-fitting approach. Recently, Laing et al. (2003) suggested a retrospective method, which does not have steady-state assumption and proposed logistic regression technique for the analysis of data. In our study, though this methodology was followed, we could not adopt logistic regression technique for determining decay rate, as the number of zeroes (disappearance of dung piles) was not sufficient over the study period. Therefore, a different analytical approach had to be followed for estimating the dung decay rate. We found Kaplan-Meier product limit function as an appropriate method. In this technique, time is treated as a major outcome variable and is nonparametric and therefore avoids many of the statistical problems of other techniques. Further it takes into account the survival time information available on censored observations unlike other techniques. To be more precise, the survival analysis takes into consideration the dung piles that are contributed to the number at risk along the time horizon until they are lost to follow up (Table 19.7).

Dung decay rate highly varies across sites. Only a very few experiments have been conducted in India. The estimated decay rate in our study is 0.0102 for all the vegetation types in Wayanad Elephant Reserve. The dung decay experiments conducted earlier in Wayanad Wildlife Sanctuary reported the decay rate of 0.0191 during dry season and 0.0360 during wet season (Easa 1999). Sale et al. (1990) reported an overall decay rate of 0.0136 during dry season in the same area and decay rate of 0.0146 during early dry season in Parambikulam Wildlife Sanctuary. Varman et al. (1995) reported a dung decay rate of 0.010 in dry season and 0.013–0.007 in wet season in Mudumalai Sanctuary. For African elephants, Barnes and Barnes (1992) obtained an elephant dung decay rate of 0.0026 based on exponential curve and 0.022 based on modeling variable age-specific mortality of dung piles. The decay rate obtained by White (1995) varied from 0.013 to 0.007.

Dung decay rate is highly affected by a number of factors such as rainfall (Barnes et al. 1997; Barnes and Dunn 2002), relative humidity, canopy cover, sunlight, amorphous dung, boli volume, and content of dung (grass fragments, leaf fragments, fruit fibers, and hard seeds) along with activities of mammals and birds (Nchanji and Plumptre 2001). The statistical technique adopted by Nchanji and Plumptre (2001) for the analysis of data was multiple linear regression technique with the duration of dung piles as dependent variable. In the recent past, the proportional hazard model is used for analyzing survival data that can also be tried in large-scale dung decay experiments where censoring occurs (Hosmer and Lemeshow 1999). It does not assume the nature or shape of the underlying survival distribution. It only assumes that the underlying hazard rate is a function of independent variables (covariates). Thus it may be considered as a nonparametric. Nonetheless, the model has two assumptions: (1) multiplicative relationship between the underlying hazard function and the log-linear function of the covariates (proportionality assumption) and (2) log-linear relationship exists between the independent variables and the underlying hazard function. If the covariates can have different values at different points in time for the same case (e.g., relative humidity), then covariate is dependent on time, and hence Cox regression with time-dependent covariates is suggested to use. The advantage of Kaplan-Meier over the Cox's proportional hazard model is that the latter is a model dependent and needs a mathematical function to express.

It must be noted that the dung decay rate used in 2005 was based on dung decay experiments conducted during 2005 census in Wayanad Elephant Reserve alone. The same decay rate was again used in 2007. In order to improve the estimates, it is necessary that the dung decay experiments are conducted in all the Elephant Reserves representing different habitat conditions. The sampling effort in terms of number of blocks chosen for the dung survey was increased significantly from 34% in 2005 to 50% in 2007. Despite methodological and coverage issues, results of 2005 and 2007 surveys were compared. The estimated elephant population of the State using dung survey for the census year 2007 was 6068. This is higher than the estimated population of 5135 elephants in the year 2005. An increasing trend in elephant population was seen in all the Elephant Reserves.

Acknowledgments The authors are grateful to Kerala Forests and Wildlife Department for their excellent support.

References

Barnes RFW, Barnes KL (1992) Estimating decay rates of elephant dung-piles in forest. Afr J Ecol 30:316–321

Barnes RFW, Dunn A (2002) Estimating forest elephant density in Sapo National Park (Liberia) with a rainfall model. Afr J Eco 40:159–163

Barnes RFW, Jenson KL (1987) How to count elephants in forests. IUCN Afr Elephant Rhino Specialist Group 1:1–6

Barnes RFW, Asamoah-Boateng B, Naada Majam J, Agyei-Ohemeng J (1997) Rainfall and the population dynamics of elephant dung-piles in the forests of southern Ghana. Afr J Ecol 35:39–52

Buckland ST, Anderson DR, Burnham KP, Laake JL, Borchers DL, Thomas L (2001) Introduction to distance sampling estimating abundance of biological populations. Oxford University Press, New York

Burnham KP, Anderson DR, Laake JL (1980) Estimation of density from line transect sampling of biological populations. Wildlife monographs No. 72: 202p

Easa PS (1999) Status, habitat utilisation and movement pattern of larger mammals in Wayanad wildlife sanctuary. KFRI Research Report No. 173. Kerala Forest Research Institute, Peechi

Easa PS, Sivaram M, Jayson EA (2002) Population estimation of major mammals in the forests of Kerala-2002. KFRI Consultancy Report No. 09. Kerala Forest Research Institute, Peechi

Hosmer DW, Lemeshow S (1999) Applied survival analysis: regression modeling of time to event data. A Wiley-Interscience Publication, Wiley, New York

Jachmann H (1991) Evaluation of four methods for estimating elephant densities. Afr J Ecol 29:188–195

Laing SE, Buckland ST, Burns RW, Lambie D, Amphlett A (2003) Dung and nest surveys: estimating decay rates. J Appl Ecol 40:1102–1111

Nchanji AC, Plumptre AJ (2001) Seasonality in elephant dung decay and implications for censusing and population monitoring in south-western Cameroon. Afr J Ecol 39:24–32

Sale JB, Johnsingh AJT, Dawson, S (1990) Preliminary trials with an indirect method of estimating Asian elephant numbers. Report presented to the IUCN/SSC Asian Elephant Specialist Group

Sivaram M, Ramachandran KK, Nair PV, Jayson EA (2005) Population estimation of wild elephants in the elephant reserves of Kerala State-2005. KFRI Extension Report No. 18. Kerala Forest Research Institute, Peechi

Sivaram M, Ramachandran KK, Nair PV, Jayson EA (2007) Population estimation of wild elephants in the elephant reserves of Kerala State-2007. KFRI Extension Report No. 27. Kerala Forest Research Institute, Peechi

SPSS (2006) SPSS Version 14.0 (Help Menu) SPSS Inc. Headquarters, 233 S.Wacker Drive, Chicago, Illinois

Staines BW, Ratcliffe PR (1987) Estimating the abundance of red deer (*Cervus elaphus* L.) and roe deer (*Capreolus capreolus* L.) and their current status in Great Britain. Symp Zool Soc Lond 50:131–152

Thomas L, Laake JL, Strinberg S, Marques FFC, Buckland ST, Borchers DL, Anderson DR, Burnham KP, Hedley SL, Pollard JH, Bishop JR, Marques TA (2006) Distance 5.0. Release 2. Research unit for wildlife population assessment. University of St. Andrews, St. Andrews. http://www.ruwpa.st-and.ac.uk/distance/

Varman KS, Ramakrishnan U, Sukumar R (1995) Direct and indirect methods of counting elephants: a comparison of results from Mudumalai Sanctuary. In: Daniel JC, Datye SH (eds) A week with elephants: proceedings of the international seminar on Asian elephants. Oxford University Press, New Delhi

Watve M (1992) Ecology of host-parasite interaction in a wild mammalian host community in Mundumalai, Southern India. Ph. D. thesis, Indian Institute of Science, Bangalore

White LJT (1995) Factors affecting the duration of elephant dung piles in rain forest in the Lope Reserve, Garbon. Afr J Ecol 33:142–150

MIX
Papier aus verantwortungsvollen Quellen
Paper from responsible sources
FSC® C105338
FSC
www.fsc.org

If you have any concerns about our products,
you can contact us on
ProductSafety@springernature.com

In case Publisher is established outside the EU,
the EU authorized representative is:
Springer Nature Customer Service Center GmbH
Europaplatz 3, 69115 Heidelberg, Germany

Printed by Libri Plureos GmbH
in Hamburg, Germany